Mechanical Behaviour of Engineering Materials

J. Rösler · H. Harders · M. Bäker

Mechanical Behaviour of Engineering Materials

Metals, Ceramics, Polymers, and Composites

With 320 Figures and 32 Tables

 Springer

Prof. Dr. Joachim Rösler
TU Braunschweig
Institut für Werkstoffe
Langer Kamp 8
38106 Braunschweig, Germany
j.roesler@tu-bs.de

Dr.-Ing. Harald Harders
Gartenstraße 28
45468 Mülheim
Germany
h.harders@tu-bs.de

Priv.-Doz. Dr. Martin Bäker
TU Braunschweig
Institut für Werkstoffe
Langer Kamp 8
38106 Braunschweig, Germany
martin.baeker@tu-bs.de

ISBN 978-3-642-09252-7   e-ISBN 978-3-540-73448-2

Springer is a part of Springer Science+Business Media
springer.com
© Springer-Verlag Berlin Heidelberg 2010

Cover design: wmx Design GmbH, Heidelberg

By the authors

Prof. Dr. rer. nat. Joachim Rösler, born in 1959, studied materials science at the University Stuttgart, Germany, from 1979 to 1985. After earning a Ph. D. at the Max-Planck Institute for Metals Research, Stuttgart, Germany, and a post-doctoral fellowship at the University of California, Santa Barbara, USA, he worked at Asea Brown Boveri AG, Switzerland, from 1991 to 1996, being finally responsible for the material laboratory of ABB Power Generation Ltd., Switzerland. Since 1996, he has been professor for materials science and director of the Institute for Materials Science at the Technical University Braunschweig, Germany. His main research interest lies in high-temperature materials, the mechanical behaviour of materials, and in materials development.

Dr.-Ing. Harald Harders, born in 1972, studied mechanical engineering, with a focus one mechanics and materials, at the Technical University Braunschweig, Germany. In 1999, he worked as research scientist at the German Aerospace Center (DLR). From 1999 to 2004, he worked as research scientist at the Institute for Materials Science at the Technical University Braunschweig, finishing with a Ph. D. thesis (2005) on fatigue of metal foams. Since 2004, he has been working in the field of life time prediction and modelling of superalloys and coating systems at Siemens Power Generation in Mülheim an der Ruhr, Germany.

Priv.-Doz. Dr. rer. nat. Martin Bäker, born in 1966, studied physics at the University Hamburg, Germany, from 1987 to 1993 and finished his Ph. D. at the II. Institute for Theoretical Physics of the University Hamburg in 1995, where he also worked as Post-Doc for a year. Since 1996, he has been working as research scientist at the Institute for Materials Science at the Technical University Braunschweig, Germany, focusing on continuum mechanics simulation of materials. In 2004, he finished his 'habilitation' (lecturer qualification) in the field of materials science.

Preface

Components used in mechanical engineering usually have to bear high mechanical loads. It is, thus, of considerable importance for students of mechanical engineering and materials science to thoroughly study the mechanical behaviour of materials. There are different approaches to this subject: The engineer is mainly interested in design rules to dimension components, whereas materials science usually focuses on the physical processes in the material occurring during mechanical loading. Ultimately, however, both aspects are important in practice. Without a clear understanding of the mechanisms of deformation in the material, the engineer might uncritically apply design rules and thus cause 'unexpected' failure of components. On the other hand, all theoretical knowledge is practically useless if the gap to practical application is not closed.

Our objective in writing this book is to help in solving this problem. For this reason, the topics covered range from the treatment of the mechanisms of deformation under mechanical loads to the engineering practice in dimensioning components. To meet the needs of modern engineering, which is more than ever characterised by the use of all classes of materials, we also needed to discuss the peculiarities of metals, ceramics, polymers, and composites. This is reflected in the structure of the book. On the one hand, there are some chapters dealing with the different types of mechanical loading common to several classes of materials (Chapter 2, elastic behaviour; Chapter 3, plasticity and failure; Chapter 4, notches; Chapter 5, fracture mechanics; Chapter 10, fatigue; Chapter 11, creep). The specifics of the mechanical behaviour of the different material classes that are due to their structure and the resulting microstructural processes are treated in separate chapters (Chapter 6, metals; Chapter 7, ceramics; Chapter 8, polymers; Chapter 9, composites).

In this book, we thus aim to comprehensively cover the mechanical behaviour of materials. It addresses students of mechanical engineering and materials science as well as practising engineers working on the design of components. Although the book contains an in-depth treatment of the mechanical behaviour and is thus not to be considered as an introduction, all topics can be understood without much previous knowledge of material physics and mechanics. To make it more accessible, the book starts with an introductory chapter on the structure of materials and contains appendices on tensors, crystal orientation, and thermodynamics.

> In many cases, we thought it desirable to cover some topics in greater depth for those readers with a special interest in the subject matter. These sections can be skipped without compromising the understanding of other subjects. These advanced sections are indented, as here, or, in the case of longer sections, marked with a * on the section number.

At the end of the main part, the reader can find some exercises with complete solutions. They serve as numerical examples for the topics covered in the text and enable the reader to check their understanding of the subject.

This book has evolved from lectures at the Technical University of Braunschweig on the mechanical behaviour of materials, aimed at graduate students, and was first published in German by the Teubner Verlag, Wiesbaden. Due to its success and many encouraging remarks from readers, it seemed worthwhile to prepare an English edition of the book. In doing so, the nomenclature and some of the references were adapted to improve the usability of the book for English readers.

We wish to thank Günter Lange who provided valuable help in preparing this book. Furthermore, we want to thank Jürgen Huber (CeramTec AG), Dr. Peter Neumann (Max-Planck-Institut für Eisenforschung GmbH), Volker Saß (ThyssenKrupp Nirosta GmbH), Johannes Stoiber (Allianz-Zentrum für Technik GmbH), the Lufthansa Technik AG, the Institut für Werkstofftechnik of the Universität Gh Kassel, the Institut für Füge- und Schweißtechnik of the Technische Universität Braunschweig, the Institut für Baustoffe, Massivbau und Brandschutz of the Technische Universität Braunschweig, and all members of the Institut für Werkstoffe. Steffen Müller has made a significant contribution to the lecture notes that were the starting point for writing this book. Furthermore, we want to thank Allister James and Gary Merrill who proofread parts of the manuscript. We are also indebted to many readers who sent book evaluations to the Teubner Verlag that have been helpful in preparing the second German edition [123]. The Teubner Verlag kindly gave the permission to publish an English translation. We finally want to thank the Springer publishing company for the cooperation in preparing this edition.

Braunschweig,   *Joachim Rösler*
Mülheim an der Ruhr,   *Harald Harders*
May 2007   *Martin Bäker*

Contents

[1] Sections with a title marked by a * contain advanced information which can be skipped without impairing the understanding of subsequent topics.

1

The structure of materials

There is a vast multitude of materials with strongly differing properties. A copper wire, for instance, can be bent easily into a new shape, whereas a rubber band will snap back to its initial form after deformation, while the attempt to bend a glass tube ends with fracture of the tube. The strongly differing properties are reflected in the application of engineering materials – you would neither want to build cars of glass nor rubber bridges. The multitude of materials enables the engineer to select the best-suited one for any particular component. For this, however, it is frequently necessary not only to know the mechanical properties of the materials, but also to understand the physical phenomena causing them.

The mechanical properties of materials are determined by their atomic structure. To understand these properties, some knowledge of the structure of materials is therefore required. This is the topic covered in this chapter. The structure of materials is investigated by solid state physics, but to understand the mechanical properties, it is not necessary to understand the more arcane aspects of this discipline as they can usually be explained with rather simple models.

This chapter starts with a short explanation of the basic principles of atomic structure and the nature of the chemical bond. Afterwards, the three main groups of materials, *metals, ceramics,* and *polymers,* are discussed. The most important characteristics of their interatomic bonds are covered, and the microscopic structure of the different groups is also treated.

For a more thorough introduction into the structure of materials the books by *Beiser* [17] and *Podesta* [110] are recommended.

1.1 Atomic structure and the chemical bond

Atoms consist of a positively charged *nucleus* surrounded by negatively charged *electrons*. Almost the complete mass of the atom is concentrated in the nucleus because it comprises heavy elementary particles, the *protons*

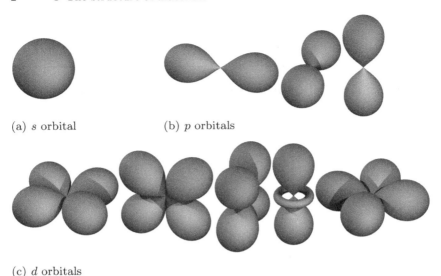

(a) *s* orbital   (b) *p* orbitals

(c) *d* orbitals

Fig. 1.1. Sketch of selected electron orbitals

and *neutrons.* The number of positively charged protons within the nucleus determines the *atomic number* and thus the chemical element. Thus *hydrogen,* containing one proton in the nucleus, has an atomic number of 1, *oxygen* an atomic number of 8, and *iron* of 26. The nucleus is not involved in chemical reactions which are governed by the electrons surrounding it.

The electrons of an atom are not arranged in an arbitrary configuration. Instead, they are confined to so-called *electron shells* that are arranged in increasing distance around the nucleus and that can only contain a limited number of electrons. The further away an electron shell is from the nucleus, the higher is the energy of the electrons in this shell so that electrons on the outer shells are more weakly bound to the nucleus than those on the inner ones.

In general, it is not possible to localise electrons at a certain point i. e., their position is not defined. It is only possible to know the probability that an electron is situated at a certain point if one tries to find it there. This probability varies in space, so there are some regions near the nucleus where the electron will be located preferentially, whereas it avoids others. The region where the electron can be found is called the *orbital.* Figure 1.1 shows some examples of such orbitals. As can be seen from the figure, orbitals can be spherically symmetric or directed. An electron shell usually comprises several orbitals. Each orbital can be occupied by no more than two electrons (*Pauli exclusion principle*).

The basic structure of all electron shells is the same in all atoms. The innermost shell, called K shell, can contain at most two electrons because there is only one, spherically symmetric, orbital (the *s* orbital) in it. The next

Table 1.1. Electron configurations of selected elements

		K	L		M			N			
		$1s$	$2s$	$2p$	$3s$	$3p$	$3d$	$4s$	$4p$	$4d$	$4f$
1	H	1									
2	He	2									
3	Li	2	1								
4	Be	2	2								
5	B	2	2	1							
6	C	2	2	2							
7	C	2	2	3							
8	O	2	2	4							
9	F	2	2	5							
10	Ne	2	2	6							
11	Na	2	2	6	1						
17	Cl	2	2	6	2	5					
19	K	2	2	6	2	6		1			
20	Ca	2	2	6	2	6		2			
21	Sc	2	2	6	2	6	1	2			
22	Ti	2	2	6	2	6	2	2			
26	Fe	2	2	6	2	6	6	2			
28	Ni	2	2	6	2	6	8	2			
29	Cu	2	2	6	2	6	10	1			
30	Zn	2	2	6	2	6	10	2			

shell, the L shell, can be occupied by up to eight electrons. Two of these are situated in a spherically symmetric s orbital, whereas the other six occupy directed orbitals, the three p orbitals. The subsequent M shell offers space to 18 electrons in s, p, and d orbitals.[1] As nature tends to states of lowest energy, these shells will be filled in the atoms starting with the innermost one, until the number of electrons equals the atomic number so that the atom is electrically neutral. Table 1.1 shows the electron configurations of several atoms.

As the energy of the electrons is higher on the outer shells than on the inner ones, it is only the electrons on these shells that are involved in chemical reactions. The binding energy of the weakest bound electron is called the *ionisation energy* because when the electron is removed a positively charged ion remains. Thus, the ionisation energy is a measure of the binding strength of an electron in the outermost shell.

The ionisation energy of an atom is particularly high if the outermost shell is fully occupied.[2] Fully occupied electron shells are energetically favourable so that atoms tend to attain configurations with completely filled outermost

---

[1] In general the number k of electrons in the nth shell is given by $k = 2n^2$.

[2] Due to their higher binding energy, the inner electrons are never involved in chemical reactions. They, however, do play a role in the generation of X rays.

shell. This explains why noble gases are almost completely chemically inert, why fluorine, lacking only one electron to fill its outer shell, has a high *electron affinity,* and why, on the other hand, an element like sodium, with only one electron on the outer shell, has a low ionisation energy.

A chemical bond between atoms is formed by several atoms 'sharing' their electrons, or by one atom completely transferring electrons to another to achieve a favourable electron configuration. Hydrogen, for instance, with only one electron on the K shell needs another electron to fill this shell. Therefore, two hydrogen atoms can bond with each other and share their electrons. A hydrogen molecule H_2 is formed. In this, rather simplified, picture of the chemical bond, each atom can form as many bonds as there are electrons missing on the outermost shell. This type of bond is called *covalent* and will be described in section 1.3.1. The number of bonds formed by an atom is called its *valency.* So fluorine has a valency of 1, oxygen of 2, and carbon of 4.[3]

The valency model of the elements can explain many chemical compounds, but not all of them. A simple example shows the limitations of the model: If a hydrogen molecule is ionised, the resulting molecule has the chemical formula H_2^+. Both hydrogen nuclei share a single electron although neither of them obtains a full outer shell in this way. Nevertheless, the H_2^+ molecule has a rather large binding energy and does not dissociate into a proton and a hydrogen atom. This is caused by a special property of electrons: electrons tend to occupy states in which they can spread over a region with the largest possible extension. The more an electron is confined to a small region, the higher its energy becomes. For the electron, it is therefore favourable to stay simultaneously at both hydrogen nuclei, for this reduces its energy.

This property of the electrons also explains why electrons do not fall into the nucleus. According to the rules of classical physics, it should be expected that an electron orbiting a proton minimises its energy by being as close to the proton as possible because both particles attract each other strongly. However, the closer the electron is to the proton, the more does its energy increase because it is more and more confined. These two effects with opposing signs lead to a minimisation of the electron energy at a certain distance to the nucleus. As we will see in the next section, this principle determines the physical properties of metals.

The chemical bond between two atoms causes an attraction between them. If they get too close, the electrostatic repulsion of the electron shells causes a repulsive force. Another repulsive effect comes about because the size of the orbitals reduces when they approach, which, as explained, is energetically unfavourable. An equilibrium distance is reached where the energy is minimised

---

[3] The valency of elements whose outer shell is less than half occupied is given not by the number of missing electrons but by the number of electrons present. Thus, sodium has a valency of 1, magnesium of 2. The situation is more complicated with the transition metals. Iron, for instance, can react with oxygen to form either FeO (valency 2) or Fe_2O_3 (valency 3).

and there is no net force on the atoms (see section 2.3). Typically, atomic distances of covalent bonds are between 0.1 nm and 0.3 nm.

Depending on the elements forming the bond, different types of bonds with distinctly different properties can be formed. These types will be discussed in the next sections together with those material classes they are typical of.

1.2 Metals

Metals are an especially important class of materials. They are distinguished by several special properties, namely their high thermal and electrical conductivity, their ductility (i. e., their ability to be heavily deformed without breaking), and the characteristic lustre of their surfaces. Their ductility, together with the high strength[4] that can be achieved by alloying, renders metals particularly attractive as engineering materials.

In nature, metals occur only seldom as they possess a high tendency for oxidation. If one looks at the pure elements, more than two thirds of them are in a metallic state. Many elements are soluble in metals in the solid state and thus allow to form a metallic alloy. For instance, steels can be produced by alloying iron with carbon. The large number of metallic elements offers a broad range of possible alloys. Of most technical importance are alloys based on iron (steels and cast irons), aluminium, copper (bronzes and brasses), nickel, titanium, and magnesium.

In this section, we start by explaining the nature of the chemical bond of metals. We will see that metals usually arrange themselves in a regular, crystalline order. Therefore, we will afterwards discuss the structure of crystals and, finally, explain how a metallic material is composed of such crystals.

1.2.1 Metallic bond

A look at the periodic table shows that metals are distinguished by possessing rather few electrons on their outer shell (figure 1.2) and thus would need a large number of electrons to fill this shell. On the other hand, they have the possibility to achieve a fully occupied outer shell by dispensing with their outer electrons. The ionisation energy of metals is, therefore, rather small.

Due to the small number of outer electrons, the metallic bond cannot be based on several atoms sharing their electrons to achieve a full outer shell. That, nevertheless, a bond forms is due to the property of electrons to tend to spread over as large a region as possible, as discussed above in the context of the H_2^+ molecule.

How this can lead to the formation of a metal can be explained most easily using an example: Lithium is an alkali metal with only one electron on the

---

[4] The strength of a material is defined by the load it can withstand without failure. This will be discussed in section 3.2

main   group elements

																2
1 H hcp																	**2** He hcp
3 Li bcc	**4** Be hcp	transition metals										**5** B tet	**6** C dia	**7** N hcp	**8** O cub	**9** F mon	**10** Ne fcc
11 Na bcc	**12** Mg hcp											**13** Al fcc	**14** Si dia	**15** P cub	**16** S ort	**17** Cl ort	**18** Ar fcc
19 K bcc	**20** Ca fcc	**21** Sc hcp	**22** Ti hcp	**23** V bcc	**24** Cr bcc	**25** Mn cub	**26** Fe bcc	**27** Co hcp	**28** Ni fcc	**29** Cu fcc	**30** Zn hcp	**31** Ga ort	**32** Ge dia	**33** As rho	**34** Se hcp	**35** Br ort	**36** Kr fcc
37 Rb bcc	**38** Sr fcc	**39** Y hcp	**40** Zr hcp	**41** Nb bcc	**42** Mo bcc	**43** Tc hcp	**44** Ru hcp	**45** Rh fcc	**46** Pd fcc	**47** Ag fcc	**48** Cd hcp	**49** In tet	**50** Sn dia	**51** Sb rho	**52** Te hcp	**53** I ort	**54** Xe fcc
55 Cs bcc	**56** Ba bcc	**57** La hcp	**72** Hf hcp	**73** Ta bcc	**74** W bcc	**75** Re hcp	**76** Os hcp	**77** Ir fcc	**78** Pt fcc	**79** Au fcc	**80** Hg rho	**81** Tl hcp	**82** Pb fcc	**83** Bi rho	**84** Po cub	**85** At	**86** Rn (fcc)
87 Fr (bcc)	**88** Ra	**89** Ac fcc	**104** Ku														

☐ metal   ☐ semi-metal   ☐ nonmetal

hcp – hexagonal close-packed   bcc – body-centred cubic
fcc – face-centred cubic   cub – cubic
ort – orthorhombic   tet – tetragonal
rho – rhombohedral   dia – diamond lattice

Fig. 1.2. Periodic table of the elements excluding lanthanides (atomic numbers 58 to 71) and actinides (atomic numbers 90 to 103). The crystal structures will be explained below.
Semi-metals have bonds of a mixed covalent-metallic type. Some materials exhibit different crystal structures depending on the temperature [10, 84]

outermost shell, thus offering seven unoccupied sites for other electrons. If two lithium atoms approach, both outer electrons, the *valence electrons,* of the atoms can occupy the space around both atoms and can thus reduce their energy. This is similar to the formation of the H_2^+ molecule discussed above. If a third lithium atom is added, this atom can also spread out its electron over all three atoms, thus forming a Li_3 molecule. A further lithium atom can also add its electron to the mix. Finally, a structure is formed in which each lithium atom is surrounded by eight nearest neighbours and shares its electrons with them. Each bond between two lithium atoms contains on average one quarter of an electron. The bond between the electrons is caused by the spreading of the electrons.

This spreading of the electrons makes it impossible to assign the electrons to the atoms they originally belonged to. The electrons spread over the whole material so that on average one electron is always close to any lithium atom,[5] but this electron is not stationary and can move about freely. This is the reason why it is often said that the atoms release their electrons to a common

---

[5] The inner electrons of course always stay close to their lithium atoms and are not considered in this discussion.

electron gas, resulting in positively charged metallic ions surrounded by a 'gas' of negatively charged electrons.[6]

The mobility of the electrons within the electron gas explains many of the physical properties of metals because the excellent electrical and thermal conductivity are based on it. The shininess of metals is also caused by it, for the electrons can easily vibrate in an oscillating electromagnetical field (e.g., light) and thus bar it from entering the metal [47, 110].

As the metallic bond does not result in a fully occupied shell of the single atoms, it is weaker than other types of bond. The binding energy of a metallic bond between any two atoms takes values between approximately 0.1 eV and 0.3 eV.[7] On the other hand, each atom in a metal has a relatively large number of nearest neighbours so that in total relatively large binding energies result, for example 1.1 eV for sodium and 3.5 eV for copper. As the binding energies are lower than in ceramics, which possess fully occupied outer shells, the melting temperature of metals is usually lower as well.

The distribution of the electrons over a large region leads to a slow decrease of the interatomic force with the distance of the atoms compared to other types of bonds. Because it is thus possible to displace single atoms with a rather small amount of energy, metals can be easily deformed plastically. If some metal atoms are replaced by those of another metallic element, the metallic bond is usually not destroyed because, for the bond, it is mainly relevant that electrons are released to the electron gas. This explains why it is possible to alloy metals in many different compositions.

How exactly the mechanical properties of metals are determined by the metallic bond will be discussed in detail in chapters 2 and 6.

1.2.2 Crystal structures

As we learned in the previous section, atoms in a metallic solid arrange themselves so that their electrons can spread over many atoms. This spreading is most easy if the atoms are arranged in a dense and regular manner. Therefore, metals form *crystals* which are distinguished by their well-ordered structure. To understand the different types of crystal structures found in nature, it is useful to think rather generally about the problem of arranging objects.

---

[6] This picture of an electron gas is suitable to describe many properties of metals correctly. Its main drawback is that in this picture the metallic bond seems to be completely different from a covalent bond. This, however, is not true as there are intermediate states between these two types, occurring in the so-called *semi-metals.*

[7] Atomic energies are frequently measured in the unit electron volt (eV). 1 eV corresponds to an energy of 1.602×10^{-19} J. In chemistry, energies are frequently calculated per mole: $1 \text{ eV} \approx 105 \text{ kJ/mol}$.

Fig. 1.3. Simple cubic crystal structure

Mathematically, a crystal can be considered as a three-dimensional arrangement of points (i. e., a *lattice* of points) that looks identical from each of the points. In a real-world crystal each of these points will be occupied by an atom[8]. The crystal has a regular, periodic structure that repeats itself exactly. It thus not only possesses a short-range order, but also a long-range order, for the structure of even a remotely distant region can be predicted exactly from each point. Figure 1.3 shows a simple cubic crystal as an example. The crystal can be visualised as consisting of cubes that all look alike. These cubes are the 'building blocks' from which the crystal can be constructed by putting them together. These building blocks are called *unit cells*. Unit cells cannot have arbitrary shapes. As the crystal has to be built from them without gaps, only such unit cells can form a crystal that can completely fill space.

Altogether, there are 14 different possibilities to arrange atoms on a lattice so that the lattice looks the same from each lattice point. These are called *Bravais lattices,* named for their discoverer, Auguste Bravais. Their unit cells are depicted in figure 1.4. For instance, the simple orthorhombic and the simple cubic lattice differ in the orthorhombic unit cell being a quadrangular prism with differing edge lengths, whereas the unit cell of the cubic lattice is a cube. The geometry of the different crystal types will be explained in more detail below.

Some of the 14 Bravais lattices are very similar. The simple cubic and the body-centred cubic lattice differ only in the additional atom that is situated in the centre of the unit cell. Such similarities can be described using the *symmetries* of a crystal. A symmetry of an object is defined as an operation that leaves the object unchanged. The simple cubic crystal structure shown in figure 1.3, for example, remains unchanged when it is rotated by 90° along one of its edges, by 120° along the cube diagonal, or if it is reflected using any of the mid-planes of the cube as mirror plane. All crystal types possessing the same symmetries with respect to rotations and reflections as this cubic crystal are grouped into the same *crystal system,* the *cubic crystal system.* Although the body-centred cubic, the face-centred cubic and the simple cubic

---

[8] Sometimes more than one atom may form a lattice point, see section 1.3.6.

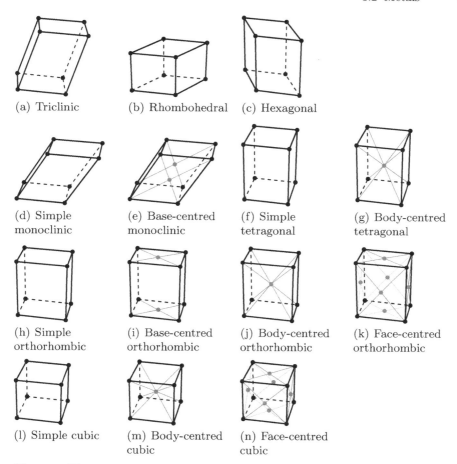

(a) Triclinic   (b) Rhombohedral   (c) Hexagonal

(d) Simple
monoclinic

(e) Base-centred
monoclinic

(f) Simple
tetragonal

(g) Body-centred
tetragonal

(h) Simple
orthorhombic

(i) Base-centred
orthorhombic

(j) Body-centred
orthorhombic

(k) Face-centred
orthorhombic

(l) Simple cubic

(m) Body-centred
cubic

(n) Face-centred
cubic

Fig. 1.4. The unit cells of the 14 Bravais lattices

lattice differ in the arrangement of their atoms, they all possess the same cubic symmetry.

The 14 Bravais lattices can be grouped into seven crystal systems according to their symmetry as listed in table 1.2. Generally, each crystal system is characterised by six numbers: three *lattice constants,* indicating the edge lengths of the three axes making up the unit cell, and the three angles between these axes. Typical values of the lattice constant in metals are between 0.2 nm and 0.6 nm.

The symmetry of a crystal type is relevant because frequently it is reflected in its material properties. A cubic crystal, for instance, has the corresponding symmetries in its mechanical properties. The lower the symmetry of a crystal, the more complicated is the anisotropy of its properties. This will be discussed in chapter 2, using the elastic properties as an example.

Table 1.2. The seven crystal systems

name	lattice-constants	lattice angle	
triclinic	$a \neq b \neq c$	$\alpha \neq \beta \neq \gamma$	
monoclinic	$a \neq b \neq c$	$\alpha = \gamma = 90° \neq \beta$	
orthorhombic	$a \neq b \neq c$	$\alpha = \beta = \gamma = 90°$	
hexagonal	$a = b \neq c$	$\alpha = \beta = 90°, \gamma = 120°$	
tetragonal	$a = b \neq c$	$\alpha = \beta = \gamma = 90°$	
rhombohedral	$a = b = c$	$\alpha = \beta = \gamma \neq 90°$	
cubic	$a = b = c$	$\alpha = \beta = \gamma = 90°$	

In metals, three lattice structures are especially frequent. Two of these are Bravais lattices with cubic symmetry:

- *face-centred cubic* (figures 1.4(n) and 1.5(a), abbreviated *fcc*),[9]
- *body-centred cubic* (figures 1.4(m) and 1.5(b), abbreviated *bcc*).

The third important crystal structure of metals is the *hexagonal close-packed structure*, abbreviated *hcp*. This structure is not a Bravais lattice as not all atoms occupy identical positions. Looking at figure 1.6, it can be seen that the atom at the front right edge of the cell has a neighbour that can be

---

[9] In the periodic table of the elements, figure 1.2, the crystal structures of the elements are listed.

(a) Face-centred cubic   (b) Body-centred cubic

Fig. 1.5. A sphere model of the cubic crystals

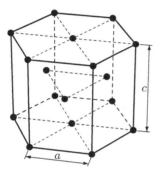

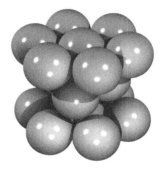

(a) Lattice representation   (b) Sphere model

Fig. 1.6. The hexagonal close-packed structure

reached by moving up by $c/2$ and to the left and back by $a/\sqrt{3}$. If this step is repeated from the atom reached in this way, there is no atom at the new position. The hexagonal close-packed lattice can be constructed by stacking two simple hexagonal lattices into each other. Such lattices are called *lattices with a basis* and will be discussed further in section 1.3.6.

A special unit cell of a crystal is the *primitive unit cell*, defined as the smallest unit cell from which the crystal can be built. As visualised in figure 1.7, the primitive unit cell is not uniquely defined but can be chosen in different ways. However, all possible primitive unit cells obviously have the same volume. One primitive unit cell of a body-centred cubic lattice is shown in figure 1.8. This cell is only part of the cube that one usually visualises when putting together the crystal lattice. As the crystal symmetries are less obvious when using this cell, frequently the cubic unit cell is used instead, called *conventional unit cell*. It is easy to determine whether a unit cell of a

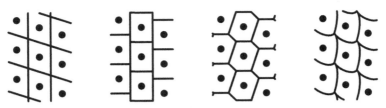

Fig. 1.7. Different unit cells of the same lattice structure in two dimensions (after [10])

Fig. 1.8. Body-centred cubic lattice, primitive unit cell (thick lines) and conventional unit cell (thin lines). Both cells are centred on one atom

Bravais lattice is primitive: If it contains only one atom, it is primitive; if it contains more, it is not. While counting the atoms it has to be kept in mind to count only the appropriate fractions of those atoms occupying more than one cell. For instance, the conventional unit cell of the body-centred cubic lattice contains two atoms and is therefore not primitive, the conventional unit cell of the face-centred cubic lattice contains four atoms and is thus not primitive either.

Two important properties of a crystal lattice are its coordination number and its relative density. As explained above, metals arrange their atoms in crystal structures because this enables them to share their electrons with many other atoms. Therefore, it is favourable if they have a large number of nearest neighbours. This number of nearest neighbours is called the *coordination number* of the crystal. The coordination number is twelve in a face-centred cubic and a hexagonal close-packed crystal, eight in a body-centred cubic crystal, and only six in a simple cubic crystal. If we imagine the atoms to be spheres touching each other, they fill up a certain fraction of space. This fraction, called the *relative density*, takes its maximum value of 74% in the face-centred cubic and the hexagonal close-packed lattice (see exercise 1).[10] Figures 1.5 and 1.6(b) use sphere models to illustrate the relative density. As can be seen, the size of the interatomic gaps is larger in the body-centred cubic than in the face-centred cubic or hexagonal close-packed lattice.

---

[10] It is impossible to pack spheres of equal size with a higher relative density than in the fcc and hcp structure. This has been conjectured by Johannes Kepler in 1611, but it was proven only in 1999 by Hales und Ferguson, using the power of modern computer algebra [136].

(a) Hexagonal close-packed lattice

(b) Face-centred cubic lattice

Fig. 1.9. Construction of the hexagonal close-packed and the face-centred cubic lattice by stacking close-packed layers of spheres. The structures differ in their stacking sequence: In the hexagonal close-packed structure spheres in the third layer are placed perpendicularly above those in the first, in the face-centred cubic lattice the spheres are offset

The hexagonal close-packed and the face-centred cubic lattice are both close-packed structures. They differ in the arrangement of atoms. This can be visualised using the stacking sequence as shown in figure 1.9. We start by arranging spheres in a close-packed way in the plane so that each sphere has six nearest neighbours. If we stack another layer of spheres onto this plane, only every second gap is occupied. When stacking a third layer onto the second, there are two different possibilities: If the spheres are placed directly above those in the first plane, the hexagonal close-packed structure results; if they are placed in the other gaps not directly over the spheres in the first plane, we get the face-centred cubic structure.

To be able to discuss the properties of crystalline materials, it is frequently necessary to uniquely identify directions within the crystal. This is done using Miller indices as explained in detail in appendix B.

More complicated crystal structures than those described so far may result if the crystal is made up of different elements. As this is most frequently the case in ceramics, it will be dealt with in section 1.3.6.

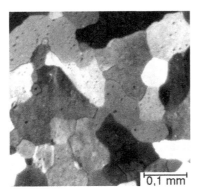

(a) Micrograph (optical microscope)

(b) Microstructure of a nickel-base alloy (scanning electron microscope picture of an intercrystalline fracture surface)

Fig. 1.10. Exemplary microstructures of metals

1.2.3 Polycrystalline metals

If a metal is cooled down from a melt and solidifies, it starts to crystallise. Depending on the cooling rate, many small nuclei of crystallisation form, small solidified regions with crystalline structure. These nuclei then grow and coalesce. As the initial nuclei develop independently, they possess no long-range order between them. Therefore, a metal does not usually consist of one single crystal with long-range order, but rather of several crystalline regions called *crystallites* or *grains.* They have a diameter of the order of a few micrometres up to a fraction of a millimetre, but can also be much larger in special cases. Grains can be made visible by polishing the surface of the metal and then etching it because the acid attacks differently oriented grains differently (see figure 1.10(a)). The structure of the grains of a metal is usually termed its *microstructure.*

The *grain boundaries* i. e., the interfaces between the grains, do not have a perfectly crystalline order as differently oriented regions adjoin here. Therefore, they can be considered as lattice imperfections. Frequently, they strongly influence the properties of a material because, for example, they may be preferred diffusion paths for corroding media. This kind of weakening of grain boundaries may then lead to failure of the material. This is called *intercrystalline fracture* and is shown in figure 1.10(b).

Technical alloys frequently consist of different phases i. e., regions with differing chemical composition or crystal structure. As we will see later (in section 6.4.4), particles of a second phase that are enclosed by a matrix of a first phase are especially important to influence mechanical properties. One example for this is iron carbide (cementite, Fe_3C) that increases the strength of steels when precipitated as fine particles.

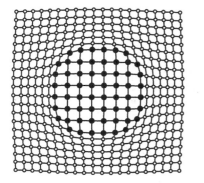 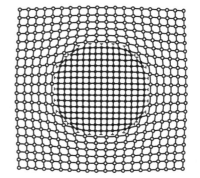

(a) Coherent. All crystal planes are continuous between matrix and particle

(b) Semi-coherent. Some of the crystal
planes are continuous between matrix
and particle

Fig. 1.11. Coherent and semi-coherent particles. The symbol ⊥ in subfigure (b) denotes inserted half-planes of the lattice. The edge where such a half-plane ends is called an edge dislocation. This will be discussed in section 6.2

Depending in the crystal structure of the two phases, the interface between them may adopt different structures: If the crystal structures and the crystal orientation of both phases are identical and the lattice constants do not differ too much, the particles of the second phase will be *coherent* i.e., the lattice planes of the matrix continue within the particle (see figure 1.11(a)). If the lattice structure and orientation are identical, but the lattice constants differ strongly, the particles will be *semi-coherent* because some lattice planes of the matrix continue inside the particle but others do not (figure 1.11(b)). Generally, the crystal lattice is distorted near to the coherent or semi-coherent particle. If the lattice structure of both phases or the lattice orientation differ, the particles are incoherent; the lattice planes of particle and matrix have no relation at all (figure 1.12).

Even within a grain, the lattice may not be perfect. Some lattice sites may not be occupied (so-called *vacancies*) or may be occupied by foreign atoms. More complicated lattice imperfections may also arise, most importantly the dislocations. As they are especially important in determining the plastic behaviour of metals, they are discussed in detail in chapter 6.

1.3 Ceramics

All non-metallic, non-organic materials are called ceramics [70].[11] Physically, the distinction between ceramics and metals can be based on their bond type –

---

[11] The classification of engineering materials is not unique, and other criteria to distinguish between the classes exist. Therefore, some materials are classified differently by different authors.

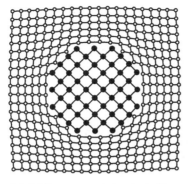
(a) Different crystal orientations

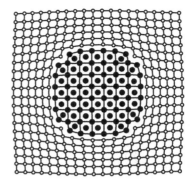
(b) Different crystal structures

Fig. 1.12. Incoherent particles

ceramics do not possess a metallic bond, but bond types that result in a completely filled outer shell.

Ceramics can be elementary i. e., they may consist of only one element (carbon, for example, can exist in two different ceramic forms, as diamond or graphite), or they can be compounds of different elements. Of technical importance are silicate ceramics, containing silicon oxide (for example, porcelain or mullite), oxide ceramics i. e., compounds of metallic elements with oxygen (for example, aluminium oxide Al_2O_3, zirconium oxide ZrO_2, or magnesium oxide MgO)[12], and non-oxide ceramics i. e., oxygen-free compounds like silicon carbide and silicon nitride.

Ceramics can be chemically bound in different ways. Rather strong bond types are the *covalent* and *ionic* bonds, weaker ones are *van der Waals, dipole,* and *hydrogen bonds.*

1.3.1 Covalent bond

The covalent bond was already discussed in section 1.1. Atoms that lack only a few electrons to achieve a fully occupied outer shell share some of their electrons. As an example, the H_2 molecule was explained. To form a solid with strong bonds between the atoms, it is insufficient if each electron lacks only one electron because in this case a two-atomic molecule will form only. An atom with a valency of four, like carbon, can form large units in which each atom has four bonded neighbours. Figure 1.13 shows the resulting carbon macro-molecule, diamond. Other elements with four valencies, like silicon and germanium, form similar structures.

---

[12] Often, metal oxides are also denoted by the name of the metal with an appended 'a' instead of 'ium oxide', e. g., alumina, zirconia, magnesia.

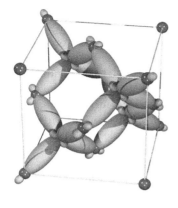

Fig. 1.13. Diamond structure with electron orbitals

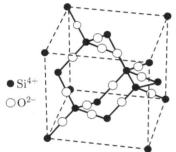

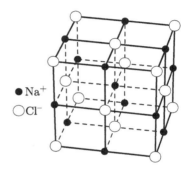

(a) Silicon oxide (high cristobalite, SiO_2)   (b) Common salt (NaCl)

Fig. 1.14. Unit cells of some ceramics

An example for a ceramic comprising different elements is silicon oxide, SiO_2, shown in figure 1.14(a). In this ceramic, each oxygen atom is linked to two silicon atoms which serve as the nodes in the three-dimensional network.

In contrast to the metallic bond, the covalent bond is directed. Thus, the electrons do not spread evenly over a wide region of the crystal, but are concentrated on the connecting line between two atoms. Therefore, it is much more difficult to move atoms in a covalent crystal against each other, resulting in brittleness and poor deformability of these ceramics.

The binding energy of the covalent bond is typically about $1\,\mathrm{eV}$ per bond, but reaches a value of $1.85\,\mathrm{eV}$ in diamond. Due to the smaller number of nearest neighbours, the difference between the overall binding energy of ceramics and metals is smaller – even in diamond the binding energy of an atom is $7.4\,\mathrm{eV}$, only twice that of copper, a metal with a rather high binding energy. In other covalent crystals, typical values are between $3\,\mathrm{eV}$ and $5\,\mathrm{eV}$, again approximately twice that of typical metals.

1.3.2 Ionic bond

Many ceramics are compounds of a metal and a non-metal. Common salt, for instance, consists of sodium and chlorine (NaCl). From this formula and the fact that common salt forms a crystal, it can be deduced that the bond cannot be covalent, for as chlorine has a valency of only one, only a diatomic molecule could form, but not a crystal. Instead, an *ionic bond* is formed.

The ionic bond is based on the high *electron affinity* (also known as *electronegativity*) of the non-metal (the chlorine in the example of common salt), whereas the metal (the sodium in the example) has only a small ionisation energy. If the outer electron of the metal is transferred to the non-metal, only a comparably small amount of energy is needed. Additional energy can be gained because the two resulting ions are electrically charged and attract each other. A diatomic molecule forms, held together by the electrostatic attraction of its ions.

> The binding energy of NaCl can be calculated rather easily (see also exercise 3): The ionisation energy of sodium is 5.1 eV, the electron affinity of chlorine i. e., the energy gained if an electron is added to a chlorine atom, is 3.6 eV. Thus, an energy of 1.5 eV is needed to transfer the electron from the sodium to the chlorine atom. In itself, this is obviously not sufficient to form an attractive bond. The ions formed by the electron transfer are, however, electrically charged and additional energy can be gained if they approach each other. If the ionic distance takes a value of 0.4 nm (a smaller distance is impossible due to the repulsion of the electron shells), this additional energy takes a value of 3.6 eV. In total, a binding energy of 2.1 eV results for a diatomic sodium chloride molecule.

In an ionic crystal, the binding energy is even higher than in a diatomic molecule because each ion is surrounded by several oppositely charged ions. Figure 1.14(b) shows the structure of a sodium chloride crystal which is a cubic crystal with alternating atom types. Each ion has six oppositely charged neighbours. If we look at atoms of each type separately, we see that they occupy the lattice points of a face-centred cubic lattice, with the two lattices being shifted by half a lattice constant. This cubic structure of a sodium chloride crystal can be observed even macroscopically – salt crystals always show rectangularly arranged faces. The resulting binding energy takes similar values to that in covalent crystals, with 3.28 eV per atom for NaCl and 4.33 eV per atom for lithium fluoride (LiF).

Similar to the covalent bond, the ionic bond is directed. Shifting the atoms would strongly increase the electrostatic repulsion of the ions. Therefore, ionic crystals are also brittle.

There is a smooth transition between covalent and ionic bonds. The ionisation energy of metals increases with increasing number of outer electrons, whereas the electron affinity of the non-metals decreases with an increasing

Fig. 1.15. Dipole bond between two carbon dioxide molecules. Differently charged atoms attract each other

valency. In between the purely covalent and the purely ionic bond, there are also intermediate states where the electron is preferentially located at one atom, but can also be found at the other. One example of this is carbon dioxide (CO_2) in which the oxygen atoms have a higher electron affinity than the carbon atom. The electrons therefore have a higher tendency of being close to the oxygen atoms so that these are partially negatively charged, whereas the carbon atom has a partially positive charge. The molecule is electrically polar and can be considered as consisting of two electric dipoles. This kind of bond is called a *polar bond.*

1.3.3 Dipole bond

If carbon dioxide (CO_2) is cooled down to $-78°C$, it forms dry ice, a solid. As the atoms of each CO_2 molecule have fully occupied shells, none of the binding mechanisms discussed so far can be responsible for the cohesion between molecules.

The bond between the carbon dioxide molecules is due to the polarity of the molecules in which electrical charges are distributed inhomogeneously (see figure 1.15). Because the molecules form electric dipoles, this type of bond is called *dipole bond.* As the atoms in the molecules do not carry complete elementary charges, but are charged rather weakly, the attractive force between the molecules is correspondingly small. Typical binding energies lie in the range of $0.2\,eV$–$0.4\,eV$ per bond.

Solids like dry ice are, according to the definition, ceramics, but due to the small binding forces they are not used as engineering materials. However, the dipole bond plays an important role in binding polymers as will be discussed below.

1.3.4 Van der Waals bond

Even completely nonpolar molecules like oxygen or the noble gases finally solidify if cooled down sufficiently. The attraction between such molecules is even smaller than that between molecular dipoles, but it is nevertheless present. This attractive force is called *van der Waals force* or, sometimes, *dispersion force.*

The van der Waals force originates in charge fluctuations in the electron shell of the atoms. Slightly simplified, it can be imagined that the charge distribution of an atom is not static because the outer electrons move about.

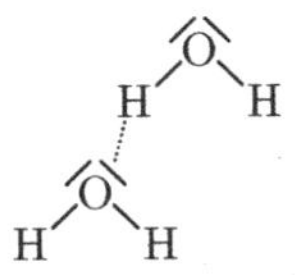

Fig. 1.16. Hydrogen bond

At any instant in time, the atom therefore forms a weak dipole, although on the average it is still electrically neutral. Neighbouring atoms possessing such dipole moments attract each other, for proximity is energetically favourable if the movement of the electrons is correlated.

A van der Waals force acts between all molecules. Because it is the weakest of all bond types, it can only play a role if no other binding mechanism is present. The strength of the van der Waals force is between $0.01\,eV$ and $0.1\,eV$ per bond. In addition, it is very short-ranged and decreases rapidly with growing distance of the molecules.[13] The van der Waals force is stronger in large atoms than in small ones because, due to their larger radius, they can produce larger dipole moments.

1.3.5 Hydrogen bond

Water has very special properties. If we compare the boiling temperature of hydrogen compounds of elements of the sixth group of the periodic table (tellurium, selenium, sulfur, and oxygen), these values are $-2°C$ for H_2Te, $-42°C$ for H_2Se, and $-60°C$ for H_2S. The decrease is due to the decreasing dipole moments with decreasing atomic radius. Therefore, we would expect water to have a very low boiling temperature. Instead, H_2O boils at $+100°C$. The binding force between the water molecules is thus much higher than expected from the comparison with other molecules.

Water is a polar molecule and as oxygen has a slightly higher electron affinity than, for example, sulfur, the larger boiling temperature may at least partially be due to this, but a detailed calculation shows that the dipole bond is far too weak to explain the large boiling temperature.

The special property of water is based on the formation of so called *hydrogen bonds*. As explained above, the hydrogen atoms are partially charged positively. To achieve an optimal electron configuration, the hydrogen atoms can arrange themselves in a way that allows them to enter those orbitals of neighbouring oxygen atoms that are not involved in the covalent bond. Thus, they enable these electrons to spread out over a larger region and in this way

---

[13] Nevertheless, the van der Waals force is strong enough to enable some lizards to walk on smooth, vertical glass panes. A large number of microscopically small and soft lamellae on the feet of these animals are pressed so closely to the ground that the van der Waals force is sufficient to carry the weight of the lizard [12].

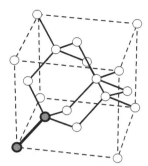

Fig. 1.17. Diamond lattice, constructed as a face-centred cubic lattice with a diatomic basis

to lower their energy. This effect makes the hydrogen bond stronger than a dipole bond. Figure 1.16 shows the formation of hydrogen bonds between different water molecules, with the hydrogen atoms acting as links between the molecules.

This type of bridge linkage can only be formed by hydrogen, for a positively charged hydrogen atom is nothing but a proton. Because of its small size and because it does not have a negatively charged outer shell, the proton can deeply penetrate the orbital of another atom and form a hydrogen bond. Binding energies are typically in the range between 0.1 eV and 0.3 eV.

Hydrogen compounds of the other elements of the sixth group do not form hydrogen bonds because their electron affinity is smaller and because they cannot approach each other as closely due to their larger size.

1.3.6 The crystal structure of ceramics

Frequently, the crystal structure of ceramics is more complex than that of metals. Even an elementary ceramic, like diamond, does not crystallise in the cubic or hexagonal structure typical of metals. Because carbon in diamond is covalently bound with a valency of 4, each carbon atom has four nearest neighbours. A unit cell of the forming three-dimensional network is shown in figure 1.13. As can be seen, the structure of the diamond lattice is cubical, but it is not a Bravais lattice because it does not look the same from each atomic site.

Such types of lattices are called *lattices with a basis*. The diamond lattice can be constructed by placing not one, but two atoms (a *diatomic basis*) on each site of a face-centred cubic lattice (figure 1.17). Another example of a lattice with a basis, the hexagonal close-packed structure, was already discussed in section 1.2.2. It can also be constructed by placing a diatomic basis on each site of a Bravais lattice, in this case a simple hexagonal lattice.[14]

---

[14] Alternatively, it can be visualised as consisting of two lattices stacked into each other.

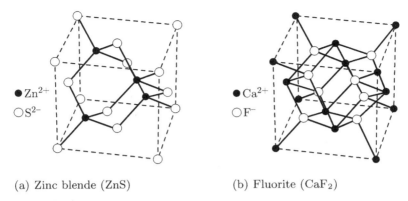

(a) Zinc blende (ZnS)   (b) Fluorite (CaF$_2$)

Fig. 1.18. Unit cells of some ceramics

Crystals comprising different elements always have to be described as lattice with a basis because the atoms are non-identical. Common salt (NaCl), figure 1.14(b), crystallises in a simple cubic structure, where the lattice sites are occupied alternatingly with sodium and chlorine ions, and can also be described as a face-centred cubic lattice with a diatomic basis. Zinc blende (ZnS, figure 1.18(a)) crystallises in a diamond lattice in which the sites are again occupied by the alternating ion types. A similar structure, this time with a three-atomic basis, is found in high cristobalite (SiO$_2$, figure 1.14(a)). Another crystal structure based on the face-centred cubic lattice is found in fluorite (CaF$_2$, figure 1.18(b)). Many even more complex structures are possible according to the stoichiometric ratio of the crystal-forming elements.

Similar to metals, ceramics are usually not single-crystalline but consist of grains. Figure 1.19 shows the microstructure of aluminium oxide as an example.

1.3.7 Amorphous ceramics

Ceramics are frequently not used in a crystalline form, but in an *amorphous* structure. In this case, they are called *glasses*. An amorphous structure is characterised by not possessing a long-range order. Figure 1.20 shows a two-dimensional image of this kind of structure. Although the valencies of each atom are saturated, no ordered structure is formed. The arrangement of the atoms is similar to that in a melt, and glasses can indeed be considered as *undercooled melts*. In many cases, glasses are transparent because there are no grain boundaries to refract light.

Frequently, glasses are based on silicon oxide, SiO$_2$. One common example is window glass, consisting of approximately 70 % SiO$_2$, 15 % Na$_2$O, and 10 % CaO. Another important glassy material is enamel as coating for metals. It has a low melting temperature that is used because of its high impact strength and corrosion resistance.

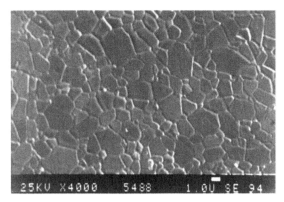

Fig. 1.19. Scanning electron microscope micrograph of the microstructure of aluminium oxide (Al$_2$O$_3$). The horizontal scale bar has a length of 1 μm. Courtesy of CeramTec AG, Plochingen, Germany

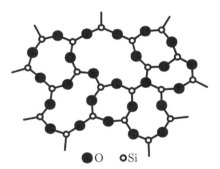

Fig. 1.20. Amorphous structure of a glass (after [9, 19]). Due to the two-dimensional representation, only three bonds per silicon atom are drawn

In principle, metals can also exist in amorphous structure. They are then called *metallic glasses*. Due to the characteristics of the metallic bond, the metal atoms tend to have a larger number of nearest neighbours than covalently bound ceramics, making it more difficult to enforce an amorphous structure. Metallic glasses can thus only be formed if the metal is cooled with extremely high cooling rates of up to 10^5 K/s. Using special alloys, it is nowadays possible to reduce these rates. Metallic glasses simultaneously exhibit high strength and high ductility.

1.4 Polymers

Polymers (plastics) consist of macromolecules, frequently in the form of large molecular chains in which the atoms are held together by covalent bonds, whereas the bonds between the different chains are much weaker. For this reason, chain molecules can be considered as the basic building units of a polymer.

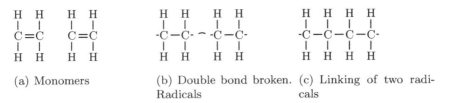

(a) Monomers   (b) Double bond broken.   (c) Linking of two radi-
                           Radicals   cals

Fig. 1.21. Chemical reaction to produce polyethylene (PE)

Contrary to metals and ceramics, polymers are thus composed not of point-like particles (atoms), but of linear components. Therefore, their structure is more complicated than that of the other classes of materials.

1.4.1 The chemical structure of polymers

The individual chain molecules within a polymer are usually organic compounds. These chain molecules consist of numerous identical units, called *monomers*. Typically, the number of monomers in a molecular chain is of the order of 10^3 to 10^5, resulting in an overall molecular length of up to a few micrometres. The average number of monomers in the chain molecules of a polymer is called the *degree of polymerisation.*

All molecules that can link in a chemical reaction to form a chain are suitable monomers.[15] One example for such a reaction is the formation of polyethylene from ethylene. Ethylene consists of two carbon atoms linked by a double bond, with the free valencies of the carbon atoms being saturated by hydrogen. Two ethylene molecules can react by using electrons from the double bond to create a link between the molecules as shown in figure 1.21. The remaining free electrons at the ends are not paired, resulting in an extremely reactive C_4H_8 molecule that can dissociate further double bonds of other molecules. A chain of carbon atoms is formed, in which each atom is linked by a single bond to two other carbon atoms along the chain. The remaining valencies of the carbon atoms are occupied by hydrogen (figure 1.22). To stop the reaction, special chemicals can be added to terminate the reaction by saturating the free electrons of the radicals.

All molecules that can link in such a chain reaction can be used to synthesise polymers. Therefore, there exists a wide spectrum of polymers with strongly varying chemical and physical properties. A selection of technically important polymers will be presented in the next section.

In between the molecular chains, there are no strong chemical bonds. Depending on the molecular structure, the strongly temperature dependent dipole, hydrogen, or van der Waals bonds are formed.

---

[15] Polymers form by two different types of *polymerisation reactions, addition polymerisation* and *condensation polymerisation*. These reactions are explained in *Jastrzebski* [78].

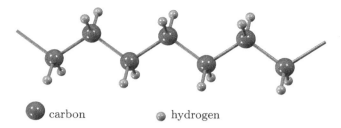

carbon   hydrogen

Fig. 1.22. Spatial structure of polyethylene. The binding angle along the chain has a value of 109°

Examples of polymers

The mechanical properties of polymers are mainly determined by the mobility of the chain molecules and will be discussed in detail in chapter 8. The mobility depends on the chemical structure of the polymer. A polymer with a carbon chain with single bonds, for instance, is flexible at each of the carbon atoms because a single bond between two carbon atoms can rotate freely. Double bonds, on the other hand, are rigid. The mobility is also affected by the presence of side groups. In this section, we will exemplify the structure of some polymers.

The simplest possible monomer that can form a polymer chain is ethylene, as already discussed above. The resulting polymer consists of a chain with a carbon atom backbone. Symbolically, this is written as $[C_2H_4]_n$, with the index 'n' denoting the number of repeat units, the degree of polymerisation. Starting with ethylene as basic unit, a large number of different polymers can be created by replacing one or more of the hydrogen atoms by varying side groups. Examples of this are polyvinyl chloride, where one hydrogen atom is replaced by chlorine, or polystyrene, in which a benzene ring substitutes a hydrogen atom. Table 1.3 and figure 1.23 provide more examples.

It is, of course, not necessary to use a derivative of ethylene as a monomer. Nylon (polyamide) consists of monomers containing an amino group (NCHO); in polydimethylsiloxane the chain itself consists of alternating silicon and oxygen atoms, with two methyl groups being linked to the silicon atoms.

1.4.2 The structure of polymers

While metals and ceramics can be fully crystalline, this is generally not possible for polymers. In principle, the molecular chains can be arranged in parallel and thus create a regular structure, but due to their length, it is highly improbable that the molecules are linear or regularly folded up when cooling the polymer from a liquid state. Statistically, it is much more likely that a chain molecule is highly twisted and entangled with other molecules. Polymers thus always possess an at least partially amorphous structure.

Table 1.3. Survey of some polymers. T_g and T_m are the glass transition temperature and the melting temperature explained in chapter 8, respectively. As these values depend on the degree of polymerisation and the amount of additives in the polymer, they are to be understood as gross estimates. *am* means 'amorphous'. If *am* and a number are given, the polymer can be either amorphous or semi-crystalline (after [13, 44, 98])

name	application example	$T_g/°C$	$T_m/°C$
	thermoplastics		
low-density poly-ethylene, LDPE	foils, electr. insulations	$-110\ldots-20$	$100\ldots110$
high-density poly-ethylene, HDPE	tubes, bottles, household articles	$-100\ldots-20$	$125\ldots135$
polypropylene, PP	tubes, food packages, electr. insulation	$-20\ldots0$	$160\ldots175$
polystyrene, PS	toys, acoustic or thermal insulation, packages	100	$am\,/\,270$
polyvinyl chloride, PVC	tubes, packages, floor coverings, window frames	$70\ldots90$	$am\,/\,212$
polymethylmetha-crylate, PMMA	windows (e. g., in airplanes), lighting technology	100	am
polyamide, PA	gearwheels, ball bearing cages, bearings	$40\ldots150$	$170\ldots300$
polycarbonate, PC	casings, gearwheels, valves, tapes, packages	150	$am\,/\,220\ldots260$
polytetrafluor ethy-lene, PTFE	gaskets, bearings, food industry	126^a	327
polyethylene-terephtalate, PET	glues, connectors, roofings, tanks	80	$am\,/\,240\ldots250$
	elastomers		
polybutadiene	car tyres	$-100\ldots-15$	$-$
	duromers		
polyester	glass-fibre laminates	$-$	$-$
aromatic polyamides (aramid)	fibres for composites	$-$	$-$
polyimide, PI	piston rings, bearings, gaskets, electr. insulation	$-\,/\,310\ldots365^b$	$-\,/\,am$

[a] Literature values for the glass transition temperature of PTFE vary strongly as measuring them is difficult [79]. The value given is taken from [13].

[b] Usually, polyimide is a duromer, but it can also form a thermoplastic. The glass transition temperature is valid for the latter state.

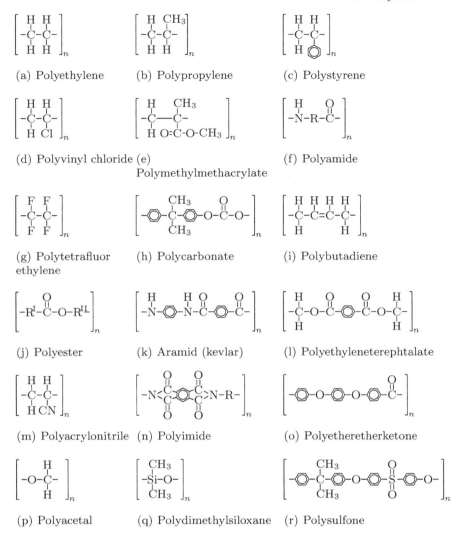

Fig. 1.23. Chemical structure of some polymers. The index 'n' denotes the repeat of the monomer according to the degree of polymerisation. 'R' denotes an arbitrary molecular chain ('Remainder')

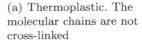

(a) Thermoplastic. The molecular chains are not cross-linked

(b) Elastomer. A few cross-links exist between the chains

(c) Duromer. Many cross-links exist between the chains

Fig. 1.24. Schematic sketch of the cross-linking of different polymers

As explained above, linear chains are the constituting units of polymers. However, it is possible to covalently cross-link the chains, forming a molecular network. These cross-links are crucial in determining the mechanical properties of the polymer because they fix the chains relative to each other and thus render it impossible to draw out single chain molecules. Therefore, a distinction is drawn between *thermoplastics* with no cross-linkage, *elastomers* (or *rubbers*) with a small number of cross-links, and *duromers* (also called *thermosetting polymers, thermosets,* or *resins,* the latter name being due to the fact that they are formed by hardening a resin component) with many cross-links.[16] In figures 1.24(a), 1.24(b), and 1.24(c), the different structures are sketched. The *cross-linking density* can be quantified in the following way: If we consider a diamond crystal as composed of parallel carbon-chain molecules in which each carbon atom is linked to a neighbouring chain, the cross-linking density takes the maximum value possible. To this a value of 1 is assigned. With this definition, elastomers have a cross-linking density, relative to diamond, of 10^{-4} to 10^{-3}, whereas the cross-linking density of duromers is much higher, with values of 10^{-2} to 10^{-1}.

Elastomers and duromers are always completely amorphous because the chemical bonds make a regular arrangement of the chain molecules impossible. Thermoplastics, on the other hand, can be semi-crystalline i. e., contain a mixture of crystalline and amorphous regions. The volume fraction of the crystalline regions in a semi-crystalline thermoplastic is called its *crystallinity.*

In a semi-crystalline thermoplastic, the crystalline regions do not consist of straight chain molecules aligned in parallel, but rather of regularly folded molecules (see figure 1.25). The crystalline regions typically have a thickness of approximately 10 nm and a length between 1 µm and 10 µm. In between

---

[16] In some duromers, the molecular network is formed not by cross-linking the chains but directly from the monomers. Strictly speaking, in this case it is not possible to talk of cross-linked chains.

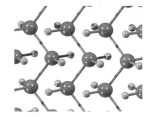

(a) Crystalline region (after [9])

(b) Alignment of polymer chains in the crystalline region

Fig. 1.25. Schematic drawing of the crystalline regions in a polymer

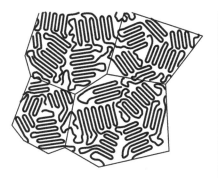

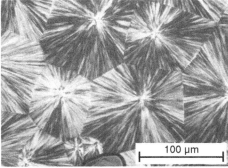

(a) Schematic structure (after [19])

(b) Micrograph. Courtesy of Institut für Baustoffe, Massivbau und Brandschutz, Technische Universität Braunschweig, Germany

Fig. 1.26. Structure of spherulites. The crystalline regions in a spherulite are arranged radially, starting from a centre point, with the folded chain molecules being oriented tangentially. In between the crystalline regions the material is amorphous

them are amorphous regions. The crystalline regions themselves are frequently arranged radially with gaps filled by amorphous material, forming so-called *spherulites* (figure 1.26) that are analogous to the crystallites in a metal. Their extension is about 0.01 mm to 0.1 mm.

2

Elasticity

2.1 Deformation modes

If a material is loaded with a force, the atoms within the material are displaced – the material responds with a deformation. This deformation determines the *mechanical behaviour* of the material. Different types of deformation exist which are not only caused by different physical mechanisms, but are also used in different engineering applications. In particular, we distinguish *reversible* deformations, with the deformation disappearing after unloading, and *irreversible* deformations that preserve the deformation after unloading. Reversible deformations are used in springs and vibrating chords; irreversible deformations are employed to produce components, e. g. by forging, or to absorb energy in crash elements. Generally, reversible deformations are called *elastic,* irreversible deformations are called *plastic.*

Different types of deformation can also be distinguished in another way, for they can be either time-dependent or time-independent. A deformation is time-dependent if the material responds with a delay to changes of the load. If – in contrast – the deformation coincides with the change of the load, the deformation is time-independent. Time-dependent deformations are denoted by the prefix *visco-*. Altogether, four different deformation types exist since elastic as well as plastic deformations can be time-dependent or time-independent.

In this chapter, we will start by discussing how external forces and the resulting material deformations can be described. Subsequently, the *time-independent elastic* behaviour of materials will be described. Often, it is simply called *'the elastic behaviour',* although this is not completely correct.

Time-independent plastic deformation will be described in chapters 3, 6, and 8, the time-dependent plastic behaviour is subject of chapters 8 and 11. Time-dependent elastic behaviour is mainly observed in polymers, described in chapter 8.

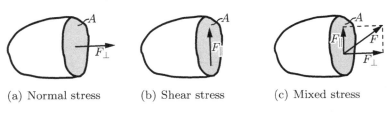

(a) Normal stress   (b) Shear stress   (c) Mixed stress

Fig. 2.1. Different stress measures

2.2 Stress and strain

Components used in engineering have strongly varying dimensions and often also a complicated geometry, resulting in loads that vary strongly throughout the component. To dimension components, characteristic parameters for each material are required that describe its mechanical behaviour. These parameters have to be independent of the geometry and dimension of the components so that they can be determined in experiments using standardised specimens. This can be achieved by normalising the load and the deformation on the dimension (area and length, respectively). To describe the varying conditions within a component, the load and deformation measures are specified for small volume elements. Usually, a continuum mechanical approach is used: The investigated scale is large in comparison to the atomic distance. The matter is considered to be distributed continuously, which results in all variables being continuous.

2.2.1 Stress

Components are usually loaded with certain forces or moments. How strong the material is stressed depends on the area loaded. If the area is increased, the stress decreases. The *stress* σ is thus defined as the force divided by the area the force is acting on. Stresses can be distinguished by the relative orientation of the force and the area. If the force F is perpendicular to the area A, the stress

$$\sigma = \frac{F_\perp}{A} \tag{2.1}$$

is called a *normal stress* (sometimes also *direct stress*, see figure 2.1(a)). If the force is parallel to the area (figure 2.1(b)), the stress is a *shear stress*

$$\tau = \frac{F_\parallel}{A}. \tag{2.2}$$

In all other cases, the force can be decomposed into a normal and a parallel component and normal and shear stresses act simultaneously (figure 2.1(c)).

To describe the loading in a certain point of a material, we imagine it to be cut apart at this point along a *cutting plane*. The stress that was transferred

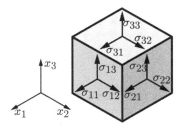

Fig. 2.2. Numbering of the components of the stress tensor $\underline{\sigma}$

through this plane by the material cut away now has to be replaced by an external stress vector, the so-called *surface traction*, to retain the equilibrium of force in the material. The value of the surface traction depends on the orientation of the cutting plane. If, for example, we cut a rod loaded with a uniaxial stress σ along a plane perpendicular to the applied force, the surface traction is a vector in the direction of the force with magnitude σ. If the cutting plane is parallel to the force vector, the surface traction vanishes i. e., we don't need to apply a surface traction vector to preserve the equilibrium. The stress state in three dimensions can be determined by cutting along three cutting planes that are preferentially chosen parallel to the coordinate axes. The nomenclature of the stresses is chosen as follows: The first index denotes the normal vector of the cutting plane considered (figure 2.2), the second index denotes the direction of the stress: $\sigma_{ij} = F_j/A_i$.[1] The shear stress on each of the three cutting planes is decomposed into its two components parallel to the coordinate axes. These 9 components of the stress are collected in a component matrix (σ_{ij}) that forms the *stress tensor* of second order $\underline{\sigma}$.

In a so-called *classical continuum*, an infinitesimal small material element cannot transfer moments.[2] From this, it can be shown that

$$\sigma_{ij} = \sigma_{ji} \quad \text{for } i,j = 1 \dots 3 \tag{2.3}$$

holds i. e., the stress tensor is symmetric [67]. It has only 6 independent components, 3 on the diagonal and 3 off-diagonal ones.

If we change the coordinate system, the components of the stress tensor $\underline{\sigma}$ (its matrix representation) change, but it still describes the same state of stress. The transformation rules are detailed in appendix A.5.

For any stress tensor $\underline{\sigma}$, there is a coordinate system where only the diagonal components of the tensor are non-vanishing, whereas all off-diagonal parts are zero. In this coordinate system, all stresses are thus normal stresses. These stresses are called *principal stresses* of the stress tensor (see appendix A.7); the axes of the coordinate system are called the *principal axes*. Principal stresses are denoted with Roman numerals when they are sorted: $\sigma_{\text{I}} \geq \sigma_{\text{II}} \geq \sigma_{\text{III}}$;

---

[1] For shear stresses, τ_{ij} (with $i \neq j$) is frequently used instead.

[2] This assumption can be relinquished, resulting in the theory of a *Cosserat continuum*. In this case, infinitesimal material elements can transfer moments, resulting in an asymmetric stress tensor.

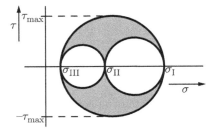

Fig. 2.3. Mohr's circle. Only those stress pairs of the surface traction lying in the grey region can occur.

if they are unsorted, Arabic numerals are used: σ_1, σ_2, σ_3. In its principal coordinate system, the stress tensor is thus simply

$$
\underline{\underline{\sigma}} = \begin{pmatrix} \sigma_1 & 0 & 0 \\ 0 & \sigma_2 & 0 \\ 0 & 0 & \sigma_3 \end{pmatrix} .
$$

In many cases (for example when we consider plastic yielding of materials, see section 3.3.2), it is necessary to calculate the shear stresses that can occur in arbitrarily oriented coordinate systems from the known principal stresses. This can be done geometrically with a construction known as *Mohr's circle* [58, 81], see figure 2.3. We draw a diagram with the normal stresses on the abscissa and the shear stresses on the ordinate. The three principal stresses are marked in the diagram and three circles are drawn, each of them bounded by two of the principal stresses. If we cut the material at the point considered, each cutting plane has a certain surface traction which can be decomposed into a pair of a normal (σ) and a shear (τ) component. If we mark all such pairs of σ-τ values for all possible orientations of the cutting plane in the diagram, they form the shaded area in figure 2.3. For instance, there is a cutting plane of maximum shear stress, with a shear stress value of $\tau_{\mathrm{max}} = (\sigma_{\mathrm{I}} - \sigma_{\mathrm{III}})/2$ and a normal stress given by the average of the largest and smallest principal stress, $(\sigma_{\mathrm{I}} + \sigma_{\mathrm{III}})/2$.

If two principal stresses take the same value, a simple circle without any open area results; if all three principal stresses are identical, the circle degenerates to a point, and the stress state is isotropic.

2.2.2 Strain

If a component is stressed, points within it are displaced. There are different kinds of displacements: The component can be displaced as a whole in a *rigid-body displacement* or it can be rotated rigidly (*rigid-body rotation*). In these cases, distances and angles between points in the material remain unchanged; the component itself is thus still undeformed. To describe the deformation of a component, considering the displacements only is therefore not too helpful. Instead, *changes* of distances and angles between points have to be looked at. This can be done by calculating the change of the displacement with position.

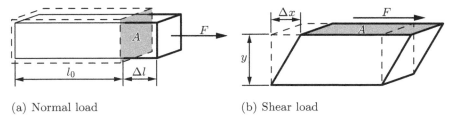

(a) Normal load   (b) Shear load

Fig. 2.4. Simple load cases

All deformations, also called *strains*, can be composed from changes in lengths and angles (shearing of the material). To describe changes in length, the *normal* or *direct strain* ε is defined as the difference Δl between the final length l_1 and the initial length l_0 (figure 2.4(a)):

$$\varepsilon = \frac{l_1 - l_0}{l_0} = \frac{\Delta l}{l_0} . \tag{2.4}$$

Changes in the angles are described by the *shear* γ, corresponding to the change in an initially right angle. For small deformations Δx (see figure 2.4(b)), it is defined as

$$\gamma = \frac{\Delta x}{y} , \tag{2.5}$$

with Δx and y being perpendicular.

An arbitrary deformation with small strains[3] of a material element can be described – analogous to the stress – by a tensor, the *strain tensor* of second order $\underline{\underline{\varepsilon}}$. To calculate the strain tensor, we chose a coordinate system that is fixed in space and consider the displacement of material points in this system as sketched in figure 2.5. This position-dependent displacement is described by a vector field $\underline{u}(\underline{x})$. To understand how the strain is calculated from the displacement, we first consider some special cases.

A pure strain in normal direction, for example in the x_1 direction, causes the displacement u_1 to increase with increasing x_1. If we consider two neighbouring points $x_1^{(1)}$ and $x_1^{(2)}$, with an initial, infinitesimal distance $\Delta x_1 \to 0$, that are displaced by $u_1^{(1)}$ and $u_1^{(2)}$, respectively, the resulting strain is

$$\varepsilon_{11} = \lim_{\Delta x_1 \to 0} \frac{u_1^{(2)} - u_1^{(1)}}{\Delta x_1} = \frac{\partial u_1}{\partial x_1} .$$

Transferring this result to the other spatial directions, we get for the normal strains

$$\varepsilon_{\underline{\underline{ii}}} = \frac{\partial u_i}{\partial x_i} . \tag{2.6}$$

---

[3] Arbitrary deformations with large strains will be discussed in section 3.1.

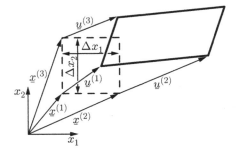

Fig. 2.5. Two-dimensional displacement field in a material. The coordinate system x_i remains fixed in space; the displacements $\underline{u}^{(j)}$ of material elements with the original coordinates $\underline{x}^{(j)}$ refer to the original position

The indices are underscored to denote that the Einstein summation convention is *not* to be used for the repeated index (see appendix A) i. e., they are not summed over.

If the material is sheared, the region considered is distorted and initially right angles are made obtuse or acute. The rotation of the edge parallel to the x_1 axis and of the other edge both contribute to this angular change (cf. figure 2.5). For small rotations and in the limit $\Delta x_1 \to 0$ and $\Delta x_2 \to 0$, the resulting shear is

$$\gamma_{12} = \lim_{\Delta x_1 \to 0} \frac{u_2^{(2)} - u_2^{(1)}}{\Delta x_1} + \lim_{\Delta x_2 \to 0} \frac{u_1^{(3)} - u_1^{(1)}}{\Delta x_2} = \frac{\partial u_2}{\partial x_1} + \frac{\partial u_1}{\partial x_2}.$$

Generalising to all coordinate axes yields

$$\gamma_{ij} = \frac{\partial u_i}{\partial x_j} + \frac{\partial u_j}{\partial x_i} \quad \text{for } i \neq j. \tag{2.7}$$

This definition implies $\gamma_{ji} = \gamma_{ij}$.

Using equations (2.6) and (2.7), all strains can be calculated if they are assumed to be small. However, they cannot be used as components of a tensor, for they do not transform correctly as tensors should. A correct transformation behaviour can be achieved when the shear strain γ_{ij} is replaced by half of its value: $\varepsilon_{ij} = \gamma_{ij}/2$. An additional advantage of this formulation is that equations (2.6) and (2.7) do not have to be written separately for the components, but can be collected in one equation:

$$\varepsilon_{ij} = \frac{1}{2} \left(\frac{\partial u_i}{\partial x_j} + \frac{\partial u_j}{\partial x_i} \right). \tag{2.8}$$

This definition ensures $\varepsilon_{ij} = \varepsilon_{ji}$, rendering the strain tensor symmetric. Similar to the stress tensor, only 6 of its components are independent.

If the material is displaced relative to the coordinate system in a rigid-body translation, the displacement vectors are the same at any material point, $\underline{u}(\underline{x}) = \text{const}$. This yields $\partial u_i / \partial x_j = 0$ and thus $\varepsilon_{ij} = 0$ as should be expected. This result is intuitively obvious, for a rigid-body translation does not cause strains.

A rigid-body rotation is more problematic. For small rotations around the x_3 axis with an angle α, we find $\partial u_1 / \partial x_1 = \cos \alpha - 1 \approx 0$, $\partial u_2 / \partial x_2 = \cos \alpha - 1 \approx 0$, $\partial u_1 / \partial x_2 = -\sin \alpha \approx -\alpha$ and $\partial u_2 / \partial x_1 = \sin \alpha \approx \alpha$. If we insert this into equation (2.8), the mixed terms $\partial u_1 / \partial x_2$ and $\partial u_2 / \partial x_1$ cancel, resulting in $\varepsilon_{ij} = 0$. However, for large rotations, the approximations are not valid and definition (2.8) is not applicable anymore. Suitable definitions of the strain need more involved tensor calculations and will be discussed in more detail in section 3.1.

2.3 Atomic interactions

In the previous chapter, we saw that different material classes have different types of chemical bonds. The atoms in the materials attract each other by different physical mechanisms. If there were only an attractive force between the atoms, their distance would quickly reduce to zero. However, in addition to the attractive interaction of the atoms, there also is a repulsive one. The repulsive interaction is – in a slightly simplified picture – based on the repulsion of the electron orbitals that cannot penetrate each other. The repulsive interaction is short-ranged i. e., it is only relevant if the distances are small, but for very small distances it becomes much larger than the attractive force.

The distance r between neighbouring atoms (e. g., in a solid) takes a value that minimises the potential energy of the total interaction between the atoms. If we superimpose the repulsive potential $U_R(r)$ and the attractive potential $U_A(r)$, the total potential is

$$U(r) = U_A(r) + U_R(r).$$ (2.9)

It is minimised at a stable atomic distance r_0 as sketched in figure 2.6. Usually, atomic distances are between $0.1\,\text{nm}$ and $0.5\,\text{nm}$ [17]. Due to the shape of the potential, the term *potential well* is frequently used to describe it.

The *interaction force* (or *binding force*) $F_i(r)$ between the atoms can be calculated by differentiating the potential:

$$F_i(r) = -\frac{\mathrm{d}U(r)}{\mathrm{d}r}.$$ (2.10)

In equilibrium, $F_i(r_0) = 0$. If an external force is added to the interaction forces, the stable atomic distance changes and the material deforms.

Because the first derivative of the potential (the negative force) vanishes in equilibrium, the potential energy can be approximated by a spring model

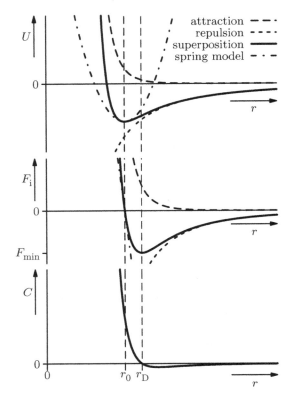

Fig. 2.6. Interaction between two atoms (potential U, binding force $F_i = -\mathrm{d}U/\mathrm{d}r$, stiffness $C = \mathrm{d}^2U/\mathrm{d}r^2$)

(a harmonic law) with a spring stiffness k if we are sufficiently close to the equilibrium position r_0:[4]

$$U(r) \approx U_0 + \frac{1}{2}k(r - r_0)^2 \,,$$
$$F_i(r) \approx -k(r - r_0)\,. \tag{2.11}$$

The external load on a single bond is equal to the negative internal (binding) force:

$$F \approx k(r - r_0)\,. \tag{2.12}$$

Thus, for small displacements, the force is proportional to the displacement.

If the external force is so large that the distance of the atoms attains the value r_D ('D' for 'debonding') shown in figure 2.6 where the restoring force is

---

[4] Mathematically, this is a Taylor series cut off at the second-order term.

maximal, a further increase in the external load cannot be borne by the bond. The bond, and thus the material, breaks.

This is also reflected in the stiffness. It decreases from its initial value k at r_0 to zero at r_D and then becomes negative, rendering the bond unstable. If we use the simplifying assumption of a harmonic law to describe the spring, we assume a constant spring stiffness. This is a valid assumption for small displacements, typical for the elastic deformation of metals and ceramics.

2.4 Hooke's law

For small displacements from the equilibrium position, the force between the atoms is proportional to the displacement (see equation (2.12)). This is true not only for a single bond, but also for larger atomic compounds and thus for macroscopic solids. This *linear-elastic behaviour* is described mathematically by *Hooke's law*. It is valid only for small strains. In metals and ceramics, this is not an important constraint because the elastic part of any deformation is small.

For uniaxial loads (figure 2.4(a)), Hooke's law is

$$\sigma = E\varepsilon \tag{2.13}$$

with *Young's modulus E*, also sometimes called the *elastic modulus*. Young's modulus quantifies the stiffness of a material: the larger Young's modulus is, the smaller is the elastic deformation for a given load.

If a component is strained by a strain ε, strains in perpendicular directions also develop. Usually, a positive strain causes a contraction in the transverse direction, justifying the name *transversal contraction* for this phenomenon. It is measured by *Poisson's ratio ν*, defined as

$$\varepsilon_{\text{trans}} = -\nu\varepsilon \,. \tag{2.14}$$

In many metals, Poisson's ratio is approximately $\nu \approx 0.33$; if the material is incompressible so its volume remains constant, $\nu = 0.5$ holds.

For pure shear (figure 2.4(b)), Hooke's law is

$$\tau = G\gamma \,,$$

where G is the *shear modulus*. Similar to Young's modulus, the shear modulus quantifies the stiffness of the material in shear.

In elastically isotropic materials, the elastic properties are the same in all spatial directions. In this case, the elastic constants are related as follows:

$$G = \frac{E}{2(1+\nu)} \,. \tag{2.15}$$

This equation will be discussed in section 2.4.3.

Table 2.1. Young's modulus of selected materials [8]. For polymers, a more detailed compilation is given in table 8.2

material	E/GPa
metals	$\approx \mathbf{15\ldots500}$
tungsten	411
nickel alloys	$180\ldots234$
ferritic steels	$200\ldots207$
austenitic steels	$190\ldots200$
cast iron	$170\ldots190$
copper alloys	$120\ldots150$
titanium alloys	$80\ldots130$
brasses and bronzes	$103\ldots124$
aluminium alloys	$69\ldots79$
magnesium alloys	$41\ldots45$
ceramics	$\approx \mathbf{40\ldots1000}$
diamond	1000
tungsten carbide, WC	$450\ldots650$
silicon carbide, SiC	450
aluminium oxide, Al_2O_3	390
titanium carbide, TiC	379
magnesium oxide, MgO	250
zirconium monoxide, ZrO	$160\ldots241$
zirconium dioxide, ZrO_2	145
concrete	$45\ldots50$
silicon	107
silica glass, SiO_2	94
window glass	69
polymers	$\approx \mathbf{0.1\ldots5.0}$
polyester	$1.0\ldots5.0$
nylon	$2.0\ldots4.0$
polymethylmethacrylate	$3.0\ldots3.4$
epoxy resins	3.0
polypropylene	0.9
polyethylene	$0.2\ldots0.7$
composites	
carbon-fibre reinforced polymers	$70\ldots200$
glass-fibre reinforced polymers	$7\ldots45$
wood, $\parallel$ to the fibres	$9\ldots16$
wood, $\perp$ to the fibres	0.6

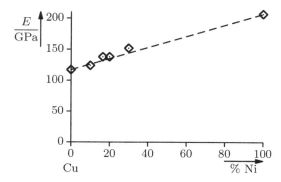

Fig. 2.7. Dependence of Young's modulus on the amount of alloyed nickel in copper [33]

In table 2.1, a survey of the Young's moduli of several engineering materials is given. The elastic stiffness of ceramics slightly exceeds that of metals, but is of the same order of magnitude. Young's modulus of most polymers is much smaller.[5] This should be expected, for the stiffness is determined by the strength of the atomic bonds, which is larger in ceramics than in metals. In polymers, the weaker inter-molecular bonds determine the stiffness. How Young's modulus can be measured will be described in section 3.2.

From table 2.1, it can also be seen that alloying does not significantly change the stiffness of materials. For example, Young's modulus of different aluminium alloys varies only by about 10%, whereas their strength (see chapter 6) can be raised considerably by alloying.

> If two different metals are alloyed, the resulting Young's modulus is not necessarily the weighted average of their two moduli because the binding energy U_{AB} between the atoms A and B is usually not the average of the single-type energies U_{AA} and U_{BB}. Depending on the alloying elements, Young's modulus may even be larger than those of both constituent elements. A rule of thumb is that adding a material with a high melting point (e. g., tungsten to nickel) increases the elastic modulus.
>
> There are a few alloy systems where Young's modulus can be increased considerably. This is the case when both the solubility of the elements and the difference in Young's modulus are large. For example, nickel ($E_{Ni} = 207\,\mathrm{GPa}$) and copper ($E_{Cu} = 121\,\mathrm{GPa}$) are completely soluble, and their Young's moduli differ almost by a factor of two. Therefore, Young's modulus of copper-nickel alloys (nickel bronze) can be strongly increased by raising the nickel content (figure 2.7).
>
> Usually, though, these effects are small because the solubility of alloying elements is usually small ($< 10\%$) in technical alloys. Therefore,

---

[5] Polymer fibres are an exception, see section 8.5.2.

Young's modulus of most engineering alloys differs only by less than $\pm 10\%$ from that of the un-alloyed matrix. In contrast, the strength, a measure of the maximum load the material can bear, can be strongly increased by alloying and may widely exceed the strength of all alloying elements (see section 6.4).

A particularly efficient way of increasing Young's modulus is to use composites, containing, for example, fibres with large stiffness in a matrix of another material. Composites are the subject of chapter 9.

So far, Hooke's law has only been stated for loads that were either normal or shear loads. In real-world applications, components are usually loaded in a multiaxial state where normal and shear stresses are combined. This case will be considered in section 2.4.2. Afterwards, different cases of special symmetries are considered that allow simplifications of Hooke's law. Prior to this, we will discuss the energy stored in elastic deformations.

2.4.1 Elastic strain energy

Any elastic deformation of a material stores energy as can be easily understood by considering the spring model from section 2.3. To calculate this energy, we consider an (infinitesimal) brick-shaped volume element of length l and cross section A to which a load F is applied. The resulting stress is $\sigma = F/A$. If we increase the stress by an amount $\mathrm{d}\sigma$, the external force must increase by $\mathrm{d}F = \mathrm{d}\sigma A$. The material lengthens by an amount $\mathrm{d}l$.

The work done is $\mathrm{d}W = F\mathrm{d}l$.[6] If we insert $\sigma = F/A$ and the definition of strain, $\mathrm{d}\varepsilon = \mathrm{d}l/l$, we find for the work done

$$\mathrm{d}W = F\mathrm{d}l = \sigma A \mathrm{d}\varepsilon\, l = \sigma \mathrm{d}\varepsilon\, V\,, \tag{2.16}$$

where $V = Al$ is the volume of the brick. If we normalise the work to the volume, thus switching to the *energy density* $\mathrm{d}w = \mathrm{d}W/V$, we find $\mathrm{d}w = \sigma \mathrm{d}\varepsilon$.

The total work done per unit volume in a material strained up to $\varepsilon_{\max}$ is the integral over $\mathrm{d}w$:

$$w = \int_0^{\varepsilon_{\max}} \sigma \mathrm{d}\varepsilon\,. \tag{2.17}$$

This equation is valid for arbitrary uniaxial deformations. If the deformation is irreversible, part of the work is transformed to heat and cannot be recovered on unloading. In elastic (reversible) deformations, the energy is stored in

---

[6] Here we use the force at the beginning of the strain increment. As we can neglect second-order terms in this infinitesimal calculation, this does not make a difference: $\mathrm{d}W = (F + \mathrm{d}F)\mathrm{d}l = F\mathrm{d}l + \mathrm{d}F\mathrm{d}l = F\mathrm{d}l$.

the strained atomic bonds and can be recovered.[7] Because the work done is stored as potential energy of the atomic bonds, the name *elastic potential* is frequently used to describe the stored energy (cf. section 2.3).

This calculation was valid for uniaxial stresses and strains only. For arbitrary stresses and strains, we have to generalise by switching to tensors:

$$w = \int_0^{\underline{\underline{\varepsilon}}_{max}} \underline{\underline{\sigma}} \cdot\cdot \, \mathrm{d}\underline{\underline{\varepsilon}} \, . \tag{2.18}$$

The product of the stress and the strain increment in this equation is the so-called *double contraction* explained in appendix A.4.

In a linear-elastic material under uniaxial loads, stress and strain are related by Hooke's law, $\sigma = E\varepsilon$. In this case, the integral in equation (2.17) can easily be solved:

$$w^{(\mathrm{el})} = \int_0^{\varepsilon_{max}} E\varepsilon\mathrm{d}\varepsilon = \frac{1}{2}E\varepsilon_{max}^2 = \frac{1}{2E}\sigma_{max}^2 \, . \tag{2.19}$$

The elastic strain energy increases quadratically with the stress or the strain (see also exercise 6).

* 2.4.2 Elastic deformation under multiaxial loads[8]

We already saw in section 2.2.2 that a load that causes a normal strain in its direction also causes transversal normal strains. For example, a stress in x_1 direction, σ_{11}, causes the following strains, according to equations (2.13) and (2.14): $\varepsilon_{11} = \sigma_{11}/E$, $\varepsilon_{22} = \varepsilon_{33} = -\nu\sigma_{11}/E$. One component of the stress tensor $\underline{\underline{\sigma}}$ thus acts on several components of the strain tensor $\underline{\underline{\varepsilon}}$. Similarly, a prescribed strain in one direction may change the stresses in other directions. If we restrict ourselves to small deformations, the relation between stress and strain is linear. Mathematically, an arbitrary linear relation between two tensors of second order can be described using a double contraction:

$$\sigma_{ij} = C_{ijkl}\,\varepsilon_{kl} \quad \text{or} \quad \underline{\underline{\sigma}} = \underset{4}{\underline{C}} \cdot\cdot \underline{\underline{\varepsilon}} \tag{2.20}$$

The *elasticity tensor* $\underset{4}{\underline{C}}$ is a tensor of fourth order. It can be considered as a four-dimensional 'matrix' with three components in each of its 4 directions. Its $3^4 = 81$ components C_{ijkl} are the material parameters that completely describe the (linear) elastic behaviour.

Because the stress and the strain tensor contain only 6 independent components each, due to their symmetry, the elasticity tensor $\underset{4}{\underline{C}}$ needs only $6^2 = 36$ independent parameters.

---

[7] The storage and dissipation of energy is also discussed in exercise 26.

[8] Sections with a title marked by a * contain advanced information which can be skipped without impairing the understanding of subsequent topics.

That not all 81 components of the elasticity tensor are needed can be most easily understood using an example. For σ_{12}, we find from equation (2.20)

$$
\begin{aligned}
\sigma_{12} = &\; C_{1211}\,\varepsilon_{11} + C_{1212}\,\varepsilon_{12} + C_{1213}\,\varepsilon_{13} \\
&+ C_{1221}\,\varepsilon_{21} + C_{1222}\,\varepsilon_{22} + C_{1223}\,\varepsilon_{23} \\
&+ C_{1231}\,\varepsilon_{31} + C_{1232}\,\varepsilon_{32} + C_{1233}\,\varepsilon_{33}\,.
\end{aligned}
$$

Using the symmetry condition $\varepsilon_{ij} = \varepsilon_{ji}$, we can collect terms as follows:

$$
\begin{aligned}
\sigma_{12} = &\; C_{1211}\,\varepsilon_{11} + C_{1222}\,\varepsilon_{22} + C_{1233}\,\varepsilon_{33} \\
&+ (C_{1212} + C_{1221})\,\varepsilon_{12} \\
&+ (C_{1213} + C_{1231})\,\varepsilon_{13} \\
&+ (C_{1223} + C_{1232})\,\varepsilon_{23}\,.
\end{aligned}
$$

The components C_{ijkl} and C_{ijlk} always appear together and thus represent only one independent parameter. This can be implemented by using the condition $C_{ijkl} = C_{ijlk}$. Thus, the 9 components C_{12kl} reduce to only 6 independent components C_{1211}, C_{1222}, C_{1233}, C_{1212}, C_{1213}, and C_{1223}.

Furthermore, because $\sigma_{12} = \sigma_{21}$, we can also set $C_{ijkl} = C_{jikl}$. The two symmetry conditions $C_{ijkl} = C_{jikl}$ and $C_{ijkl} = C_{ijlk}$ reduce the number of independent components of the elasticity tensor to 36.

The reduced number of components enables us to use a simplified matrix notation (*Voigt notation*), rewriting the tensors of second order as column matrices and the tensor of fourth order as a quadratic matrix: $(\sigma_{ij}) \longrightarrow (\sigma_\alpha)$, $(\varepsilon_{ij}) \longrightarrow (\varepsilon_\alpha)$, and $(C_{ijkl}) \longrightarrow (C_{\alpha\beta})$. The new Greek indices α and β take values from 1 to 6. Writing down the components explicitly, we have

$$
\begin{aligned}
(\sigma_\alpha) &= \left(\begin{array}{cccccc} \sigma_{11} & \sigma_{22} & \sigma_{33} & \sigma_{23} & \sigma_{13} & \sigma_{12} \end{array}\right)^{\mathrm{T}}, \\
(\varepsilon_\alpha) &= \left(\begin{array}{cccccc} \varepsilon_{11} & \varepsilon_{22} & \varepsilon_{33} & \gamma_{23} & \gamma_{13} & \gamma_{12} \end{array}\right)^{\mathrm{T}}
\end{aligned}
$$

with $\gamma_{ij} = 2\varepsilon_{ij}$. The factors of 2 for the mixed components are due to the re-writing of the tensor components.

This can again be understood most easily using an example. The stress component σ_{11} is, according to equation (2.20),

$$
\begin{aligned}
\sigma_{11} = &\; C_{1111}\varepsilon_{11} + C_{1112}\varepsilon_{12} + C_{1113}\varepsilon_{13} \\
&+ C_{1121}\varepsilon_{21} + C_{1122}\varepsilon_{22} + C_{1123}\varepsilon_{23} \\
&+ C_{1131}\varepsilon_{31} + C_{1132}\varepsilon_{32} + C_{1133}\varepsilon_{33}\,.
\end{aligned}
$$

With help of the symmetry conditions $\varepsilon_{21} = \varepsilon_{12}$, $\varepsilon_{31} = \varepsilon_{13}$, $\varepsilon_{32} = \varepsilon_{23}$, $C_{1121} = C_{1112}$, $C_{1131} = C_{1113}$, and $C_{1132} = C_{1123}$, we find

$$\sigma_{11} = C_{1111}\varepsilon_{11} + C_{1122}\varepsilon_{22} + C_{1133}\varepsilon_{33}$$
$$+ 2C_{1123}\varepsilon_{23} + 2C_{1113}\varepsilon_{13} + 2C_{1112}\varepsilon_{12} \,.$$

The sequence of the mixed terms is not universally agreed upon, but a consistent convention has to be used in any calculation.[9]

The elasticity tensor possesses further symmetries due to the existence of an elastic potential [108]. The elasticity matrix $(C_{\alpha\beta})$ is symmetric because of this and the number of independent components reduces further to 21 (6 diagonal and 15 off-diagonal ones).

The elastic potential was already introduced in equation (2.18). Writing it in differential form yields $\mathrm{d}w = \underline{\underline{\sigma}} \cdot\cdot\, \mathrm{d}\underline{\underline{\varepsilon}}$, or, after re-writing,

$$\sigma_{ij} = \frac{\mathrm{d}w}{\mathrm{d}\varepsilon_{ij}} \quad \text{or} \quad \underline{\underline{\sigma}} = \frac{\mathrm{d}w}{\mathrm{d}\underline{\underline{\varepsilon}}} \,.$$

Thus, the stress tensor can be calculated by differentiating the elastic potential with respect to the strains.

Hooke's law, equation (2.20), can also be written in differential form:

$$C_{ijkl} = \frac{\partial \sigma_{ij}}{\partial \varepsilon_{kl}} \quad \text{or} \quad \underset{\widetilde{4}}{C} = \frac{\partial \underline{\underline{\sigma}}}{\partial \underline{\underline{\varepsilon}}} \,.$$

The elasticity tensor is thus the derivative of the stress with respect to the strain.

Inserting the stress from the previous equation, we find

$$C_{ijkl} = \frac{\partial^2 w}{\partial \varepsilon_{ij} \partial \varepsilon_{kl}} \quad \text{or} \quad \underset{\widetilde{4}}{C} = \frac{\partial^2 w}{\partial \underline{\underline{\varepsilon}} \partial \underline{\underline{\varepsilon}}} \,.$$

Because the sequence of taking the derivatives is arbitrary, we find the symmetry condition $C_{ijkl} = C_{klij}$ for the elasticity tensor, or, for the elasticity matrix, $(C_{\alpha\beta}) = (C_{\beta\alpha})$.

Altogether, the three symmetry conditions $C_{ijkl} = C_{jikl} = C_{ijlk} = C_{klij}$ reduce the number of independent components to 21 even in anisotropic materials.

Writing out all components, Hooke's law looks like this:

---

[9] When working with material parameters, the convention in use has to be checked carefully.

$$
\begin{pmatrix} \sigma_{11} \\ \sigma_{22} \\ \sigma_{33} \\ \sigma_{23} \\ \sigma_{13} \\ \sigma_{12} \end{pmatrix} = \begin{pmatrix} C_{11} & C_{12} & C_{13} & C_{14} & C_{15} & C_{16} \\ C_{12} & C_{22} & C_{23} & C_{24} & C_{25} & C_{26} \\ C_{13} & C_{23} & C_{33} & C_{34} & C_{35} & C_{36} \\ C_{14} & C_{24} & C_{34} & C_{44} & C_{45} & C_{46} \\ C_{15} & C_{25} & C_{35} & C_{45} & C_{55} & C_{56} \\ C_{16} & C_{26} & C_{36} & C_{46} & C_{56} & C_{66} \end{pmatrix} \begin{pmatrix} \varepsilon_{11} \\ \varepsilon_{22} \\ \varepsilon_{33} \\ \gamma_{23} \\ \gamma_{13} \\ \gamma_{12} \end{pmatrix}. \tag{2.21}
$$

This notation is easier to handle than the tensor notation. Its disadvantage is that coordinate transformations cannot be performed; in this case, the tensor notation must be used.

The arrangement of atoms in a crystal lattice causes further symmetry conditions that will be discussed in the next sections.

* 2.4.3 Isotropic material

A material is mechanically *isotropic* if all of its mechanical properties are the same in all spatial directions. The elasticity tensor must thus remain unchanged by arbitrary rotations of the material or the coordinate system. Its components must be invariant with respect to rotations.

This invariance property can be used to show that the elasticity matrix has the simple form:

$$
(C_{\alpha\beta}) = \begin{pmatrix} C_{11} & C_{12} & C_{12} & & & \\ C_{12} & C_{11} & C_{12} & & & \\ C_{12} & C_{12} & C_{11} & & & \\ & & & C_{44} & & \\ & & & & C_{44} & \\ & & & & & C_{44} \end{pmatrix} \tag{2.22}
$$

with the additional relation

$$
C_{44} = \frac{C_{11} - C_{12}}{2}. \tag{2.23}
$$

All components not specified vanish, so there are only two independent parameters, C_{11} and C_{12}.

The following relations between these parameters and the more familiar Young's modulus E, Poisson's ratio ν, and shear modulus G hold:

$$
\begin{aligned}
C_{11} &= \frac{E(1-\nu)}{(1+\nu)(1-2\nu)}, \\
C_{12} &= \frac{E\nu}{(1+\nu)(1-2\nu)}, \\
C_{44} &= G = \frac{E}{2(1+\nu)}.
\end{aligned} \tag{2.24}
$$

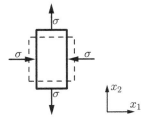
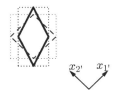

(a) Initial coordinate system   (b) Rotated coordinate system

Fig. 2.8. Example to demonstrate the isotropy equation (2.23): Illustration of the loading and the resulting deformation. Both figures show the same deformation, merely viewed in different coordinate systems

Thus, the σ_{11} component is

$$\sigma_{11} = \frac{E}{(1+\nu)(1-2\nu)} \Big((1-\nu)\varepsilon_{11} + \nu(\varepsilon_{22} + \varepsilon_{33}) \Big) \qquad (2.25a)$$

and σ_{12} is given by

$$\sigma_{12} = G\gamma_{12} . \qquad (2.25b)$$

Apart from E, G, and ν, the so-called *Lamé's elastic constants* λ and μ are sometimes used. Their relation to the other elastic constants is as follows [16, 112]:

$$\lambda = C_{12} = \frac{E\nu}{(1+\nu)(1-2\nu)} ,$$
$$\mu = C_{44} = \frac{E}{2(1+\nu)} .$$

From equation (2.23), we find $C_{11} = \lambda + 2\mu$.

The validity of the condition (2.23) can be illustrated using the following example.[10] A material is deformed in plane strain with the following strain tensor

$$(\varepsilon_{ij}) = \begin{pmatrix} -\varepsilon & 0 & 0 \\ 0 & \varepsilon & 0 \\ 0 & 0 & 0 \end{pmatrix} ,$$

written in the x_i coordinate system (see figure 2.8(a)). Using Hooke's law (2.21) and the elasticity matrix from equation (2.22), we find for the required stress

---

[10] The calculation is further elaborated in exercise 5.

$$(\sigma_{ij}) = \begin{pmatrix} -\varepsilon(C_{11} - C_{12}) & 0 & 0 \\ 0 & \varepsilon(C_{11} - C_{12}) & 0 \\ 0 & 0 & 0 \end{pmatrix}. \tag{2.26}$$

If we consider the same deformation in a coordinate system $x_{i'}$ that is rotated by 45° relative to the x_i system, the coordinate transformation results in the following strain tensor:

$$(\varepsilon_{i'j'}) = \begin{pmatrix} 0 & \varepsilon & 0 \\ \varepsilon & 0 & 0 \\ 0 & 0 & 0 \end{pmatrix}. \tag{2.27}$$

This corresponds to pure shear with $\gamma_{12} = 2\varepsilon$, see figure 2.8(b). If we ignore the isotropy of the elasticity tensor for a moment, we have to assume that its components are different in different coordinate systems. In the primed coordinate system, $\sigma_{\alpha'} = C_{\alpha'\beta'}\,\varepsilon_{\beta'}$ leads to

$$(\sigma_{i'j'}) = \begin{pmatrix} 0 & 2\varepsilon C_{4'4'} & 0 \\ 2\varepsilon C_{4'4'} & 0 & 0 \\ 0 & 0 & 0 \end{pmatrix}. \tag{2.28}$$

The stresses (σ_{ij}) and $(\sigma_{i'j'})$ describe the same state of stress, Therefore, a coordinate transformation must transform (σ_{ij}) to $(\sigma_{i'j'})$:

$$(\sigma_{i'j'}) = \begin{pmatrix} 0 & 2\varepsilon(C_{11} - C_{12}) & 0 \\ 2\varepsilon(C_{11} - C_{12}) & 0 & 0 \\ 0 & 0 & 0 \end{pmatrix}. \tag{2.29}$$

Comparing the components in equation (2.28) and (2.29), we find

$$C_{4'4'} = \frac{C_{11} - C_{12}}{2}. \tag{2.30}$$

Because the material is isotropic, $C_{\alpha'\beta'} = C_{\alpha\beta}$ and, especially, $C_{4'4'} = C_{44}$. Thus, equation (2.30) is the same as (2.23).

Frequently, Hooke's law is not needed to calculate the stress components from a given strain, as in equation (2.20), but to determine the strains from the stresses. We can rearrange equation (2.20) as follows:

$$\varepsilon_{ij} = S_{ijkl}\,\sigma_{kl}. \tag{2.31}$$

$\underset{\sim}{S}$ is the *compliance tensor*, the inverse of the elasticity tensor $\underset{\sim}{C}$.[11] Because inverting a matrix is an awkward calculation, the components of the compliance matrix are written explicitly here:

---

[11] We can also invert the elasticity matrix in the Voigt notation instead: $(S_{\alpha\beta}) = (C_{\alpha\beta})^{-1}$.

$$(S_{\alpha\beta}) = \begin{pmatrix} {}^{1}\!/E & -\,{}^{\nu}\!/E & -\,{}^{\nu}\!/E & & & \\ -\,{}^{\nu}\!/E & {}^{1}\!/E & -\,{}^{\nu}\!/E & & & \\ -\,{}^{\nu}\!/E & -\,{}^{\nu}\!/E & {}^{1}\!/E & & & \\ & & & {}^{1}\!/G & & \\ & & & & {}^{1}\!/G & \\ & & & & & {}^{1}\!/G \end{pmatrix}. \tag{2.32}$$

Again, there is an additional condition, $S_{44} = 2(S_{11} - S_{12})$, from which we can derive equation (2.15), $G = E/2(1 + \nu)$. Inserting equation (2.32) into Hooke's law, we find for the ε_{11} component, for instance,

$$\varepsilon_{11} = \frac{1}{E}\big(\sigma_{11} - \nu(\sigma_{22} + \sigma_{33})\big) \tag{2.33a}$$

and for the γ_{12} component

$$\gamma_{12} = \frac{1}{G}\sigma_{12}. \tag{2.33b}$$

The other components are analogous.

If we take a closer look at the elasticity matrix $(C_{\alpha\beta})$, equation (2.22), and the compliance matrix $(S_{\alpha\beta})$, equation (2.32), we realise the following pattern: Both are of the form

where a ● marks a number and unoccupied spaces mark zero values. The upper right and lower left sub-matrices describe the relation between shear stresses and normal strains and between normal stresses and shear strains. As they are vanishing, there is no coupling between those components. Therefore, in a fixed coordinate system, normal stresses cannot cause shear strains and shear stresses cannot cause normal strains in an isotropic material.

The lower right sub-matrix, relating shear stresses and shear strains, is diagonal. Shear stresses thus can only cause shear strains of the same orientation.

The upper left sub-matrix, which relates normal stresses and normal strains, is fully occupied. Therefore, a normal stress induces not only a strain in the same direction, but also transverse normal strains, the transverse contraction. Similarly, a normal strain causes stresses in transverse directions.

The consequences of these couplings between the different components can be illustrated using an example. We want to calculate the stiffness in x_1 direction of a component for two different cases. In the first case, the component can deform freely in the x_2 and x_3 direction, so the resulting

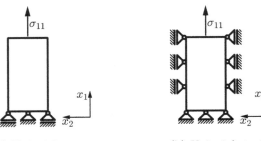

(a) Uniaxial stress   (b) Uniaxial strain

Fig. 2.9. Two different constraints on the same component

stress state is uniaxial and $\sigma_{22} = \sigma_{33} = 0$ (figure 2.9(a)). In the second case, transversal contractions are suppressed, $\varepsilon_{22} = \varepsilon_{33} = 0$, and the state is one of uniaxial strain (figure 2.9(b)).

For uniaxial stresses, it is easiest to calculate the strains with equation (2.31). This yields

$$(\varepsilon_{ij}) = \begin{pmatrix} \sigma_{11}/E & 0 & 0 \\ 0 & -\sigma_{11}\nu/E & 0 \\ 0 & 0 & -\sigma_{11}\nu/E \end{pmatrix}.$$

In x_1 direction, we thus find the uniaxial Hooke's law (2.13), $\sigma_{11} = E\varepsilon_{11}$.

In the case of uniaxial strain, equation (2.20) can be employed, resulting in

$$(\sigma_{ij}) = \begin{pmatrix} C_{11}\varepsilon_{11} & 0 & 0 \\ 0 & C_{12}\varepsilon_{11} & 0 \\ 0 & 0 & C_{12}\varepsilon_{11} \end{pmatrix}.$$

In x_1 direction, we find by using equation (2.24)

$$\sigma_{11} = \frac{E(1-\nu)}{(1+\nu)(1-2\nu)}\varepsilon_{11}.$$

If we assume a Poisson's ratio of $\nu = 1/3$, we get

$$\sigma_{11} = \frac{3}{2}E\varepsilon_{11}.$$

By suppressing transverse contractions, the stiffness of the component increases by 50% compared to the uniaxial stress state. This example also illustrates that the simple relation $\sigma = E\varepsilon$ must not be used inconsiderately, even if only the stresses and strains in one direction are of interest.

∗ 2.4.4 Cubic lattice

In a cubic crystal, the material properties are anisotropic, but there are a number of rotational symmetries. For example, rotations by multiples of 90°

around the $\langle 100 \rangle$ axes[12] do not change the crystal relative to the coordinate system. Further symmetries are rotations by multiples of 120° around the $\langle 111 \rangle$ axes and by multiples of 180° around the $\langle 110 \rangle$ axes. All these rotations must leave the elasticity tensor and the compliance tensor invariant. Using tensor algebra, the elasticity matrix can be shown to have the following form in a coordinate system parallel to the edges of the unit cell:

$$
(S_{\alpha\beta}) = \begin{pmatrix} S_{11} & S_{12} & S_{12} & & & \\ S_{12} & S_{11} & S_{12} & & & \\ S_{12} & S_{12} & S_{11} & & & \\ & & & S_{44} & & \\ & & & & S_{44} & \\ & & & & & S_{44} \end{pmatrix} . \tag{2.34}
$$

Unspecified components vanish. Thus, the three independent constants S_{11}, S_{12}, and S_{44} remain. If the coordinate system is not parallel to the edges of the unit cell, a coordinate transformation of the elasticity tensor has to be used to find the components. In this case, the elasticity matrix takes a shape different from that in equation (2.34).

Because the material properties are direction-dependent in a cubic crystal, they have to be stated together with the corresponding direction. According to the definition, the load direction has to be stated for Young's modulus: E_i. Because the shear stress τ_{ij} and shear strain γ_{ij} have two indices, two indices are needed for the shear modulus G_{ij}. Poisson's ratio relates strains in two directions. Here the second index 'j' denotes the direction of the strain that causes the transversal contraction in the direction marked by the first index 'i': $\varepsilon_{\underline{ii}} = -\nu_{ij}\varepsilon_{\underline{jj}}$.[13] If the coordinate system is aligned with the axes of the unit cell, the directions can be characterised using Miller indices, for example $E_{\langle 100 \rangle}$. The following relations between the components S_{ij} and E, G, and ν hold:

$$
\begin{aligned}
S_{11} &= \frac{1}{E_{\langle 100 \rangle}} , \\
S_{12} &= -\frac{\nu_{\langle 010 \rangle \langle 100 \rangle}}{E_{\langle 100 \rangle}} = -\frac{\nu_{\langle 001 \rangle \langle 100 \rangle}}{E_{\langle 100 \rangle}} , \\
S_{44} &= \frac{1}{G_{\langle 010 \rangle \langle 100 \rangle}} = \frac{1}{G_{\langle 001 \rangle \langle 100 \rangle}} .
\end{aligned} \tag{2.35}
$$

$\langle 100 \rangle$ is the set of all directions that are parallel to the edges of the unit cell. In cubic crystals, it is rather unusual to work with E, G, and ν. Instead, the components S_{11}, S_{12}, and S_{44} of the compliance matrix or C_{11}, C_{12}, and C_{44} of the elasticity matrix are used.

---

[12] Directions and planes in crystals are described using *Miller indices*, explained in appendix B.

[13] As before, underscoring the indices indicates that no summation over this repeated index is done, see appendix A.4.

There is no equation similar to (2.23) in a cubic crystal; S_{11}, S_{12}, and S_{44} (or C_{11}, C_{12}, and C_{44}) are not related. This can be seen from the example from section 2.4.3 on page 47. Up to equation (2.30), $C_{4'4'} = (C_{11} - C_{12})/2$, the calculation remains unchanged. If the material is anisotropic, as in the case of a cubic crystal, $C_{4'4'} \neq C_{44}$, so

$$C_{44} \neq \frac{C_{11} - C_{12}}{2} .$$

It is sufficient to know the elastic constants in one coordinate system (for example, S_{11}, S_{12}, and S_{44}) to calculate the properties in any other coordinate system.

To do this, we have to transform $\underset{\sim}{C}$ or $\underset{\sim}{S}$ to the desired coordinate system. The transformation has to be done using the tensors, not the matrices $\underline{C}$ or $\underline{S}$ in the simplified Voigt notation, because these matrices do not transform correctly.

Young's modulus in arbitrary directions $[hkl]$, for instance, follows the relation

$$\frac{1}{E_{[hkl]}} = S_{11} - [2(S_{11} - S_{12}) - S_{44}](\alpha^2\beta^2 + \alpha^2\gamma^2 + \beta^2\gamma^2) \tag{2.36}$$

with $\alpha = \cos([hkl], [100])$, $\beta = \cos([hkl], [010])$, and $\gamma = \cos([hkl], [001])$.

The *anisotropy factor* A quantifies the difference of the mechanical behaviour relative to an isotropic material. It is defined as

$$A = \frac{2(S_{11} - S_{12})}{S_{44}} . \tag{2.37}$$

If $A = 1$, the material is isotropic, otherwise it is anisotropic.

In the elasticity matrix $(C_{\alpha\beta})$, the same components are occupied as in the compliance matrix $(S_{\alpha\beta})$.[14] Both matrices can be converted using the following equations which are also valid for an isotropic material:

$$C_{11} = \frac{S_{11} + S_{12}}{(S_{11} - S_{12})(S_{11} + 2S_{12})} , \tag{2.38a}$$

$$C_{12} = -\frac{S_{12}}{(S_{11} - S_{12})(S_{11} + 2S_{12})} , \tag{2.38b}$$

$$C_{44} = \frac{1}{S_{44}} \tag{2.38c}$$

and

$$S_{11} = \frac{C_{11} + C_{12}}{(C_{11} - C_{12})(C_{11} + 2C_{12})} , \tag{2.39a}$$

---

[14] As long as the coordinate system is parallel to the edges of the unit cell.

$$S_{12} = -\frac{C_{12}}{(C_{11} - C_{12})(C_{11} + 2C_{12})}, \tag{2.39b}$$

$$S_{44} = \frac{1}{C_{44}}. \tag{2.39c}$$

The considerations concerning the coupling between different stress and strain components from the end of section 2.4.3 apply also to cubic crystals.

*2.4.5 Orthorhombic crystals and orthotropic elasticity

The unit cell of the orthorhombic crystal is brick-shaped. The elastic properties are therefore symmetric with respect to three perpendicular planes. In a coordinate system that is parallel to the edges of the unit cell, the compliance matrix (equation (2.31)) takes the form

$$(S_{\alpha\beta}) = \begin{pmatrix} S_{11} & S_{12} & S_{13} & & & \\ S_{12} & S_{22} & S_{23} & & & \\ S_{13} & S_{23} & S_{33} & & & \\ & & & S_{44} & & \\ & & & & S_{55} & \\ & & & & & S_{66} \end{pmatrix}$$

$$= \begin{pmatrix} 1/E_1 & -\nu_{12}/E_2 & -\nu_{13}/E_3 & & & \\ -\nu_{21}/E_1 & 1/E_2 & -\nu_{23}/E_3 & & & \\ -\nu_{31}/E_1 & -\nu_{32}/E_2 & 1/E_3 & & & \\ & & & 1/G_{23} & & \\ & & & & 1/G_{13} & \\ & & & & & 1/G_{12} \end{pmatrix}. \tag{2.40}$$

Again, the unspecified components vanish. Altogether, there are nine independent elastic constants. It has to be noted that the compliance tensor is symmetric, so some parameters are related, for example $-\nu_{21}/E_1 = -\nu_{12}/E_2$. Nevertheless, it is useful to discriminate between ν_{12} and ν_{21}, for they are defined by transversal contraction.

In a coordinate system parallel to the edges of the unit cell, normal stresses can only cause normal strains, and shear stresses only shear strains. This is not valid anymore if the coordinate system is arbitrarily oriented, so normal strain and shear are coupled.

The orthorhombic crystal lattice itself is not too important technically because there are only a small number of materials crystallising in this structure. Composites (chapter 9), however, frequently have the same symmetry because they may contain aligned fibres. Materials with the same symmetry as an orthorhombic crystal are called *orthotropic*.

Table 2.2. Number of independent elastic constants for different lattice types (cf. table 1.2). Names specifying a symmetry rather than a lattice are printed in italics (e. g., *isotropic*)

lattice type	number of elastic constants
isotropic	2
cubic	3
hexagonal, *transversally isotropic*	5
tetragonal	6
orthorhombic / *orthotropic*	9
monoclinic	13
triclinic	21

* 2.4.6 Transversally isotropic elasticity

In a *transversally isotropic* material, there is a plane in which all properties are isotropic. Perpendicular to this plane, the properties differ. One example for such a material is a hexagonal crystal which is transversally isotropic with respect to its mechanical properties.[15] Other technically important materials may also be transversally isotropic, for example directionally solidified metals in which the grains have a preferential orientation (see also section 2.5), or composites (chapter 9) with fibres oriented in one direction, but aligned arbitrarily (or hexagonally) in the perpendicular plane.

In a coordinate system where the 3 axis is the axis of symmetry, the compliance matrix (equation (2.31)) looks like this:

$$
(S_{\alpha\beta}) = \begin{pmatrix}
S_{11} & S_{12} & S_{13} & & & \\
S_{12} & S_{11} & S_{13} & & & \\
S_{13} & S_{13} & S_{33} & & & \\
& & & S_{44} & & \\
& & & & S_{44} & \\
& & & & & 2(S_{11}-S_{12})
\end{pmatrix}
$$

$$
= \begin{pmatrix}
1/E_1 & -\nu_{21}/E_1 & -\nu_{13}/E_3 & & & \\
-\nu_{21}/E_1 & 1/E_1 & -\nu_{13}/E_3 & & & \\
-\nu_{31}/E_1 & -\nu_{31}/E_1 & 1/E_3 & & & \\
& & & 1/G_{13} & & \\
& & & & 1/G_{13} & \\
& & & & & 2(1+\nu_{21})/E_1
\end{pmatrix}.
$$
$$(2.41)$$

In this case, we have five independent elastic parameters since there is a relation between the ν_{ij} due to the symmetry of the compliance matrix, similar to that for orthotropic materials: $\nu_{21} = \nu_{12}$ and $\nu_{31}/E_1 = \nu_{13}/E_3$.

---

[15] The crystal lattice itself, however, is only symmetric when rotated by multiples of 60°.

Table 2.3. Elastic constants of different single crystals [35,98,105]. $E_{\text{isotr.}}$ is Young's modulus of a nearly isotropic polycrystal

material	$E_{\text{isotr.}}$ GPa	$E_{\langle 100 \rangle}$ GPa	$E_{\langle 111 \rangle}$ GPa	A	C_{11} GPa	C_{12} GPa	C_{44} GPa
cubic materials							
metals and semi-metals							
Al	70	64	76	1.23	108	61	29
Au	78	43	117	1.89	186	157	42
Cu	121	67	192	3.22	168	121	75
α-Fe	209	129	276	2.13	233	124	117
Ni	207	137	305	2.50	247	147	125
Si	–	130	188	1.57	166	64	80
W	411	411	411	1.00	501	198	151
ceramics							
diamond	–	1050	1200	1.20	1076	125	576
MgO	310	247	343	1.54	291	90	155
NaCl	37	44	32	0.72	49	13	13
TiC	–	476	429	0.88	512	110	117

material	$E_{\text{isotr.}}$ GPa	C_{11} GPa	C_{33} GPa	C_{44} GPa	C_{12} GPa	C_{13} GPa
hexagonal materials						
Mg	44	60	62	16	26	22
Ti	112	162	181	47	92	69
Zn	103	164	64	39	36	53

∗ 2.4.7 Other crystal lattices

The number of independent elastic parameters can also be determined for the other crystal lattices and is listed in table 2.2. Generally, couplings between shear stresses and normal strains and normal stresses and shear strains can occur when the number of independent parameters is larger than three. In this case, a uniaxial stress can cause not only normal strains, but also shear strains as we already saw for the example of the orthorhombic crystal.

∗ 2.4.8 Examples

Table 2.3 contains an overview of the elastic constants for some metals and ceramics. As can be seen, the anisotropy factor of tungsten is 1.0, so it is (almost) isotropic even as a single crystal. For most other materials, almost isotropic properties can only be found in a polycrystalline state. The direction dependence of Young's modulus for selected materials is plotted in figure 2.10.

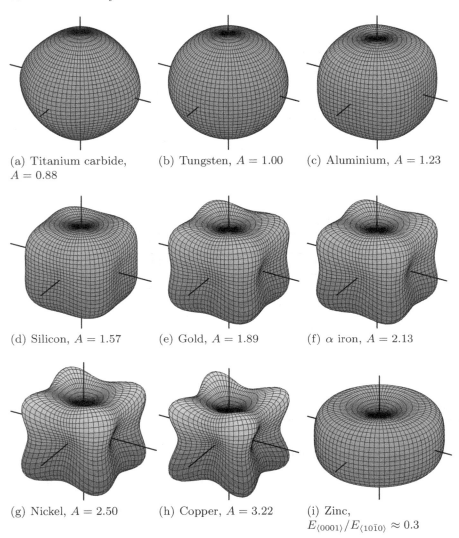

(a) Titanium carbide, $A = 0.88$

(b) Tungsten, $A = 1.00$

(c) Aluminium, $A = 1.23$

(d) Silicon, $A = 1.57$

(e) Gold, $A = 1.89$

(f) α iron, $A = 2.13$

(g) Nickel, $A = 2.50$

(h) Copper, $A = 3.22$

(i) Zinc, $E_{\langle 0001 \rangle}/E_{\langle 10\bar{1}0 \rangle} \approx 0.3$

Fig. 2.10. Orientation dependence of Young's modulus for some materials of table 2.3. In each spatial direction, the distance of the surface from the origin is a measure of Young's modulus

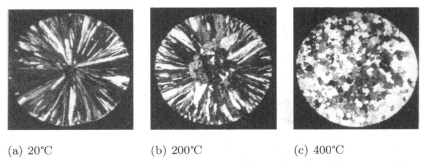

(a) 20°C   (b) 200°C   (c) 400°C

Fig. 2.11. Microstructure of technical pure aluminium for different mould temperatures (cast temperature 900°C). The resulting preferential crystal orientation is the more pronounced, the colder the mould is

* 2.5 Isotropy and anisotropy of macroscopic components

Single crystals are usually mechanically anisotropic as we saw in the preceding sections. In a polycrystalline material, the grains are frequently oriented randomly, and the mechanically anisotropic effects are evened out macroscopically. The material is thus approximately isotropic.

However, there are some cases where a macroscopic component can be anisotropic:

- The component consists of a single crystal. One example are turbine blades used at extreme thermal loads (see also page 58).
- The grains are not small compared to the dimensions of the component itself, so there is insufficient averaging.
- The material is a composite with preferred orientation of the reinforcing phase. Fibre composites are the most important example (see chapter 9).
- During solidification or recrystallisation, a *texture* is formed in the material i.e., the grains have a preferential orientation. This may be due to thermal gradients during solidification of an alloy (see figure 2.11): Solidification starts at the coldest point with the formation of a large number of small nuclei that grow in the direction of the temperature gradient. The speed of crystal growth depends on the crystal orientation, resulting in some grains overtaking the others. The final crystal structure is transversally isotropic. This process can be exploited technically to manufacture directionally solidified materials. One example are turbine blades containing very long grains oriented in the longitudinal direction of the blade (see figure 2.12). Why this is done is explained in the next section.
- The grains have rotated due to large plastic deformations ($> 50\%$), producing a textured material. The reason for this orientation is that crystals

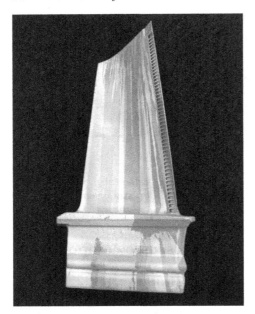

Fig. 2.12. Directionally solidified gas turbine blade. The grains are made visible by etching. Some grains extend over the total length of the blade (385 mm). Courtesy of Siemens AG, Power Generation, Mülheim, Germany

can only deform plastically in certain planes and directions.[16] Deformations of this magnitude are frequently encountered in metal working, for example drawing or rolling.

* How to exploit the elastic anisotropy: Gas turbine blades

Gas turbine blades (figure 2.13(a)) are facing extreme conditions: They have to withstand large mechanical loads due to centrifugal forces at high temperature. To at least partly protect the material from the extreme gas temperatures of 1200°C or more, the blades are cooled from the inside with air of about 500°C. If the wall of the turbine blade has a thickness of about 2 mm and is exposed to the process gas with a temperature of 1200°C, its surface temperature will be about $T_{out} = 1000°C$, whereas on the inside it is only $T_{in} = 600°C$ (figure 2.13(b)). Due to thermal expansion, the material would expand on the outside, but is partly constrained by the cooler inside wall. Thus, large compressive thermal stresses form on the outside and tensile stresses on the inside. In the middle of the wall, there will be a neutral axis at about $T_{m} =$ ·

---

[16] We will discuss this in chapter 6.

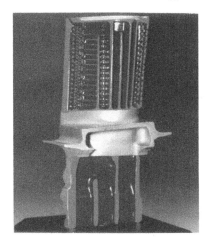

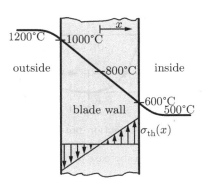

(a) Cut-away view. Courtesy of Lufthansa Technik AG, Hamburg, Germany

(b) Temperature distribution in the wall

Fig. 2.13. Gas turbine blade of a jet engine. Cooling channels inside the blade are used to air-cool the blade

800°C where thermal stresses vanish. The thermal stress σ_{th} at any point x can be calculated approximately by

$$\sigma_{th}(x) = E\,\varepsilon_{th} = E\,\alpha\big(T_m - T(x)\big)\,. \tag{2.42}$$

Here $T(x)$ is the local temperature. The thermal stress is thus proportional to the *coefficient of thermal expansion* α and to Young's modulus E. If we can reduce Young's modulus in the direction of the thermal stresses, the stresses are reduced, thus either increasing the stress tolerance or allowing to raise the temperature and thus the efficiency of the turbine. In this context, it is irrelevant that the elastic deformations due to centrifugal loads increase when E is reduced, for they are small enough not to compromise the component in any case.

If we assume, as an example, a turbine blade made of a polycrystalline, isotropic nickel-base superalloy with Young's modulus $E_{isotr.} = 200\,000\,\text{MPa}$ and a coefficient of thermal expansion of $\alpha = 15 \times 10^{-6}\,\text{K}^{-1}$, we can estimate the stresses at the outside to $\sigma_{th,out} = -600\,\text{MPa}$ and those at the inside to $\sigma_{th,in} = 600\,\text{MPa}$.

Now we manufacture the turbine blade from a single crystal or a directionally solidified material oriented in the $\langle 100 \rangle$ direction with Young's modulus of $E_{\langle 100 \rangle} = 135\,000\,\text{MPa}$. The thermal stresses at the same temperature are now $\sigma_{th,out,\langle 100 \rangle} = -405\,\text{MPa}$, $\sigma_{th,in,\langle 100 \rangle} = 405\,\text{MPa}$. If we assume that the maximum stress the material can bear is 600 MPa, we can raise the surface temperature to almost 1100°C without having to change the material.

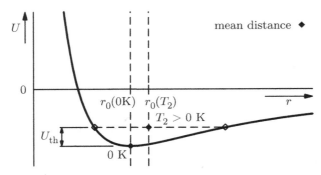

Fig. 2.14. Interaction potential between two atoms. When the temperature is increased, additional thermal energy U_{th} is available. The asymmetry of the potential well causes an increase of the average atomic distance

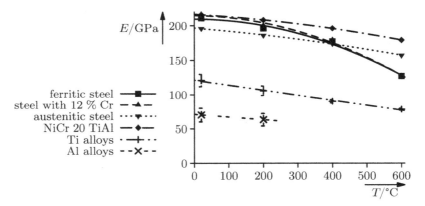

Fig. 2.15. Temperature dependence of Young's modulus for some metals [18]

2.6 Temperature dependence of Young's modulus

In this section, we will discuss the temperature dependence of Young's modulus in metals and ceramics; polymer elasticity will be dealt with in chapter 8.

At typical service temperatures, which are usually smaller than half the melting temperature T_m measured in kelvin ($T < 0.5\,T_m$, $[T] = K$), some rules-of-thumb can be stated for the temperature dependence of Young's modulus. In metals, the temperature dependence of Young's modulus E_M is rather large:

$$E_M(T) \approx E_M(0\,K) \cdot \left(1 - 0.5\frac{T}{T_m}\right). \tag{2.43}$$

Here, $E_M(0\,K)$ is Young's modulus at $0\,K$. Some experimentally determined values are shown in figure 2.15. The temperature dependence of Young's modulus of ceramics is smaller [51]:

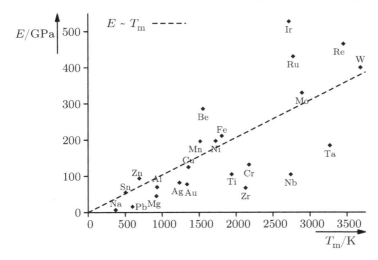

Fig. 2.16. Young's modulus versus melting temperature in some metals [18, 143]

$$E_{\mathrm{K}}(T) \approx E_{\mathrm{K}}(0\,\mathrm{K}) \cdot \left(1 - 0.3\frac{T}{T_{\mathrm{m}}}\right) . \tag{2.44}$$

We can understand the temperature dependence from the properties of the atomic bond discussed in section 2.3. Raising the temperature increases the energy of the atoms by a temperature-dependent amount U_{th}. The atoms start to oscillate around their equilibrium position. The amplitude of the oscillations can be estimated by adding the thermal energy to the energy in the minimum of the potential well as sketched in figure 2.14. Because the repulsive interaction is shorter-ranged than the attractive interaction, the slope is larger on the left side of the well. The mean distance of the atoms thus grows when the temperature is raised. This explains the phenomenon of *thermal expansion*.

Due to thermal expansion, the mean equilibrium position of the atom is at a position in the potential well where the slope of the force curve and thus the stiffness is smaller – Young's modulus is reduced.

This simple model relates thermal expansion and the reduction of the elastic modulus with increasing temperature. It is confirmed by the fact that metals have a larger temperature dependence of Young's modulus than ceramics and also a larger coefficient of thermal expansion.

The reason for this is the larger bond length of the metallic bond. Because it is based on electrons in a widely spread electron gas, the interaction energy does not decrease as strongly with increasing distance as in a covalent bond that involves only two atoms. The range of the ionic bond is also rather small because the electric field is shielded by the neighbouring ions of different charges.

As a rule-of-thumb, we can state that within each class of materials, Young's modulus is roughly proportional to the melting temperature:

$$E \sim T_\mathrm{m} \, . \tag{2.45}$$

This relation can also be explained with the help of figure 2.14. The energy needed to melt the material is roughly proportional to the depth of the potential well because the bonds have to be sufficiently dissolved to allow free movement of the atoms. The deeper the potential well is, the steeper are its sides, for the range of the attractive and repulsive forces are roughly the same for all materials within a certain class. As the second derivative of the energy determines the elastic properties, materials with a larger bond energy have to have a larger elastic modulus. In figure 2.16, the relation between melting temperature and Young's modulus is sketched.

3

Plasticity and failure

In this chapter, we will discuss plasticity phenomenologically, without discussing the particular mechanisms of different materials. These will be treated in chapters 6 to 9.

In contrast to elastic deformations (chapter 2), *plastic deformations* are irreversible. Upon unloading, a plastically deformed material will not return to its original state. In reality, an elastic deformation is superimposed to any plastic deformation so that the elastic part of the deformation will revert, but the plastic part remains. For this reason, one important problem in investigating the deformation of materials is to distinguish between elastic and plastic parts of the strain.

Similar to elastic deformations, plastic deformations can be time-dependent or time-independent. In this book, the term *plasticity* always implies time-independent deformation. Time-dependent plastic deformation will be denoted as *viscoplasticity* or *creep*. This will be discussed in chapter 11 for the case of metals and ceramics, and in chapter 8 for polymers.

Plastic deformation allows to form components or semi-finished parts during manufacturing, with processes like rolling, deep drawing, or forging. During service, plastic deformation is usually to be avoided because the deformations are normally large. Thus, the occurrence of plastic deformations can be used as a failure criterion in designing components. On the other hand, plastic deformations can increase the safety of a component, for they may be detected before the material fails completely, thus leaving room for countermeasures.

Because large strains can occur during plastic deformation, we will start the chapter by discussing the notion of strain if strains become large. The plastic behaviour of materials is usually measured during a *tensile test,* discussed in detail in section 3.2. Next, we will consider the methods of continuum mechanics to describe the limit between elastic and plastic behaviour, plastic deformation, and hardening effects observed in plasticity (section 3.3). Another important material parameter, hardness, will be discussed in section 3.4. Finally, we will discuss different failure mechanisms leading to catastrophic rupture of materials.

3.1 Nominal and true strain

In section 2.2.2, the strain ε was defined as quotient of the change in length Δl and the initial length l_0:

$$\varepsilon = \frac{l_1 - l_0}{l_0} = \frac{\Delta l}{l_0}. \tag{3.1}$$

This is a sensible definition in the case of elastic deformations because the initial, undeformed state (length l_0) is a reference state the material returns to upon unloading.

During plastic deformation, atoms within the material rearrange, and the initial state of the material is not retained. Therefore, it is not helpful to relate all deformations to the initial state. Instead, strains should be calculated relative to the current state of the material. If, for example, a specimen is plastically lengthened and compressed to its original length, it seems to be in the original state macroscopically, but, usually, not all of the atoms have returned to their initial positions. This shows that the current state of the material depends not only on the current strain, but also on the deformation history.

> In metals, for example, the states before and after the plastic deformation can usually be distinguished macroscopically, for the stress needed for further plastic deformation (called the yield strength, see section 3.2) usually increases. To describe the deformation history, a plastic equivalent strain $\varepsilon_{\text{eq}}^{(\text{pl})}$ is defined, which always increases during plastic deformation. This equivalent strain is non-zero after the deformation. It will be discussed in section 3.3.5.

A further reason not to relate plastic deformations to the initial length is that they are usually large. Especially in the case of several deformation steps, this would lead to incorrect results for the strains (see the example on page 65).

If the strain ε is calculated relative to the initial state of the material, it is called *nominal strain* to distinguish it from the so-called *true strain* φ.

To calculate the *true strain* φ, we assume that the strain is applied incrementally in infinitesimally small steps. In each step, the infinitesimal length change $\mathrm{d}l$ is related to the current length l. At each time, the increment in the true strain $\mathrm{d}\varphi$ is, analogous to equation (3.1):

$$\mathrm{d}\varphi = \frac{(l + \mathrm{d}l) - l}{l} = \frac{\mathrm{d}l}{l}. \tag{3.2}$$

For very small (infinitesimal) strains, this is identical to the nominal strain. To calculate the total true strain φ, the increments $\mathrm{d}\varphi$ have to be integrated:

$$\varphi = \int_{l_0}^{l_1} \frac{\mathrm{d}l}{l} = \ln \frac{l_1}{l_0} = \ln \left(1 + \frac{\Delta l}{l_0} \right) = \ln(1 + \varepsilon). \tag{3.3}$$

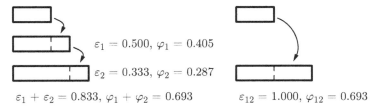

$\varepsilon_1 = 0.500, \varphi_1 = 0.405$

$\varepsilon_2 = 0.333, \varphi_2 = 0.287$

$\varepsilon_1 + \varepsilon_2 = 0.833, \varphi_1 + \varphi_2 = 0.693$

$\varepsilon_{12} = 1.000, \varphi_{12} = 0.693$

(a) Deformation in two steps   (b) Deformation in one step

Fig. 3.1. Comparison of nominal and true strain for a deformation of a tensile specimen in one or two steps, respectively. In total, the length is doubled during the deformation. The nominal strain differs ($\varepsilon_1 + \varepsilon_2 \neq \varepsilon_{12}$) while the true strain is identical ($\varphi_1 + \varphi_2 = \varphi_{12}$) for both deformation sequences

The difference between true and nominal strain can be explained using the example of two identical specimens, deformed in different steps as shown in figure 3.1.

The first specimen is lengthened from its initial length l_0 to a length $l_1 = 1.5\, l_0$ (i.e., $\Delta l = 0.5\, l_0$), corresponding to a nominal strain of

$$\varepsilon_1 = \frac{\Delta l}{l_0} = \frac{0.5\, l_0}{l_0} = 0.5$$

and a true strain of

$$\varphi_1 = \ln \frac{1.5\, l_0}{l_0} = 0.405\,.$$

During further deformation, it is lengthened again by $\Delta l = 0.5\, l_0$ to a total length of $l_2 = 2\, l_0$. This deformation is described by the following strains:

$$\varepsilon_2 = \frac{\Delta l}{1.5\, l_0} = \frac{0.5\, l_0}{1.5\, l_0} = 0.333\,,$$

$$\varphi_2 = \ln \frac{2\, l_0}{1.5\, l_0} = 0.287\,.$$

If we assume that subsequent strains can be added, the following total strains result (see also figure 3.1(a)):

$$\varepsilon_1 + \varepsilon_2 = 0.833\,,$$

$$\varphi_1 + \varphi_2 = 0.693\,.$$

The second specimen is lengthened in a single step from l_0 to $2\, l_0$ (i.e., $\Delta l = l_0$), resulting in the strains (figure 3.1(b)):

$$\varepsilon_{12} = \frac{l_0}{l_0} = 1.000\,,$$

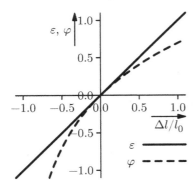

Fig. 3.2. Comparison of nominal and true strain

$$\varphi_{12} = \ln \frac{2\,l_0}{l_0} = 0.693\,.$$

If we compare the calculated strains of both specimens, which have the same length initially and eventually, the true strains are identical ($\varphi = 0.693$), but the nominal strains are not ($0.833 \neq 1.000$). Only by using the true strains can identical strain values be calculated independent of the deformation history. The distinction between true and nominal strain is also discussed in exercise 7.

Mathematical considerations

If we expand the relation between nominal and true strain, equation (3.3), in a Taylor series and cut off after the second term, we find

$$\varphi = \ln(1 + \varepsilon) \approx \varepsilon - \frac{1}{2}\varepsilon^2\,.$$

For small strains $\varepsilon \ll 1$, we can neglect ε^2 and we find $\varphi \approx \varepsilon$. In this case, the nominal strain ε is a good approximation to the true strain φ, see also figure 3.2. In general, the true strain is smaller than the nominal strain in the tensile region (positive strains) and larger in the compressive region (negative strains).

* Multiaxial large deformations

In the form defined in this section, the true strain φ can only be used for normal strains, but it cannot describe large shear deformations. To describe large and arbitrary deformations in more than one dimension, several approaches can be used [16,67,80]. They are all based on a matrix, called *deformation gradient* $\underline{F}$. The deformation gradient is similar to a coordinate transformation from the undeformed to the deformed state.

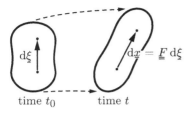

Fig. 3.3. Movement of the connecting line between two material points during a deformation (after [16])

To define the deformation gradient, we define a coordinate system for the undeformed material (at time t_0) that associates each point in the material with a coordinate ξ. Using this coordinate system, the vector between two neighbouring points at time t_0 is given by $d\xi$. This coordinate system remains fixed in space during the deformation i. e., neither its size, orientation, or origin change. Before performing the deformation, we define a second coordinate system, described by $\underline{x}$, that is identical to the ξ system at time t_0. This second system moves and deforms together with the material (see figure 3.3) and thus is, for example, not orthogonal after the deformation. The transformation from one system to the other can be done by a position-dependent matrix, the deformation gradient $\underline{\underline{F}}(\underline{\xi})$. For the vectors connecting two points, $d\underline{\xi}$ and $d\underline{x}$ in the two coordinate systems, we find

$$d\underline{x} = \underline{\underline{F}}(\underline{\xi})\,d\underline{\xi} \qquad \text{or} \qquad dx_i = F_{ij}(\underline{\xi})\,d\xi_j \,.$$

The position-dependence of the deformation gradient is given by

$$F_{ij}(\underline{\xi}) = \frac{\partial x_i(\underline{\xi}, t)}{\partial \xi_j} \,. \tag{3.4}$$

The deformation gradient $\underline{\underline{F}}$ contains not only information about the deformation, but also about rigid-body rotations of the material. These, however, do not contribute to the deformation itself, and the two contributions thus have to be separated. This can be done by considering the deformation gradient as a composition of a deformation $\underline{\underline{U}}$, called the *right stretch tensor* (or, sometimes, material stretch tensor) and a subsequent rotation $\underline{\underline{R}}$. These two are multiplied using the tensor product:

$$\underline{\underline{F}} = \underline{\underline{R}}\,\underline{\underline{U}} \tag{3.5}$$

The right stretch tensor $\underline{\underline{U}}$ can be calculated from $\underline{\underline{F}}$ by

$$\underline{\underline{U}}^2 = \underline{\underline{F}}^{\mathrm{T}}\underline{\underline{F}} \,. \tag{3.6}$$

Although the deformation is described by $\underline{\underline{U}}$, other measures of strain can be useful. One example is provided by Green's strain tensor $\underline{\underline{G}}$, defined as

$$\underline{\underline{G}} = \frac{1}{2}(\underline{\underline{U}}^2 - \underline{\underline{1}}) = \frac{1}{2}(\underline{\underline{F}}^{\mathrm{T}}\underline{\underline{F}} - \underline{\underline{1}}) \,. \tag{3.7}$$

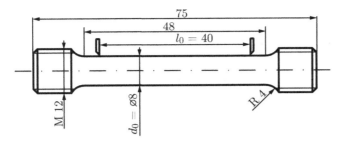

Fig. 3.4. Tensile test specimen (type B) with circular cross section with nominal diameter $d_0 = 8\,\mathrm{mm}$ and original gauge length $L_0 = 40\,\mathrm{mm}$ (designation: Tensile test piece DIN 50 125-B 8×40)

Green's strain tensor vanishes in an undeformed system: $\underline{\underline{G}} = \underline{\underline{0}}$. For small deformations, it converges to the strain tensor $\underline{\underline{\varepsilon}}$, defined in section 2.2.2.

If we write $\underline{\underline{G}}$ element-wise for the displacements $\underline{u}$, we find (the calculation can be found in *Bathe* [15], for example)

$$G_{ij} = \frac{1}{2}\left(\frac{\partial u_i}{\partial \xi_j} + \frac{\partial u_j}{\partial \xi_i}\right) + \frac{1}{2}\frac{\partial u_k}{\partial \xi_i}\frac{\partial u_k}{\partial \xi_j}\,. \tag{3.8}$$

For small deformations, the terms of the form $\partial u_k/\partial \xi_i$ are small. The product in the second term thus becomes small and can be neglected. For small deformations, Green's strain tensor thus converges to $\underline{\underline{\varepsilon}}$.

3.2 Stress-strain diagrams

3.2.1 Types of stress-strain diagrams

The elastic-plastic behaviour of materials is frequently described by *stress-strain curves* measured in tensile tests. Tensile tests are used to determine material parameters that are then listed in tables, for example in ISO standards. These parameters are used in selecting materials and in component design. Even if the shape of the component is complex and the stress state multiaxial, criteria can be used that allow to employ the parameters determined in tensile tests. We will see in section 3.3 how this is done in the case of plasticity.

In a tensile test, the specimen is lengthened at constant speed, and the extension ΔL and the required force are measured. To make results comparable, standardised specimens are used (see figure 3.4 for an example). Most common specimens have a spherical cross section, with a diameter that is constant along the gauge length. To avoid failure at the clampings, the specimen is made thicker there, and the transition between the different diameters is smooth to avoid abrupt changes in cross section because these would cause stress concentrations and might induce localised failure (see chapter 4).

To transform the measured quantities (force F and extension ΔL) to material parameters, the *nominal stress* σ and the nominal strain ε are calculated:

$$\sigma(\Delta L) = \frac{F(\Delta L)}{S_0} \, , \tag{3.9}$$

$$\varepsilon(\Delta L) = \frac{\Delta L}{L_0} \, . \tag{3.10}$$

S_0 is the original cross-sectional area, and L_0 is the initial gauge length. [1] The stress measure σ in equation (3.9) is not the true stress in the material, for it is always calculated based on the original cross-sectional area S_0, whereas the cross section of the specimen changes during the test. This is the reason for the term *nominal stress*. The actual stress in the material in a tensile test is called the *true stress* σ_t.

The shape of the stress-strain curves obtained in tensile tests differs between the different material classes. The characteristic shapes are summarised in figure 3.5.

The stress-strain curves of most metals are of the type shown in figure 3.5(a). The specimen behaviour is almost purely elastic behaviour initially, with the slope of the stress-strain curve being equal to Young's modulus E. With increasing deformation, plastic deformation (also called *yielding*) gradually starts, but it is not possible to determine the exact point when yielding sets in. In most engineering applications, it is assumed that a plastic – and thus permanent – deformation of 0.2 % can be tolerated. Thus, a plastic deformation of 0.2 % is used to define the stress where plastic deformation starts. This stress is called the *yield strength* $R_{p0.2}$ (YS, sometimes σ_Y) of the material.[2]

To determine the plastic strain, the elastic part has to be subtracted from the total strain. As we saw in section 2.3, elastic deformations are due to a change in the bond length between the atoms. Upon unloading, the atoms return to their equilibrium positions, and the elastic strain becomes zero. Since the change in bond length is independent of the rearrangement of the atoms during plastic deformation, the slope of the stress-strain curve upon unloading is given by Young's modulus.[3] However, the rearrangement of the atoms shifts

---

[1] Terms are taken from EN 10 002-1.

[2] According to EN 10 002-1, this stress is called 0.2 % *proof strength*, but in ASTM E 8M it is called *yield strength (offset* 0.2 %*)*. Since we will in many cases not need to distinguish between materials where yielding starts gradually and those where there is an apparent yield point (described below), we will use the term 'yield strength' in the following.

[3] In reality, there is a slight deviation in the slope due to the reduction in the cross-sectional area of the specimen that is caused by plastic deformation (see section 3.2.2).

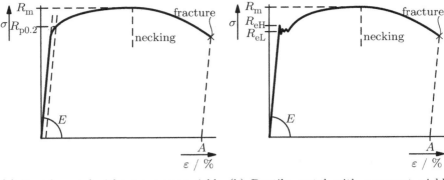

(a) Ductile metal without apparent yield point (cf. EN 10 002-1)

(b) Ductile metal with apparent yield point (cf. EN 10 002-1)

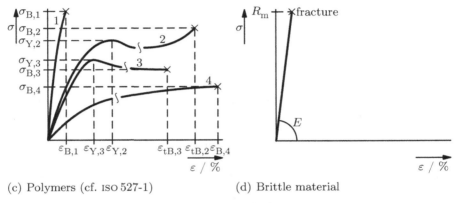

(c) Polymers (cf. ISO 527-1)

(d) Brittle material

Fig. 3.5. Types of stress-strain curves (after [48])

the elastic line to the right because the equilibrium positions are not the same as prior to the deformation. To determine the stress required for a permanent plastic strain of 0.2%, we have to draw a line parallel to the straight part of the loading curve at a distance of 0.2% strain (see figure 3.6). Its intersection with the measured stress-strain curve determines the yield strength.

Upon further elongation of the specimen, the nominal stress σ increases and reaches a maximum value. This value is called the *tensile strength* R_m of the material (sometimes also known as 'ultimate tensile strength', UTS, σ_{UTS}). At this point, the specimen starts to neck i. e., the reduction of its cross section localises in some part of the gauge length where the material is weaker than elsewhere (e. g., because of a tool mark, a slight deviation in the initial cross section, or a cavity). Locally, the stress increases and the cross section of the specimen reduces further in the necking region because plastic deformation is concentrated here. Although the true stress in the necking region still increases, the external force and thus the nominal stress σ decrease, for

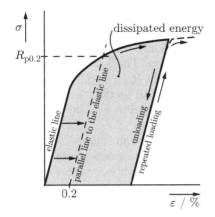

Fig. 3.6. Unloading during a tensile test. The slope of the unloading curve is nearly identical to the initial elastic slope at the beginning of the test. The yield strength $R_{\mathrm{p}0.2}$ is determined by drawing a straight line parallel to the initial elastic line at a distance of $\varepsilon = 0.2\%$ and by identifying the intersection point with the stress-strain curve

the latter is calculated using the original cross-sectional area.[4] To ensure stable deformation of the specimen even after the force decreases, tensile tests are usually strain-controlled. The part of the specimen outside of the necking region is unloaded due to the decrease in the force and does not deform plastically anymore.

The tensile test finally stops when the specimen ruptures. The elongation of the specimen at this point is measured relative to the original gauge length. Thus, the *elongation after fracture A* is defined as $A = \Delta L / L_0$. Frequently, A is characterised by a subscript k, defined as $k = L_0 / \sqrt{S_0}$. For the two common specimen dimensions $L_0 = 5\,d_0$ and $L_0 = 10\,d_0$, k takes the values $k = 5.65$ and $k = 11.3$, and the elongation after fracture is denoted as $A_{5.65}$ and $A_{11.3}$, respectively. The standard EN 10 002-1 permits to use the shorthand A for $A_{5.65}$.

The ability of a material to deform plastically before fracture is called its *ductility*. The larger the elongation after fracture, the more ductile is the tested material. Materials with low ductility are called *brittle*.

The stress level that the material can withstand before failure is called its strength. Whether the yield strength $R_{\mathrm{p}0.2}$ or the tensile strength R_{m} of the material is used as a failure criterion depends on the application in question. In chapters 5, 10, and 11, further failure mechanisms and criteria to measure material strength will be defined.

The plastic behaviour of some metals, especially plain carbon steels, deviates from that described so far. Their stress-strain curve shows a so-called *apparent yield point* (also known as *yield point phenomenon*) (figure 3.5(b)):

---

[4] This will be investigated in more detail in section 3.2.2.

They are almost completely elastic until the *upper yield strength* R_{eH} (UYS) is reached. At this stress, plastic deformation sets in rather suddenly, which is localised in so-called *Lüders bands* or flow lines. While the stress oscillates, these lines extend until they cover the whole specimen. The lowest stress occurring during this process is called *lower yield strength* R_{eL} (LYS).[5] Why this localised plastic deformation occurs, will be explained in section 6.4.3. After the specimen has plastified completely, it behaves identical to a metal without apparent yield point.

Polymers can also exhibit different types of stress-strain curves (see figure 3.5(c)). Some polymers have an initial maximum of the stress-strain curve (curves 2 and 3), (*apparent yield point*) called yield strength σ_{Y}, with a corresponding yield strain ε_{Y}. The stress then drops to a smaller value. Fracture occurs at the *rupture strength* σ_{B} and the rupture strain $\varepsilon_{\mathrm{tB}}$, with varying shapes of the curve in between. The maximum stress in the stress-strain curve is used to define the tensile strength σ_{M}. For curve 2, we thus find $\sigma_{\mathrm{M}} = \sigma_{\mathrm{B}}$, in curve 3, $\sigma_{\mathrm{M}} = \sigma_{\mathrm{Y}}$ holds. The strain at the tensile stress is called ε_{M} if the tensile strength is identical to the yield strength (curve 3), and $\varepsilon_{\mathrm{tM}}$, if the tensile strength is larger than the yield strength (curve 2). If there is no initial maximum of the stress-strain curve (as in curve 1 for a brittle and curve 4 for a ductile polymer), the stress increases monotonically until fracture. The rupture stress is identical to the tensile strength in this case, $\sigma_{\mathrm{B}} = \sigma_{\mathrm{M}}$. The strain at rupture is denoted as $\varepsilon_{\mathrm{B}} = \varepsilon_{\mathrm{M}}$. Contrary to metals, the strains (ε_{Y}, ε_{B}, $\varepsilon_{\mathrm{tM}}$, ε_{M}) are not calculated for the unloaded state, but for the loaded state and thus include elastic strains.[6] The mechanisms causing the different shapes of the curves will be the subject of chapter 8.

In brittle materials, especially in ceramics, there is no or at most very limited plastic deformation. They are elastic until final fracture (figure 3.5(d)). The stress at fracture defines the tensile strength R_{m}.

Energy is required for plastic deformation, which is partially stored in the material, but mostly dissipated as heat. The *specific plastic energy*, the energy required during plastic deformation per unit volume, corresponds to the area beneath the stress-strain curve (shown in grey in figure 3.6). According to equation (2.18), we find

$$w^{(\mathrm{pl})} = \int \underline{\underline{\sigma}} \cdot\cdot \, \mathrm{d}\underline{\underline{\varepsilon}}^{(\mathrm{pl})} = \int \sigma_{ij} \, \mathrm{d}\varepsilon_{ij}^{(\mathrm{pl})}, \tag{3.11}$$

with $\underline{\underline{\varepsilon}}^{(\mathrm{pl})}$ being the plastic strain.

Some exemplary stress-strain curves of different materials are compared in figure 3.7. The initial region is printed separately because otherwise the elastic part of the curves could not be discerned.

---

[5] In determining R_{eL}, the first minimum of the stress is not taken into account because it is usually caused by an overshooting effect within the testing bay.

[6] To simplify the discussions in this book, we will usually use the symbols $R_{\mathrm{p0.2}}$ and R_{m} even for polymers, although the standard prescribes σ_{Y} and σ_{M}.

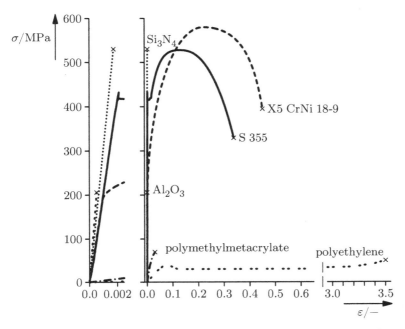

Fig. 3.7. Stress-strain curves of various materials (after [44, 91, 101]); the left plot shows the initial region in more detail. Al_2O_3 and Si_3N_4 are ceramics. S 355 is a plain carbon steel (old designation St 52), X 5 CrNi 18-10 is an austenitic steel. Polymethylmethacrylate (PMMA) and polyethylene (PE) are polymers

3.2.2 Analysis of a stress-strain diagram

In a tensile test, the following material parameters are determined: Young's modulus E, yield strength $R_{p0.2}$ or upper yield strength R_{eH} and lower yield strength R_{eL}, tensile strength R_m, and elongation after fracture A. In this section, we will discuss how they can be determined and interpreted, using the example of a ductile metal.

In tensile tests, force-displacement curves are measured. Using the original cross-sectional area S_0, the nominal stress $\sigma(\Delta L)$ is calculated from the force $F(\Delta L)$, and the nominal strain $\varepsilon(\Delta L)$ is calculated by dividing the extension ΔL by the initial gauge length L_0 (see equations (3.9) and (3.10)).

For small strains, the slope of the stress-strain curve corresponds to *Young's modulus* E. However, it has to be taken into account that the initial part of the curve is not straight and usually has a smaller slope (figure 3.8) because of setting processes in the clampings and other initial effects. In metals without an apparent yield point, yielding starts gradually, rendering it difficult to determine the end of the straight line marking elastic behaviour. For these reasons, it can be rather difficult to exactly determine Young's modulus from the stress-strain curve. To alleviate this problem, the specimens are sometimes

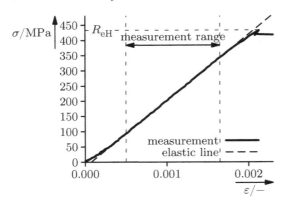

Fig. 3.8. Determination of Young's modulus $E = 217\,797\,\text{MPa}$ in a tensile test for S 355. The parts near the origin and shortly before yield must not be taken into account for calculating Young's modulus (the boundaries used are marked by vertical lines). Data: Courtesy of Institut für Werkstofftechnik, Universität Gh Kassel, Germany

loaded slightly beyond their yield strength and then unloaded. The slope of the unloading curve corresponds more closely to Young's modulus than that of the initial loading curve.

> Young's modulus can also be determined in other ways. One possibility is to measure the resonant frequency of a vibrating beam because this frequency is determined by the sonic speed $c = \sqrt{E/\varrho}$. Two advantages of this method are that very small amplitudes are sufficient and that frequencies can be measured electronically with high precision.

In metals with an apparent yield point, the transition between elastic and plastic behaviour is easy to detect by the corresponding drop in the stress (figure 3.5(b)). Otherwise, the state at which yielding starts cannot be specified precisely because yielding sets on gradually. In engineering, a practical approach is taken, using 0.2% plastic deformation to define the yield strength $R_{\text{p}0.2}$.

During plastic deformation, the specimen's cross-sectional area changes significantly. Therefore, the nominal stress differs from the *true stress* σ_{t} that is defined as the quotient between the external load $F(\Delta l)$ and the current cross-sectional area $S(\Delta L)$:

$$\sigma_{\text{t}}(\Delta L) = \frac{F(\Delta L)}{S(\Delta L)}\,. \tag{3.12}$$

During plastic deformation, the true stress in the material is the stress required to cause further plastic flow in the material. Therefore, it is called the *flow stress* σ_{F}.

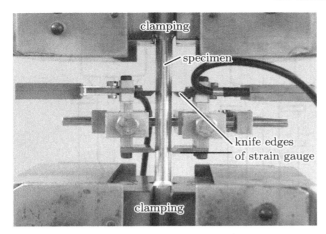

Fig. 3.9. Strain measurement in a tensile test

The current cross-sectional area $S(\Delta L)$ can be calculated approximately if we assume that the volume of the specimen remains unchanged during plastic deformation and that the volume change due to elastic deformation is small.[7] If the volume is constant, we can write

$$S(\Delta L) \cdot L(\Delta L) = S_0 \cdot L_0 = \text{const}.$$

Using equation (2.4), a relation to the nominal strain ε can be found:

$$\frac{S(\Delta L)}{S_0} = \frac{1}{1 + \varepsilon(\Delta L)}. \tag{3.13}$$

This approximation is valid if the cross-sectional area is constant over the whole gauge length L. The true stress

$$\sigma_\mathrm{t} = \sigma\,(1 + \varepsilon) \tag{3.14}$$

is thus larger than the nominal stress σ because of the reduction in area.

The specimen extension ΔL is usually measured using a strain gauge that is fixed to the specimen using two knife edges with a distance L_0 as shown in figure 3.9. All measured strain values are thus averages over the gauge length. As long as the specimen elongates uniformly on its whole length, it does not matter where and over what distance the extension and thus the strain is measured.

As soon as necking begins in the specimen, the results differ strongly depending on the region used for the measurement because plastic deformation is concentrated in the necking region. This is illustrated in figure 3.10. Figure 3.10(a) shows the unloaded specimen with three different strain gauges.

---

[7] At least in the case of metals, these assumptions are correct.

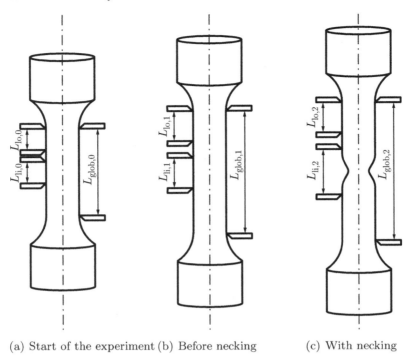

(a) Start of the experiment (b) Before necking   (c) With necking

Fig. 3.10. Sketch of a tensile test with strain gauges with different positions of the knife edges. The subscripts are used as follows: glob for the *global* strain, li for the *local* strain *inside* the necking region, and lo for the *local* strain *outside* the necking region.

As long as no necking occurs, all measured strains are equal e. g., in figure 3.10(b): $\varepsilon_{\text{glob},1} = \varepsilon_{\text{li},1} = \varepsilon_{\text{lo},1} = 0.42$. As soon as necking occurs, the obtained value depends on the position and gauge length e. g., in figure 3.10(c): $\varepsilon_{\text{glob},2} = 0.60$, $\varepsilon_{\text{li},2} = 1.19$, and $\varepsilon_{\text{lo},2} = 0.42$

In figure 3.10(b), the state is prior to necking, and all three strain gauges agree in the measured values. In figure 3.10(c), necking has started and plastic deformation is concentrated in this region. The strain ε_{lo} outside of the necking region has changed only slightly compared to figure 3.10(b). The large strain gauge (strain $\varepsilon_{\text{glob}}$) includes the plastic deformation within the necking region. However, its measurement range also includes a large part of the specimen that is almost unchanged compared to figure 3.10(b). Therefore, $\varepsilon_{\text{glob},2}$ is smaller than $\varepsilon_{\text{li},2}$ because the measurement range of the latter contains mainly the necking region. For this reason, it is necessary to state the ratio of initial gauge length L_0 and the original diameter d_0 or the cross-sectional area S_0 in

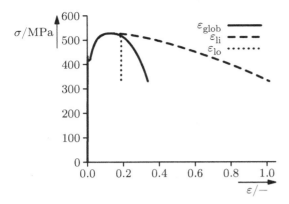

Fig. 3.11. Measured stress-strain curves of the steel S 355. Experimental data: Courtesy of Institut für Werkstofftechnik, Universität Gh Kassel, Germany

values given for the elongation after fracture. For example, $A_{11.3}$ corresponds to a ratio $L_0/\sqrt{S_0} = 11.3$ (see page 71).[8]

Figure 3.11 shows experimental values measured with the different strain gauges. The curve marked $\varepsilon_{\mathrm{glob}}$ is the stress-strain curve measured using the large strain gauge. The curve marked $\varepsilon_{\mathrm{lo}}$,[9] measured outside of the necking region, shows the unloading of the specimen in this region with no further plastic deformation. The material elastically contracts due to the load reduction so that the almost vertical curve in fact has a slope that is approximately equal to Young's modulus. The curve marked $\varepsilon_{\mathrm{li}}$ shows that plastic deformation in the necking region is much larger than the global strain suggests.[10]

In the necking region, the cross-sectional area $S(\Delta L)$ of the specimen decreases strongly, causing a decrease in the external force $F(\Delta L)$ and the nominal stress $\sigma(\Delta L)$ as well. Calculating the stress by dividing the external force by the original cross-sectional area is not meaningful anymore. To calculate the true stress σ_{t} from equation (3.12), the approximation formula (3.13) can be used initially if the measurement range is short and restricted to the necking region, for the assumption of a homogeneous cross section is then still valid. With increasing necking, the approximation becomes increasingly worse.

The true strain was defined in section 3.1 as

---

[8] If the cross section is circular, this implies $L_0/d_0 = 10$.

[9] $\varepsilon_{\mathrm{lo}}$ was defined in figure 3.10.

[10] Modern measurement methods allow to distribute a large number of measurement ranges over the whole specimen and measure simultaneously in all of them. One example of such a method is laser extensometry, where the elongation is measured using a laser.

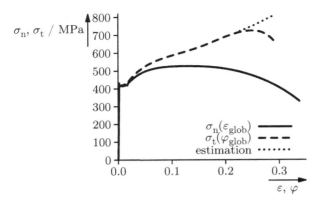

Fig. 3.12. Plots of the nominal and the true stress versus the global strain. The true stress σ_t is calculated using the approximation (3.13) for a short segment in the necking region. This equation is not valid anymore as soon as necking is strong. Thus, the true stress appearing at the thinnest position within the necking region is larger than the calculated value, given as estimation

$$\varphi_{\text{glob}} = \ln(1 + \varepsilon_{\text{glob}}) = \ln \frac{L_{\text{glob}}}{L_0} .$$

Nominal and true stress-strain curve are compared in figure 3.12. The true curve is above the nominal one because the cross section of the specimen decreases. Close to the end of the test, when necking becomes more and more pronounced, the approximated 'true' curve decreases again. Its stress values do not correspond to the real stress at the thinnest part of the specimen. The reason for this is that the approximation formula overestimates the cross-sectional area due to the strong necking and the measurement range which is too large. The stress at the narrowest part of the necking region still increases in most cases, for the flow stress of metals usually increases with increasing deformation. This is called *work hardening*. Sometimes, the flow stress decreases and the material *softens*. This will be discussed at the end of this section.

The stress state within the necking region is triaxial, with a radial stress σ_r and a circumferential stress σ_c in addition to the longitudinal stress σ_l.[11] This stress state, sketched in figure 3.13, varies over the cross section: All stress components are maximal in the centre of the specimen, with the circumferential and the radial stress, σ_c and σ_r, being identical close to the centre and differing slightly near the specimens surface [90]. The difference between the stresses $\sigma_l - \sigma_c$ and $\sigma_l - \sigma_r$ is almost constant throughout the specimen.

---

[11] This must not be confused with the stress state near notches, addressed in chapter 4. If notches are present, the stress state is usually (almost) purely elastic, but here the material yields over the whole cross section. The stress states in the two cases are completely different.

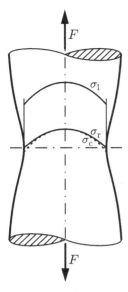

Fig. 3.13. Stress state of triaxial tension in the necking region (after [90]). The stated stress components are the longitudinal stress σ_l, circumferential stress σ_c, and the radial stress σ_r

> The reason for this is that the specimen yields everywhere in the cross section. The so-called yield criterion, defined in section 3.3.1 below, has to be fulfilled. If the Tresca yield criterion is used, the difference between smallest and largest stress has to be identical for the specimen to yield everywhere.

The plastic deformation of the specimen is largest in its centre, contrary to what might be assumed from its appearance (see figure 3.14).

The final fracture of the specimen occurs either by drawing down the material to a line or a point (tip), by a *cup-and-cone fracture,* or by a shear-face fracture [90] (figure 3.15). If the specimen is drawn down to a point, necking proceeds continuously until the cross section is reduced to zero. Very large plastic deformations can occur in this case which is thus restricted to materials with high ductility. In technical alloys, cup-and-cone fracture is predominant due to the formation of microcracks within the specimen.[12] A shear-face fracture can be favoured by different circumstances: Bending, for example due to imprecise clamping of the specimen or due to the specimen geometry (e. g., in wire ropes), can initiate shear-face fracture. Shear-face fracture can also occur if the specimen softens during plastic deformation. This may happen

---

[12] Usually, cup and cone parts can be found on both pieces of the specimen. A complete cup and cone as shown in figure 3.15(b) is rare.

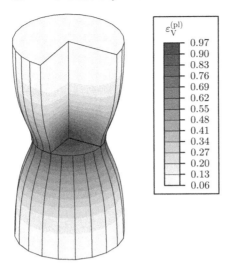

Fig. 3.14. Plastic strain in a necked tensile test specimen. The maximal deformation occurs within the necking region in the centre of the specimen. The used measure for the plastic strain is the *equivalent plastic strain* $\varepsilon_{eq}^{(pl)}$ which provides a quantitative measure for the total plastic deformation (cf. section 3.3.5)

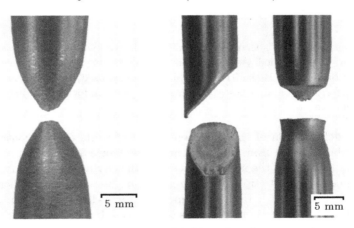

(a) Pure aluminium tensile test specimen drawn to a point

(b) Shear-face fracture and cup-and-cone fracture of AlMg 5 tensile test specimens

Fig. 3.15. Types of fracture for ductile tensile test specimens [90]

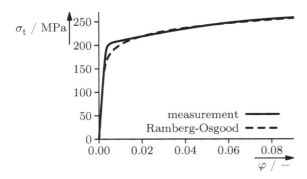

Fig. 3.16. True stress-strain curve of AlMgSi 1 with fitted Ramberg-Osgood equation. The used parameters are $E = 66.2\,\text{GPa}$, $K = 327\,\text{MPa}$, $n = 0.10$

in materials with strong temperature-dependence of the flow stress due to the heat generated in plastic deformation [90]. The mechanisms of specimen failure will be discussed in more detail in section 3.5.

3.2.3 Approximation of the stress-strain curve

To ease the calculation of plastic deformations, stress-strain curves are frequently approximated using simple equations. One commonly used example is the *Ramberg-Osgood law*

$$\varphi = \frac{\sigma_{\mathrm{t}}}{E} + \left(\frac{\sigma_{\mathrm{t}}}{K}\right)^{1/n} . \tag{3.15}$$

K is a parameter describing the absolute stress level and n is called the *strain hardening exponent*. Figure 3.16 shows the stress-strain curve of the aluminium alloy AlMgSi 1 compared with the approximation.

If plastic deformations are large, as it is the case in metal working processes like deep drawing or forging, the elastic part of the deformation can be neglected and the law simplifies to

$$\sigma_{\mathrm{t}} = K\varphi^{n} . \tag{3.16}$$

If $n = 0$, there is no hardening and the flow stress is constant at a value $\sigma_{\mathrm{t}} = K$. This is called perfectly plastic behaviour (see section 3.3.5). With increasing hardening exponent, the hardening increases. As we will see in section 6.4.1, the stress increase due to hardening, $\Delta\sigma_{\mathrm{t}}$, is similar for different materials within the same material class and does not depend strongly on the yield strength. Thus, the relative hardening, $\Delta\sigma_{\mathrm{t}}/\sigma_{\mathrm{t}}$, is larger in low-strength materials. Applying this to equation (3.16), we see that the constant K is small in low-strength materials, but n is large, whereas the situation is reversed in high-strength materials. Common values for n are in the range between 0.1

and 0.45. The constant K is a measure of the strength. It is not equal to the yield or tensile strength, but to the strength extrapolated to a strain of $\varphi = 1$.

Using this approximation, the strain φ_{neck} at which necking of the specimen sets in can be estimated.

> Necking occurs when hardening of the material is not sufficient anymore to compensate for the reduction of the cross-sectional area so that the force transferred by the cross section decreases. At this moment, the external force does not increase further: $dF_{\text{neck}}/d\varphi_{\text{neck}} = 0$. Using the true stress σ_{t}, we can write $F = \sigma_{\text{t}}S$. If we form the differential $dF/d\varphi$, the onset of necking is characterised by[13]
>
> $$\frac{dF_{\text{neck}}}{d\varphi_{\text{neck}}} = \sigma_{\text{t,neck}} \frac{dS_{\text{neck}}}{d\varphi_{\text{neck}}} + S_{\text{neck}} \frac{d\sigma_{\text{t,neck}}}{d\varphi_{\text{neck}}} = 0 \,.$$
>
> This yields
>
> $$\frac{1}{\sigma_{\text{t,neck}}} \frac{d\sigma_{\text{t,neck}}}{d\varphi_{\text{neck}}} = -\frac{1}{S_{\text{neck}}} \frac{dS_{\text{neck}}}{d\varphi_{\text{neck}}} \,. \tag{3.17}$$
>
> If we neglect elastic strains, the volume $V = S\,L$ of the specimen has to remain constant during plastic deformation (and thus also at the onset of necking):
>
> $$\frac{dV_{\text{neck}}}{d\varphi_{\text{neck}}} = S_{\text{neck}} \frac{dL_{\text{neck}}}{d\varphi_{\text{neck}}} + L_{\text{neck}} \frac{dS_{\text{neck}}}{d\varphi_{\text{neck}}} = 0 \,.$$
>
> Rearranging this equation and using the definition of the true strain increment, equation (3.2), we find
>
> $$-\frac{1}{S_{\text{neck}}} \frac{dS_{\text{neck}}}{d\varphi_{\text{neck}}} = \frac{1}{d\varphi_{\text{neck}}} \frac{dL_{\text{neck}}}{L_{\text{neck}}} = \frac{d\varphi_{\text{neck}}}{d\varphi_{\text{neck}}} = 1 \,. \tag{3.18}$$
>
> Inserting equation (3.18) into equation (3.17) yields
>
> $$\frac{d\sigma_{\text{t,neck}}}{d\varphi_{\text{neck}}} = \sigma_{\text{t,neck}} \,. \tag{3.19}$$
>
> If we differentiate equation (3.16), describing the stress-strain curve, with respect to the strain, the result
>
> $$\frac{d\sigma_{\text{t,neck}}}{d\varphi_{\text{neck}}} = n \cdot K \cdot \varphi_{\text{neck}}^{n-1} = n\frac{K\varphi_{\text{neck}}^{n}}{\varphi_{\text{neck}}} = n\frac{\sigma_{\text{t,neck}}}{\varphi_{\text{neck}}}$$
>
> can be inserted in equation (3.19) to arrive at
>
> $$\sigma_{\text{t,neck}} = n\frac{\sigma_{\text{t,neck}}}{\varphi_{\text{neck}}} \,.$$

---

[13] We assume that necking has not yet set in so that the cross section is still the same everywhere.

Table 3.1. Hardening exponents of some materials [18, 88]. This table shows that equation (3.20) is nearly fulfilled by these materials. The material parameters vary strongly depending on the applied heat treatment

material	φ_{neck}	n	$R_{\mathrm{p0.2}}, R_{\mathrm{eH}}/\mathrm{MPa}$	$R_{\mathrm{m}}/\mathrm{MPa}$
S 235 JR (St 37-2)	0.21	0.22	235	430
E 335 (St 60-2)	0.15	0.17	335	650
X 5 CrNi 18-10	0.39	0.38	185	600
AlMg 5	0.19	0.19	80	180
CuZn 36 (brass)	0.40	0.42	180	330

From this calculation, the condition for the onset of necking is

$$\varphi_{\mathrm{neck}} = n \,. \tag{3.20}$$

This equation shows that there is a relation between the maximum strain or maximum deformation of the specimen and the hardening. High-strength materials with a large yield strength, but smaller hardening exponent n than low-strength materials, are thus normally less ductile. Table 3.1 shows the material parameters of some metals.

3.3 Plasticity theory

In the previous sections, we saw that a material deforms elastically if it is loaded with increasing stress, and then it yields gradually or suddenly. During plastic deformation, the flow stress frequently increases because of hardening. All considerations so far were only valid for tensile tests with uniaxial stresses. Because real-world components are seldom subjected to uniaxial stresses only, laws to describe yielding, deformation, and hardening have to be defined for multiaxial stress states as well. This is especially important if components are designed using numerical methods, for example the method of finite elements [15, 63]. Obviously, it is not feasible to perform experiments that cover all possible stress states in a material. Therefore, the idea is to find ways to use parameters determined in tensile tests ($R_{\mathrm{p0.2}}$, for example) in the case of multiaxial loads as well.

All approaches discussed in this section have been derived phenomenologically. They are based on the theory of continuum mechanics and thus do not explicitly include the microscopic mechanisms occurring in the different material classes. Furthermore, isotropic behaviour is assumed because most materials, especially metals, are polycrystalline, and their macroscopic properties are thus averaged over many grains.

Yield criteria, the subject of sections 3.3.1 to 3.3.3, describe the transition between elastic and plastic behaviour for arbitrary stress states. Next, we will study *flow rules* that can be used to calculate how the material deforms.

How the flow stress changes during plastic deformation can be described using *hardening laws,* the topic of section 3.3.5.

3.3.1 Yield criteria

Yield criteria are mathematical tools to decide whether the stress state in a material will cause plastic deformation or not. In a polycrystalline metallic tensile test specimen, which can be assumed to be isotropic and uniaxially loaded, the material yields at a stress[14]

$$\sigma = R_{\mathrm{p}}\,. \tag{3.21}$$

Equation (3.21) is thus the yield criterion for this case.

If the stress state is multiaxial, the stress at yielding cannot be determined as easily. One possibility to do so is to calculate a scalar equivalent stress σ_{eq} from the six components of the stress tensor (see section 2.4.2) and to compare this with a critical stress σ_{crit}. The material yields if the equivalent stress reaches the critical value:

$$\sigma_{\mathrm{eq}}(\sigma_{11}, \sigma_{22}, \sigma_{33}, \sigma_{23}, \sigma_{13}, \sigma_{12}) = \sigma_{\mathrm{crit}} \tag{3.22a}$$

or, in a short-hand notation,

$$\sigma_{\mathrm{eq}}(\sigma_{ij}) = \sigma_{\mathrm{crit}}\,. \tag{3.22b}$$

Usually, yield criteria are stated in the form

$$f(\sigma_{ij}) = 0\,, \tag{3.23}$$

where f is defined by $f(\sigma_{ij}) = \sigma_{\mathrm{eq}}(\sigma_{ij}) - \sigma_{\mathrm{crit}}$. If $f(\sigma_{ij}) < 0$, the material deforms only elastically, if $f(\sigma_{ij}) = 0$, it yields. In the uniaxial case, equation (3.21) becomes

$$f(\sigma) = \sigma - R_{\mathrm{p}} = 0\,.$$

Most commonly used materials are isotropic (see section 2.5). In this case, the orientation of the load does not matter for the yield criterion. Since the principal stresses describe a stress state $\underline{\sigma}$ completely, excepting only its orientation, equation (3.23) can be rewritten using principal stresses if the material is isotropic:[15]

$$f(\sigma_1, \sigma_2, \sigma_3) = 0\,. \tag{3.24}$$

---

[14] R_{p} is the yield strength. Here, we did not specify the plastic strain used to define the yield strength. In metals, $R_{\mathrm{p0.2}}$ is the most common choice (see section 3.2).

[15] If the principal stresses are sorted by their size, a Roman subscript is used; if they are unsorted, the subscript uses Arabic numbers, see section 2.2.1.

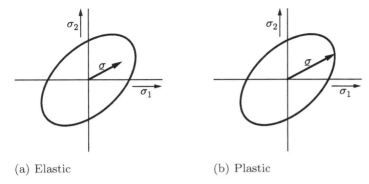

(a) Elastic   (b) Plastic

Fig. 3.17. Yield surface for a state of plane stress. If the stress state reaches the yield surface, yielding occurs

The function $f(\sigma_1, \sigma_2, \sigma_3)$ can be interpreted geometrically if we use a coordinate system with three axes σ_1, σ_2 and σ_3. These axes describe points in the *space of principal stresses* (or *stress space*, for short), not in spatial space. The function f states for each point in the space of principal stresses and thus for each possible stress state whether the material yields or not. Usually, $f < 0$ holds for points close to the origin (for small stresses), and the material deforms elastically. For points far away from the origin, $f > 0$ holds in most cases.[16] The boundary between the two cases, described by $f = 0$, forms a surface in the stress space, called the *yield surface*. For this reason, f is sometimes called the *yield function*. Within the yield surface, the material is elastic; if the stress state reaches the yield surface, the material yields.

> The yield surface can be interpreted geometrically in the form $f(\sigma_{ij})$
> as well. However, the *stress space* is six-dimensional in this case and
> $f(\sigma_{ij}) = 0$ becomes a five-dimensional hypersurface.

If we reduce the number of variable principal stresses to two, for example, by considering a plane stress state, the yield surface becomes a line. As in the general case, yielding occurs if the stress state reaches this line as sketched in figure 3.17. Figure 3.18 illustrates how the yield surface can be constructed from $f(\sigma_1, \sigma_2)$.

It is impossible to create a stress state that is outside of the yield surface; only the cases $f < 0$ and $f = 0$ can occur in reality. This can be explained using the example of a perfectly plastic material with a stress-strain diagram as in figure 3.19(a). It is easy to see that it is not possible to increase the stress beyond R_p. But what happens in a material that hardens? In such materials, stresses larger than R_p are indeed possible (see figure 3.19(b)). This, however, does not imply that the stress state leaves the yield surface. Instead, the yield

---

[16] We will see later that this can, in fact, not happen.

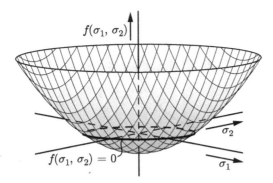

Fig. 3.18. Sketch of the yield function f for two varying principal stresses σ_1 and σ_2. The curve with $f(\sigma_1, \sigma_2) = 0$ is the yield surface for the material and has an elliptical shape. It corresponds to the von Mises yield criterion (cf. figure 3.23(b)), introduced later. For three principal stresses, f is a hypersurface in four-dimensional space which cannot be shown graphically

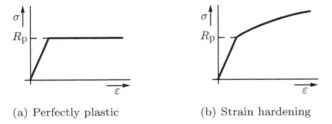

(a) Perfectly plastic   (b) Strain hardening

Fig. 3.19. Stress-strain curves for different hardening behaviours

surface itself changes during loading to ensure that the stress state always remains on it. This will be explained in more detail in section 3.3.5.

The precise form of the yield criteria depends on the material considered. We will start by discussing the most commonly used criteria for metals and afterwards we will discuss modifications pertaining to polymers.

3.3.2 Yield criteria of metals

The plastic deformation of metals is based on so-called slip processes within the grains that shift crystal planes relative to each other.[17] The material shears plastically. Slip can occur simultaneously on different planes, thus allowing for arbitrary plastic deformations. The crucial point is that plastic deformations are caused by shear processes and, therefore, by shear stresses.

---

[17] This section uses some concepts that will be explained in detail later in section 6.2.4, but can be read without reading that section first.

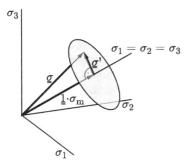

Fig. 3.20. Illustration of deviatoric stresses in the space of principal stresses

As already mentioned in section 3.2.2, a metal does not change its volume during plastic deformation, it is incompressible.[18] This is plausible, for plastic deformation corresponds to a rearrangement of atoms without changes in the interatomic distance (see section 6.2.3).

Experimentally, it has been found that *hydrostatic stresses,* characterised mathematically by $\sigma_1 = \sigma_2 = \sigma_3$, do not cause plastic deformations. Therefore, it can be assumed that the deviation of the stress state from a state of hydrostatic stress determines whether the material yields. In the space of principal stresses, the yield surface thus is a surface (cylindrical or prismatic, for example) that encloses the hydrostatic space diagonal $\sigma_1 = \sigma_2 = \sigma_3$. If we change the position on the hydrostatic axis (the space diagonal), the surface enclosing the axis changes neither its shape nor its size. Mathematically, this can be represented by subtracting the hydrostatic part of the stress[19]

$$\sigma_{\text{hyd}} = \frac{1}{3}\sigma_{ii} = \frac{1}{3}(\sigma_{11} + \sigma_{22} + \sigma_{33}) \tag{3.25}$$

of the diagonal elements of the stress tensor. The result is the so-called *deviatoric stress tensor*

$$\begin{pmatrix} \sigma'_{11} & \sigma'_{12} & \sigma'_{13} \\ \sigma'_{21} & \sigma'_{22} & \sigma'_{23} \\ \sigma'_{31} & \sigma'_{32} & \sigma'_{33} \end{pmatrix} = \begin{pmatrix} \sigma_{11} & \sigma_{12} & \sigma_{13} \\ \sigma_{21} & \sigma_{22} & \sigma_{23} \\ \sigma_{31} & \sigma_{32} & \sigma_{33} \end{pmatrix} - \begin{pmatrix} \sigma_{\text{hyd}} & 0 & 0 \\ 0 & \sigma_{\text{hyd}} & 0 \\ 0 & 0 & \sigma_{\text{hyd}} \end{pmatrix}$$

or, written in index notation,[20]

$$\sigma'_{ij} = \sigma_{ij} - \delta_{ij}\sigma_{\text{hyd}} . \tag{3.26}$$

This is illustrated in figure 3.20. The yield criterion is now defined using the deviatoric stress tensor.

---

[18] Poisson's ratio for purely plastic deformation is $\nu^{(\text{pl})} = 0.5$.

[19] The hydrostatic stress σ_{hyd} corresponds to the negative pressure: $-p = \sigma_{\text{hyd}}$.

[20] δ_{ij} is the *Kronecker delta,* $\delta_{ij} = \begin{cases} 1 & \text{for } i = j, \\ 0 & \text{else} \end{cases}$.

In isotropic materials, the yield surface must not depend on the orientation of the load. Thus, the function f describing the yield surface can only contain those parts of the deviatoric stress tensor that do not change during coordinate transformations. This is already ensured if the principal stresses are used because the hydrostatic stress σ_{hyd} is also coordinate invariant.

There are several ways to define yield criteria according to these rules. The two most important ones are discussed now.

The Tresca yield criterion

The *Tresca yield criterion* or *maximum shear stress criterion* is not directly based on the considerations of the previous sections, but it fulfils them nevertheless. It states that the maximum shear stress in the material point determines yielding. This maximum shear stress can be determined graphically using Mohr's circle, see figure 2.3 on page 34. The maximum principal stress is denoted as σ_{I}, the intermediate value as σ_{II}, and the smallest as σ_{III}. The maximum shear stress is

$$\tau_{\max} = \frac{\sigma_{\mathrm{I}} - \sigma_{\mathrm{III}}}{2}.$$

The intermediate principal stress σ_{II} is not considered.

If $\tau_{\max}$ reaches a critical value τ_{F}, the material yields. Thus, the yield criterion is

$$\frac{\sigma_{\mathrm{I}} - \sigma_{\mathrm{III}}}{2} - \tau_{\mathrm{F}} = 0. \tag{3.27}$$

The value of τ_{F} can be determined by a tensile test which is characterised by the stress state $\sigma_{\mathrm{I}} = R_{\mathrm{p}}$ and $\sigma_{\mathrm{II}} = \sigma_{\mathrm{III}} = 0$. The result is

$$\tau_{\mathrm{F}} = \frac{R_{\mathrm{p}}}{2}.$$

Equation (3.27) can thus be recast in the form

$$\sigma_{\mathrm{I}} - \sigma_{\mathrm{III}} = R_{\mathrm{p}}. \tag{3.28}$$

The term $\sigma_{\mathrm{I}} - \sigma_{\mathrm{III}}$ is the equivalent stress $\sigma_{\mathrm{eq,T}}$. The yield strength can be plotted in Mohr's circle as shown in figure 3.21. As soon as the circle touches this limit, the material yields.

If we plot the yield criterion in the space of principal stresses, as explained at the beginning of section 3.3.1, and consider a state of plane stress ($\sigma_3 = 0$), the yield surface (or rather, yield line) is hexagonal as sketched in figure 3.22(a). Again, Roman subscripts denote the principal stresses when sorted ($\sigma_{\mathrm{I}} \geq \sigma_{\mathrm{II}} \geq \sigma_{\mathrm{III}}$), Arabic subscripts are used if they are unsorted. There are six parts of the yield surface:

In section ①, both σ_1 and σ_2 are positive, and the equations $\sigma_{\mathrm{I}} = \sigma_1$, $\sigma_{\mathrm{II}} = \sigma_2$, and $\sigma_{\mathrm{III}} = 0$ hold. From the yield criterion $R_{\mathrm{p}} = \sigma_{\mathrm{I}} - \sigma_{\mathrm{III}}$, we find

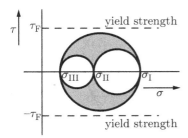

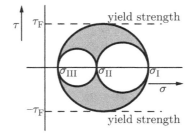

(a) The material does not yield   (b) Yielding occurs

Fig. 3.21. Yield strength in Mohr's circle. If the maximal shear stress $\tau_{\max}$, determined by the principal stresses σ_I and σ_{III}, is smaller than the shear yield strength τ_F, no yielding occurs (a). If, in contrast, $\tau_{\max}$ reaches the shear yield strength, the material yields (b)

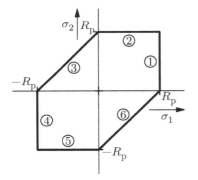

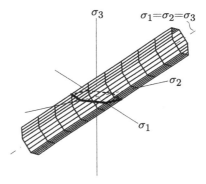

(a) Yield surface for $\sigma_3 = 0$. The vari-   (b) Yield surface for arbitrary stress
ous sections are explained in the text   states in the space of principal stresses

Fig. 3.22. Yield surface for the Tresca yield criterion

$\sigma_1 = R_p$. In section ②, the principal stresses are $\sigma_I = \sigma_2$, $\sigma_{II} = \sigma_1$, and $\sigma_{III} = 0$, resulting in $\sigma_2 = R_p$. In section ③, the signs differ: $\sigma_2 > 0$ and $\sigma_1 < 0$. Thus, the principal stresses are now $\sigma_I = \sigma_2$, $\sigma_{II} = 0$, and $\sigma_{III} = \sigma_1$. Using the yield criterion, we find $R_p = \sigma_2 - \sigma_1$, which is a linear equation $\sigma_2 = \sigma_1 + R_p$. The other sections ④, ⑤, and ⑥ are similar.

If we consider a general stress state in three dimensions, the result is a hexagonal tube centred on the space diagonal $\sigma_1 = \sigma_2 = \sigma_3$. The corresponding yield surface is shown in figure 3.22(b).

Since the Tresca yield criterion can be easily evaluated using Mohr's circle, it is often used in heuristic explanations. Using it in the calculation of plastic deformations, for example, with the method of finite elements, is problematic, though, as we will see on page 96.

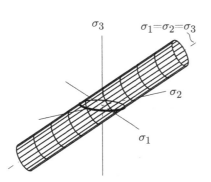
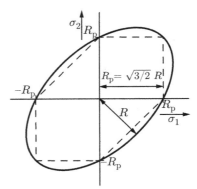

(a) Yield surface for arbitrary stress states in the space of principal stresses

(b) Yield surface for $\sigma_3 = 0$. For comparison, the Tresca yield surface is plotted as dashed line

Fig. 3.23. Yield surface for the von Mises yield criterion

The von Mises yield criterion

The yield surface for the *von Mises yield criterion* (occasionally called *distortional strain energy criterion*) is cylindrical in the space of principal stresses, with its centre coinciding with the hydrostatic axis $\sigma_1 = \sigma_2 = \sigma_3$ and a radius (see figure 3.23(a)).

$$R = \sqrt{2}\, k_{\mathrm{F}}\,.$$

k_{F} is the critical stress for the yield criterion which can be stated as follows, using principal stresses:[21]

$$\sqrt{\frac{1}{6}\left[(\sigma_1 - \sigma_2)^2 + (\sigma_2 - \sigma_3)^2 + (\sigma_1 - \sigma_3)^2\right]} = k_{\mathrm{F}}\,. \tag{3.29}$$

If we are not using principal stresses, the off-diagonal terms of the stress tensor (shear stresses) have to be considered, resulting in

$$\sqrt{\frac{1}{6}\left[(\sigma_{11} - \sigma_{22})^2 + (\sigma_{22} - \sigma_{33})^2 + (\sigma_{11} - \sigma_{33})^2\right] + \sigma_{23}^2 + \sigma_{13}^2 + \sigma_{12}^2} = k_{\mathrm{F}}\,. \tag{3.30}$$

Apart from the principal stresses, a stress tensor has another set of invariants, called the *principal invariants* J_1, J_2, and J_3 (see appendix A.7). They are defined as

---

[21] In equations (3.29) and (3.30), we use the components of the stress tensor itself, not of the deviatoric stress tensor. This does not matter because only differences of the diagonal elements occur, for which $\sigma'_{ii} - \sigma'_{jj} = (\sigma_{ii} - \sigma_{\mathrm{hyd}}) - (\sigma_{jj} - \sigma_{\mathrm{hyd}}) = \sigma_{ii} - \sigma_{jj}$ holds.

$$J_1 = \sigma_{ii}\,,$$

$$J_2 = \frac{\sigma_{ij}\sigma_{ji} - \sigma_{ii}\sigma_{jj}}{2}\,,$$

$$J_3 = \det(\sigma_{ij})\,.$$

According to equation (3.25), the hydrostatic stress is $\sigma_{\mathrm{hyd}} = J_1/3$. This yields the invariance of the hydrostatic stress with respect to coordinate transformations, which was already used above.

The von Mises yield criterion uses the principal invariants of the deviatoric stress tensor, J_1', J_2', and J_3'. Using equation (3.26), we find

$$J_1' = 0\,,$$

$$J_2' = \frac{1}{2}\left(\sigma_{ij}'\sigma_{ji}'\right) \tag{3.31}$$

$$= \frac{1}{6}\left[(\sigma_{11} - \sigma_{22})^2 + (\sigma_{22} - \sigma_{33})^2 + (\sigma_{11} - \sigma_{33})^2\right] + \sigma_{23}^2 + \sigma_{13}^2 + \sigma_{12}^2\,,$$

$$J_3' = \det(\sigma_{ij}')\,.$$

If we define the yield surface with these invariants, any yield criterion fulfils $f(J_2', J_3') = 0$.

The von Mises yield criterion, equation (3.29), results if we assume that the yield criterion depends only on the second invariant J_2' from equation (3.31):

$$f(J_2') = \frac{1}{2}\sigma_{ij}'\sigma_{ji}' - k_{\mathrm{F}}^2 = 0\,. \tag{3.32}$$

J_2' measures the distance from the hydrostatic axis in the principal stress space.

Using equation (3.29) for a uniaxial tensile test, we find, analogous to the derivation of equation (3.28),

$$k_{\mathrm{F}} = \frac{R_{\mathrm{p}}}{\sqrt{3}}\,.$$

The yield criterion can thus be rewritten as

$$\sqrt{\frac{1}{2}\left[(\sigma_1 - \sigma_2)^2 + (\sigma_1 - \sigma_3)^2 + (\sigma_2 - \sigma_3)^2\right]} = R_{\mathrm{p}} \tag{3.33}$$

or

$$\sqrt{\frac{1}{2}\left[(\sigma_{11} - \sigma_{22})^2 + (\sigma_{22} - \sigma_{33})^2 + (\sigma_{11} - \sigma_{33})^2\right] + 3\left(\sigma_{23}^2 + \sigma_{13}^2 + \sigma_{12}^2\right)} = R_{\mathrm{p}}\,. \tag{3.34}$$

The left-hand side in equations (3.33) and (3.34) is called the *equivalent stress* $\sigma_{\mathrm{eq,M}}$. Figure 3.23(b) shows the von Mises yield criterion for a state

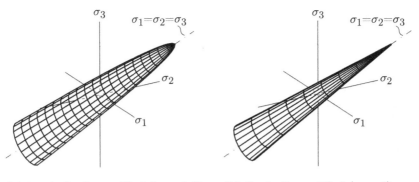

(a) Parabolically modified ($m = 2.3$)   (b) Conically modified ($m = 3$)

Fig. 3.24. Yield criteria for polymers

of plane stress compared to the Tresca yield criterion for the case that the yield strength was determined in a tensile test. The difference amounts to 15.5 % at most.

The two yield criteria can also be normalised to agree for a shear experiment with $\sigma_2 = -\sigma_1$. In this case, the von Mises ellipse is completely enclosed by the Tresca hexagon and touches it at the six linear sections.

It is not possible to prove the validity of these yield criteria theoretically. This is obvious if we remember that they are continuum-mechanical approximations of a discontinuous reality. Experiments confirm that both – especially the von Mises yield criterion – satisfactorily describe the observed material behaviour.

3.3.3 Yield criteria of polymers

In contrast to metals, the yield strength of polymers is different in compression and tension. Frequently, the yield strength in uniaxial compression is 20 % to 30 % larger than in uniaxial tension (see also section 8.4). To account for this, the von Mises yield criterion is augmented by terms that depend on the hydrostatic stress state. We will discuss two possible approaches.

The *parabolically modified yield criterion* assumes that the yield surface is a paraboloid centred on the hydrostatic axis (figure 3.24(a)). Its radius depends on the hydrostatic stress σ_{hyd} according to

$$R(\sigma_{\mathrm{hyd}}) = \sqrt{\frac{2m}{3} R_{\mathrm{p}}^2 - 2(m-1)\sigma_{\mathrm{hyd}} R_{\mathrm{p}}}, \tag{3.35}$$

where m quantifies the difference between the material behaviour in compression and tension:

$$m = \frac{R_{\mathrm{c}}}{R_{\mathrm{p}}}. \tag{3.36}$$

The yield strengths in compression and tension are denoted R_c and R_p.[22] If a negative number results in the square root, the radius is taken to be zero; the material yields.

If we insert the radius of the yield surface, $R(\sigma_{hyd})$, into equation (3.29) we find the *parabolically modified von Mises criterion* [45]

$$\sigma_{eq,pM} = \frac{m-1}{2m}(\sigma_1 + \sigma_2 + \sigma_3) \tag{3.37}$$
$$+ \sqrt{\left[\frac{m-1}{2m}(\sigma_1 + \sigma_2 + \sigma_3)\right]^2 + \frac{1}{2m}\left[(\sigma_1 - \sigma_2)^2 + (\sigma_1 - \sigma_3)^2 + (\sigma_2 - \sigma_3)^2\right]}\,.$$

The yield criterion $\sigma_{eq,pM} = R_p$ is used for compressive and tensile loads.[23]

A different approach is to assume that the yield surface is a cone (see figure 3.24(b)) with radius

$$R(\sigma_{hyd}) = \sqrt{\frac{2}{3}}\frac{1}{m+1}\left[2mR_p - 3(m-1)\sigma_{hyd}\right]\,. \tag{3.38}$$

The *conically modified von Mises criterion*, written with principal stresses, is

$$\sigma_{eq,cM} = \frac{1}{2m}\Bigg[(m-1)(\sigma_1 + \sigma_2 + \sigma_3) $$
$$+ (m+1)\sqrt{\frac{1}{2}\left[(\sigma_1 - \sigma_2)^2 + (\sigma_1 - \sigma_3)^2 + (\sigma_2 - \sigma_3)^2\right]}\Bigg] \tag{3.39}$$

where m is taken from equation (3.36), and the yield criterion is $\sigma_{eq,cM} = R_p$ (after [45]). Again, a negative value in the square root means that the radius is zero and the material yields.

In figure 3.25, the different yield criteria are compared for a plane stress state. The modified yield criteria lead to a compressive yield strength that surpasses the tensile yield strength.

3.3.4 Flow rules

As we saw in the previous sections, *yield criteria* can be used to assert for any stress state whether a material yields. How the material deforms plastically is not governed by a yield criterion. The plastic deformation itself is described using *flow rules*. We will discuss them rather briefly here, a more

---

[22] The compressive strength R_c is taken as positive although it corresponds to a compressive stress.

[23] If we assume, for example, a compressive test with $\sigma_1 = -R_c$, $\sigma_2 = \sigma_3 = 0$ and use the definition of m, we find – after some algebra – $\sigma_{eq,pM} = R_p$. Here, it has to be kept in mind that the square root in equation (3.37) is always positive. In the case $\sigma < 0$, we use $\sqrt{\sigma^2} = |\sigma| = -\sigma$.

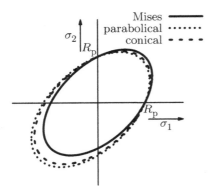

Fig. 3.25. Comparison of the yield criteria for polymers with the original von Mises yield criterion. The curves are given for identical R_p values

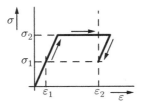

Fig. 3.26. Illustration of the fact that there is no unique relation between strains and stresses for plastic deformations, shown for the example of a perfectly plastic material. After different loading histories, different strains (ε_1 and ε_2, respectively) can occur at the same stress σ_1. Likewise, the same strain ε_2 can occur for different stresses (σ_1 and σ_2, respectively), depending on the loading history

detailed discussion can be found in books on continuum mechanics e.g., *Jirasek/Bazant* [80], *Hill* [65], *Becker/Bürger* [16], or *Kaliszky* [81].

As soon as plastic deformations have occurred, the relation between stress and strain is not unique. In a tensile test, for example, where the material is loaded until it yields and then unloaded again, two possible strains correspond to a given stress value (see figure 3.26). The strain in the material (for example, ε_1 and ε_2 at σ_1 in figure 3.26) depends on whether the material has been stressed with values larger than the current stress σ or not. Similarly, two different stresses (σ_1 or σ_2 in the figure) may cause the same strain (ε_2). Thus, the relation between stress and strain is not unique; the current state of the material depends on the deformation history.

However, the stress $\underline{\sigma}$ needed to increase the strain by a *plastic strain increment* $d\underline{\varepsilon}^{(\mathrm{pl})}$ can be stated,[24] for the yield criterion has to be fulfilled. We

---

[24] Here, we use $\underline{\varepsilon}$ instead of a quantity to characterise large deformations, like $\underline{G}$. The reason for this is that the strain increments $d\underline{\varepsilon}$ are always small. Thus, there is no error in using $d\underline{\varepsilon}$ as long it is always calculated for the current state of

can thus write down the strain increment as

$$d\underline{\underline{\varepsilon}}^{(\mathrm{pl})} = d\underline{\underline{\varepsilon}}^{(\mathrm{pl})}(\underline{\underline{\sigma}})\,.$$

The differential changes are frequently divided by the time increment dt in which they are applied. This leads to the rate equation

$$\dot{\underline{\underline{\varepsilon}}}^{(\mathrm{pl})} = \dot{\underline{\underline{\varepsilon}}}^{(\mathrm{pl})}(\underline{\underline{\sigma}})\,,$$

where $\dot{\underline{\underline{\varepsilon}}}^{(\mathrm{pl})}$ is the *plastic strain rate*.[25] The established procedure to perform the calculations is to prescribe (plastic) strain increments and calculate the required stresses, for this ensures stability of the calculations and uniqueness of the results. In a perfectly plastic material, for example, the stress can be assigned unambiguously to the strain increment (see figure 3.26), but the reverse is not true. In the tensile test described in section 3.2, we also prescribed the strains and measured the stresses for stability reasons.

A commonly used flow rule in its rate formulation is

$$\dot{\varepsilon}_{ij}^{(\mathrm{pl})} = \dot{\lambda}\sigma'_{ij}\,. \tag{3.40}$$

The proportionality constant $\dot{\lambda}$ adjusts itself at any given strain rate $\dot{\varepsilon}_{ij}^{(\mathrm{pl})}$ to ensure that the deviatoric stress tensor σ'_{ij} cannot leave the yield surface during plastic deformation.

> To derive equation (3.40), it is assumed that the power dissipated during plastic deformation, $\dot{w}^{(\mathrm{pl})}$ i.e., the plastic energy dissipation per time, is maximal. There is no theoretical proof of this assumption, which is called *Drucker's postulate,* but it can be justified by arguments from thermodynamics and is also in agreement with experiments.
>
> Mathematically, Drucker's postulate can be written as
>
> $$\dot{w}^{(\mathrm{pl})} = \sigma_{ij}\,\dot{\varepsilon}_{ij}^{(\mathrm{pl})} = \underline{\underline{\sigma}} \cdot\cdot\, \dot{\underline{\underline{\varepsilon}}}^{(\mathrm{pl})} = \max\,.$$
>
> If we take the constant volume during plastic deformation into account, we can also write
>
> $$\dot{w}^{(\mathrm{pl})} = \sigma'_{ij}\,\dot{\varepsilon}_{ij}^{(\mathrm{pl})} = \underline{\underline{\sigma}}' \cdot\cdot\, \dot{\underline{\underline{\varepsilon}}}^{(\mathrm{pl})} = \max\,. \tag{3.41}$$
>
> If we consider an arbitrary yield surface $f(\sigma_{ij}) = 0$ and prescribe a plastic strain rate (in the Voigt notation) $\dot{\underline{\varepsilon}}^{(\mathrm{pl})}$, equation (3.41) holds if the projection of $\underline{\underline{\sigma}}'$ onto $\dot{\underline{\varepsilon}}^{(\mathrm{pl})}$ becomes maximal. This is sketched in figure 3.27. To achieve a maximum power of plastic energy dissipation

---

the material. If the total deformation is to be calculated from the increments, an appropriate strain measure like $\underline{\underline{G}}$ has to be used. This may be difficult, for the contribution of rigid-body rotations has to be taken into account, which may be problematic [67, 71].

[25] This formulation does not imply that the material properties are time- or rate-dependent.

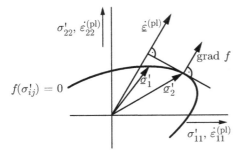

Fig. 3.27. Illustration of Drucker's postulate. For a specified plastic strain rate $\dot{\underline{\varepsilon}}^{(\mathrm{pl})}$, the dissipated power $\dot{w}^{(\mathrm{pl})} = \underline{\sigma}' \cdot\!\cdot \dot{\underline{\varepsilon}}^{(\mathrm{pl})}$ is maximal when the projection of $\underline{\sigma}'$ onto $\dot{\underline{\varepsilon}}^{(\mathrm{pl})}$ is maximal. This is the case for $\underline{\sigma}'_2$. In contrast, $\underline{\sigma}'_1$ would result in a lower $\dot{w}^{(\mathrm{pl})}$, for instance. For the stress with the maximal dissipated power $\underline{\sigma}'_2$, the following condition holds true: The gradient of the yield function grad f, which is perpendicular to the yield surface, is parallel to the plastic strain rate $\dot{\underline{\varepsilon}}^{(\mathrm{pl})}$

for a given $\dot{\underline{\varepsilon}}^{(\mathrm{pl})}$, the deviatoric stress tensor of the external stress must be $\underline{\sigma}'_2$. If it is $\underline{\sigma}'_1$, the power is smaller according to equation (3.41). If the yield surface is continuously differentiable, the position of maximum power is characterised by the gradient of the yield surface grad $f = \partial f/\partial \sigma'_{ij}$, which is perpendicular to the surface, being parallel to $\dot{\varepsilon}^{(\mathrm{pl})}_{ij}$. This yields the flow rule

$$\dot{\varepsilon}^{(\mathrm{pl})}_{ij} = \dot{\lambda}\frac{\partial f}{\partial \sigma'_{ij}} \,. \tag{3.42}$$

Here, $\dot{\lambda}$ is a proportionality factor.

Furthermore, the relation between stresses and plastic strain rates must be unique. From this, it can be seen that the yield surface must be strictly convex[26] and continuously differentiable to allow the formulation of a flow rule. The Tresca yield criterion is not continuously differentiable (there is no unique normal vector at its corners), and on the surfaces, different stress states fulfil equation (3.42) for a given $\dot{\varepsilon}^{(\mathrm{pl})}_{ij}$. Therefore, a flow rule cannot be derived using this criterion.

For the von Mises yield criterion, the yield surface is given by equation (3.32),

$$f(\sigma'_{kl}) = \frac{1}{2}\sigma'_{kl}\sigma'_{lk} - k_{\mathrm{F}}^2 \,,$$

---

[26] A (simply) convex surface may contain plane sections with a constant normal vector. If there are no plane sections and the surface is curved everywhere (without turning points), it is *strictly convex*.

resulting in

$$\frac{\partial f}{\partial \sigma'_{ij}} = \frac{1}{2} \left(\frac{\partial \sigma'_{kl}}{\partial \sigma'_{ij}} \sigma'_{lk} + \sigma'_{kl} \frac{\partial \sigma'_{lk}}{\partial \sigma'_{ij}} \right) .$$

The partial derivatives can be evaluated as

$$\frac{\partial \sigma'_{kl}}{\partial \sigma'_{ij}} = \left\{ \begin{matrix} 1 & \text{for } k = i, \, l = j \\ 0 & \text{otherwise} \end{matrix} \right\} = \delta_{ki} \delta_{lj} .$$

Thus, the equation simplifies to

$$\frac{\partial f}{\partial \sigma'_{ij}} = \frac{1}{2} (\sigma'_{ji} + \sigma'_{ij}) = \sigma'_{ij} .$$

Inserting this into equation (3.42), we find the flow rule (3.40) for the von Mises yield criterion.

The consistency of this equation can be checked by verifying that plastic deformations do not change the volume. This is rather simple. Since λ is a scalar quantity and since the trace of the deviatoric stress tensor, σ'_{ii}, vanishes, we find $\operatorname{tr} \underline{\dot{\varepsilon}} = \dot{\varepsilon}_{ii} = 0$. The change in volume over time is thus zero.

As already stated, flow rules are a tool to determine the stresses from a given strain rate. They do *not* allow to calculate how the yield surface changes by hardening (see section 3.3.1). This task is performed by hardening laws, to be discussed now.

3.3.5 Hardening

If we take another look at the stress-strain curve of a tensile test, we find that the stress increases after yielding begins (at a stress of R_{p}), seemingly moving beyond the yield surface that is determined by the yield strength. This, however, was explicitly ruled out in section 3.3.1. The apparent contradiction can be resolved by realising that the yield surface changes during plastic deformation.

That the yield surface must change during plastic deformation and that the stress state cannot move outside of it can also be seen by the fact that the material becomes elastic immediately upon unloading. Thus, the stress state must be inside the yield surface as soon as the load reduces. This is illustrated in figure 3.28. The yield surface is constant until the stress reaches the yield strength (figure 3.28(a)). During the subsequent plastic deformation and hardening, the yield surface grows together with the current flow stress (figures 3.28(b) and 3.28(c)). Upon unloading, the material becomes elastic, and the slope of the stress-strain curve is equal to Young's modulus. The yield surface remains unchanged (figure 3.28(d)).

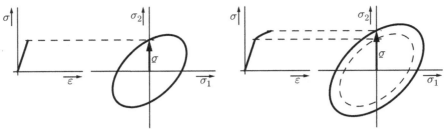

(a) Yielding begins (transition from elastic to plastic behaviour)

(b) The material yields, the yield surface grows with the flow stress

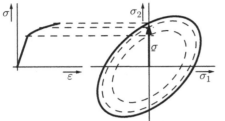

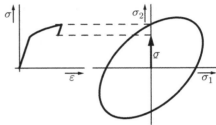

(c) The material yields, the yield surface continues to grow

(d) During unloading, the deformation immediately becomes elastic, the yield surface stays constant

Fig. 3.28. Illustration of strain hardening using a stress-strain curve. Here, we assume isotropic hardening which is explained in the text

The yield surface can change its shape, size, and position in stress space. Mathematically, this can be realised by modifying the yield criterion (3.23) with additional terms that describe the change of the yield surface (the modified yield surface is denoted as g) [115]:

$$g(\sigma_{ij}, \varepsilon_{ij}^{(\mathrm{pl})}, k_l) = 0\,. \tag{3.43}$$

Here, $\varepsilon_{ij}^{(\mathrm{pl})}$ is the current plastic deformation, and k_l is a set of *hardening parameters* which may depend on the deformation history, the strain rate, or the temperature. As in the original yield criterion, the material deforms elastically if the stress state lies with the surface $g = 0$ i. e., if $g < 0$ holds.

To take hardening into account, we need to find a quantity that can describe the deformation history of the material. This quantity has to increase during plastic deformation, regardless of the deformation orientation, for, in general, any plastic deformation causes hardening. A frequently used quantity is the so-called *equivalent plastic strain* $\varepsilon_{\mathrm{eq}}^{(\mathrm{pl})}$. To define this strain, we need the *equivalent plastic strain rate* $\dot{\varepsilon}_{\mathrm{eq}}^{(\mathrm{pl})}$ which is defined analogously to the von

Mises yield criterion:[27]

$$\dot{\varepsilon}_{\text{eq}}^{(\text{pl})} = \sqrt{\frac{2}{9}\left[\left(\dot{\varepsilon}_1^{(\text{pl})} - \dot{\varepsilon}_2^{(\text{pl})}\right)^2 + \left(\dot{\varepsilon}_1^{(\text{pl})} - \dot{\varepsilon}_3^{(\text{pl})}\right)^2 + \left(\dot{\varepsilon}_2^{(\text{pl})} - \dot{\varepsilon}_3^{(\text{pl})}\right)^2\right]} \qquad (3.44)$$

Since the equivalent plastic strain rate defined this way is positive for all plastic strain rates, the equivalent plastic strain increases for any plastic deformation, regardless of the deformation orientation.

The equivalent plastic strain $\varepsilon_{\text{eq}}^{(\text{pl})}$ is calculated from the equivalent plastic strain rate $\dot{\varepsilon}_{\text{eq}}^{(\text{pl})}$ by integrating:

$$\varepsilon_{\text{eq}}^{(\text{pl})} = \int \dot{\varepsilon}_{\text{eq}}^{(\text{pl})}\,\mathrm{d}t\,. \qquad (3.45)$$

For a uniaxial and monotonous deformation, $\varepsilon_{\text{eq}}^{(\text{pl})} = \varepsilon^{(\text{pl})}$ holds.

The hardening parameters k_l in equation (3.43) depend on the equivalent plastic strain:

$$g\big(\sigma_{ij}, \varepsilon_{ij}^{(\text{pl})} = 0, k_l(\varepsilon_{\text{eq}}^{(\text{pl})})\big) = f(\sigma_{ij})\,.$$

In the following, we will consider three important cases: the case of no hardening, and the two extreme cases of hardening behaviour, *isotropic hardening* and *kinematic hardening*. The hardening of most materials contains isotropic and kinematic parts.

No hardening

A *perfectly plastic* material does not harden so that its yield surface, equation (3.23), remains unchanged during deformation. Thus, the yield criterion is

$$g(\sigma_{ij}) = f(\sigma_{ij}) = 0\,. \qquad (3.46)$$

The corresponding stress-strain diagram is given in figure 3.29(a).

In reality, there are no perfectly plastic materials. In a tensile test, they would start to neck immediately upon yielding because the stability criterion from section 3.2.3 would be violated (in a perfectly plastic material, we find $\sigma = R_\text{p}\varphi^0$ and thus $\varphi_\text{neck} = 0$). Nevertheless, perfectly plastic material models are frequently used, especially in metal working, for they are simple and allow to solve many problems analytically, using the so-called slip-line theory [65]. If elastic deformations are allowed in the material, it is called *elastic-perfectly plastic* (figure 3.29(a)). Since the elastic part of the deformation is small when plastic deformations are large, the elastic part can be neglected, leading to the assumption of a *rigid-perfectly plastic* material (figure 3.29(b)).

---

[27] This is a reasonable definition for it ensures that the trace of the plastic strains vanishes: $\operatorname{tr}\underline{\underline{\varepsilon}}^{(\text{pl})} = 0$.

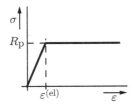

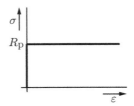

(a) Perfectly plastic and elastic   (b) Rigid-perfectly plastic

Fig. 3.29. Stress-strain curves of perfectly plastic materials

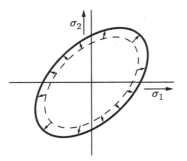

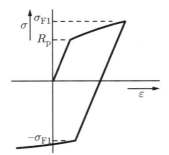

(a) Development of the yield surface during isotropic hardening   (b) Stress-strain curve with reversed load for an isotropically hardening material

Fig. 3.30. Isotropic hardening

Isotropic hardening

During *isotropic hardening,* the yield surface grows symmetrically around the origin as sketched in figure 3.30(a). Mathematically, this implies that the argument $\varepsilon_{ij}^{(\mathrm{pl})}$ in equation (3.43) plays no role, and the yield criterion is thus

$$g\big(\sigma_1, \sigma_2, \sigma_3, k_l(\varepsilon_{\mathrm{eq}}^{(\mathrm{pl})})\big) = 0\,. \tag{3.47}$$

If yielding of the material is governed by the von Mises yield criterion, we find the yield criterion for isotropic hardening, using the changing flow stress $\sigma_{\mathrm{F}}(\varepsilon_{\mathrm{eq}}^{(\mathrm{pl})})$,

$$\sqrt{\frac{1}{2}\Big[(\sigma_1 - \sigma_2)^2 + (\sigma_1 - \sigma_3)^2 + (\sigma_2 - \sigma_3)^2\Big]} = \sigma_{\mathrm{F}}(\varepsilon_{\mathrm{eq}}^{(\mathrm{pl})}) \tag{3.48}$$

with initial condition

$$\sigma_{\mathrm{F}}(\varepsilon_{\mathrm{eq}}^{(\mathrm{pl})} = 0) = R_{\mathrm{p}}\,. \tag{3.49}$$

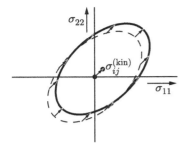

Fig. 3.31. Shift of the yield surface caused by kinematic hardening. It is not possible to draw this in the space of principal stresses because the material becomes anisotropic during hardening so that its properties depend on the spatial direction

So far, we have not specified how the yield surface changes with increasing plastic deformation. This is done by evolution equations for the hardening parameters, so-called *hardening laws.*

A simple isotropic hardening law can be written down for the case of *linear hardening,* defined by the flow stress increasing linearly with the plastic strain. Its rate formulation is [65].

$$\dot{\sigma}_{\mathrm{F}} = H \cdot \dot{\varepsilon}_{\mathrm{eq}}^{(\mathrm{pl})} \tag{3.50}$$

with the *hardening parameter H*. When hardening begins, the flow stress σ_{F} is equal to the yield strength R_{p}. The initial condition is thus (3.49). The rate equation for linear hardening can be integrated to yield

$$\sigma_{\mathrm{F}} = R_{\mathrm{p}} + H \cdot \varepsilon_{\mathrm{eq}}^{(\mathrm{pl})} \,.$$

The stress-strain diagram of a material with isotropic hardening that is deformed by uniaxial tension first and uniaxial compression afterwards can be found in figure 3.30(b). In compression, the material yields at a stress $-\sigma_{\mathrm{F}1}$, given by the absolute value of the maximum stress in tension, $\sigma_{\mathrm{F}1}$.

Kinematic hardening

In a material with *kinematic hardening,* the yield surface changes neither its shape nor its size, but moves in stress space (see figure 3.31). Mathematically, this can be realised by subtracting a *kinematic backstress* $(\sigma_{ij}^{(\mathrm{kin})})$ from the stress tensor in the yield criterion:

$$g\big(\sigma_{ij} - \sigma_{ij}^{(\mathrm{kin})}\big) = 0 \,. \tag{3.51}$$

Geometrically, $(\sigma_{ij}^{(\mathrm{kin})})$ is the shift of the yield surface from the origin, see figure 3.31.

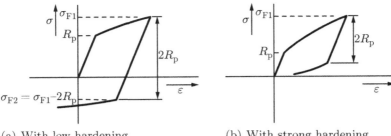

(a) With low hardening   (b) With strong hardening

Fig. 3.32. Stress-strain diagram with reversed stress for a kinematically hardening material

Since the yield surface shifts in stress space, the material becomes anisotropic because the value of the flow stress depends on the space direction. Thus, the yield criterion cannot be formulated using principal stresses. The kinematic backstress depends on the deformation history. However, because it is anisotropic, it cannot depend on the equivalent plastic stress $\varepsilon_{\mathrm{eq}}^{(\mathrm{pl})}$, only, but depends also on other variables e. g., the plastic strain $(\varepsilon_{ij}^{(\mathrm{pl})})$.

Due to the anisotropy, the von Mises yield criterion must be used in its coordinate-dependent formulation:

$$
\left(\frac{1}{2}\left[\left(\sigma_{11}^{(\mathrm{eff})} - \sigma_{22}^{(\mathrm{eff})}\right)^2 + \left(\sigma_{22}^{(\mathrm{eff})} - \sigma_{33}^{(\mathrm{eff})}\right)^2 + \left(\sigma_{11}^{(\mathrm{eff})} - \sigma_{33}^{(\mathrm{eff})}\right)^2\right]\right.
$$
$$
\left. +3\left(\sigma_{23}^{(\mathrm{eff})}\right)^2 + 3\left(\sigma_{13}^{(\mathrm{eff})}\right)^2 + 3\left(\sigma_{12}^{(\mathrm{eff})}\right)^2\right)^{1/2} = R_{\mathrm{p}}
\tag{3.52}
$$

with $\sigma_{ij}^{(\mathrm{eff})} = \sigma_{ij} - \sigma_{ij}^{(\mathrm{kin})}$ for all $i, j = 1, 2, 3$.

A simple evolution law for the backstress is [80]

$$
\underline{\dot{\sigma}}^{(\mathrm{kin})} = C \cdot \dot{\varepsilon}_{\mathrm{eq}}^{(\mathrm{pl})} \cdot \frac{\underline{\sigma} - \underline{\sigma}^{(\mathrm{kin})}}{\sigma_0} .
\tag{3.53}
$$

Here, C is a hardening parameter and σ_0 a constant reference stress which is equal to the yield strength at $\underline{\sigma}^{(\mathrm{kin})} = \underline{0}$. When plastic deformation starts, $\underline{\sigma}^{(\mathrm{kin})} = \underline{0}$ holds.

If we deform a kinematically hardening material in uniaxial tension and compression, its behaviour differs drastically from the isotropically hardening material discussed above (figure 3.32). Upon load reversal, the material yields at a stress $\sigma_{\mathrm{F2}} = \sigma_{\mathrm{F1}} - 2R_{\mathrm{p}}$ because the size of the yield surface remains unchanged. In the extreme case, this may lead to plastic deformation while the stress is still tensile (figure 3.32(b)).

A special case of pure kinematic hardening is the so-called *Masing behaviour*. If a stress-strain diagram is measured for a material with this behaviour, the material behaviour upon load reversal can be described by rotating the original stress-strain diagram by 180°, scaling both axes to twice their length, and positioning the origin of the rotated and scaled diagram at the end of the original stress-strain curve. One prerequisite for the validity of this construction is that the slope of the stress-strain curve after yielding must be the same for both loading directions.

Frequently, the term *Bauschinger effect* is used if the flow stress becomes smaller upon load reversal, like in kinematic hardening.

∗ 3.3.6 Application of a yield criterion, flow rule, and hardening rule

In this section, we will discuss the relation between elastic deformations, a yield criterion, a flow rule, and hardening, using the example of a simple material.

Consider a tensile specimen of an isotropic metal with elastic parameters $E = 210\,000\,\mathrm{MPa}$ and $\nu = 0.3$, and a yield strength $\sigma_\mathrm{F} = 210\,\mathrm{MPa}$. The material hardens linearly and isotropically according to equation (3.50), with hardening parameter $H = 10\,000\,\mathrm{MPa}$. The tensile specimen is elongated, starting with an unloaded state, at a constant strain rate of $\dot{\varepsilon}_{11} = 0.001\,\mathrm{s}^{-1}$. We want to determine the time-dependence of stresses and strains.

We start by collecting all equations required. Next, we simplify the equations and calculate the solution variables.

As the metal is not rigid, each deformation has an elastic part which follows Hooke's law (2.31):

$$\underline{\underline{\varepsilon}}^{(\mathrm{el})} = \underset{4}{\underline{S}} \cdot\cdot \, \underline{\underline{\sigma}} \, . \tag{3.54}$$

For plastic deformations, it is useful to use a rate formulation (see section 3.3.4). Equation(3.40)

$$\underline{\underline{\dot{\varepsilon}}}^{(\mathrm{pl})} = \dot{\lambda} \cdot \underline{\underline{\sigma}}' \tag{3.55}$$

holds if the yield criterion is fulfilled.

Each strain increment has an elastic and a plastic part:

$$\mathrm{d}\underline{\underline{\varepsilon}} = \mathrm{d}\underline{\underline{\varepsilon}}^{(\mathrm{el})} + \mathrm{d}\underline{\underline{\varepsilon}}^{(\mathrm{pl})} \, .$$

This is also true if we relate the increments to time:

$$\underline{\underline{\dot{\varepsilon}}} = \underline{\underline{\dot{\varepsilon}}}^{(\mathrm{el})} + \underline{\underline{\dot{\varepsilon}}}^{(\mathrm{pl})} \, . \tag{3.56}$$

Since the compliance tensor $\underset{4}{\underline{S}}$ is time-independent, equation (3.54) can be easily differentiated with respect to time, yielding

$$\underline{\underline{\dot{\varepsilon}}}^{(\mathrm{el})} = \underset{4}{\underline{S}} \cdot\cdot \, \underline{\underline{\dot{\sigma}}} \, . \tag{3.57}$$

This equation can be used together with equation (3.56).

So far, we still lack a criterion to define whether the material yields. This is provided by the von Mises yield criterion

$$\begin{aligned} \sigma_{eq} &< \sigma_F \Rightarrow \text{no yielding}, \\ \sigma_{eq} &= \sigma_F \Rightarrow \text{yielding} \end{aligned} \tag{3.58}$$

with the equivalent stress

$$\sigma_{eq} = \sqrt{\frac{1}{2}\left[(\sigma_1 - \sigma_2)^2 + (\sigma_1 - \sigma_3)^2 + (\sigma_2 - \sigma_3)^2\right]}. \tag{3.59}$$

Hardening is taken into account by equation (3.50).

We now have found all equations necessary to perform the calculations. In summary, we have to solve the following system of differential equations (using the same equation numbers as above):

$$\dot{\underline{\varepsilon}} = \dot{\underline{\varepsilon}}^{(el)} + \dot{\underline{\varepsilon}}^{(pl)}, \tag{3.56}$$

$$\dot{\underline{\varepsilon}}^{(el)} = \underset{4}{S} \cdot\cdot \, \dot{\underline{\sigma}}, \tag{3.57}$$

$$\sigma_{eq} = \sqrt{\frac{1}{2}\left[(\sigma_1 - \sigma_2)^2 + (\sigma_1 - \sigma_3)^2 + (\sigma_2 - \sigma_3)^2\right]}, \tag{3.59}$$

$$\dot{\underline{\varepsilon}}^{(pl)} = \begin{cases} 0 & \text{for } \sigma_{eq} < \sigma_F \\ \dot{\lambda} \cdot \underline{\sigma}' & \text{for } \sigma_{eq} = \sigma_F \end{cases}, \tag{3.55}, (3.58)$$

$$\dot{\varepsilon}_{eq}^{(pl)} = \sqrt{\frac{2}{9}\left[\left(\dot{\varepsilon}_1^{(pl)} - \dot{\varepsilon}_2^{(pl)}\right)^2 + \left(\dot{\varepsilon}_1^{(pl)} - \dot{\varepsilon}_3^{(pl)}\right)^2 + \left(\dot{\varepsilon}_2^{(pl)} - \dot{\varepsilon}_3^{(pl)}\right)^2\right]}. \tag{3.44}$$

$$\dot{\sigma}_F = H \cdot \dot{\varepsilon}_{eq}^{(pl)} \tag{3.50}$$

During deformation, the following quantities change: $\underline{\varepsilon}$, $\underline{\sigma}$ (and thus also $\underline{\sigma}'$, σ_{eq}), $\underline{\varepsilon}^{(el)}$, $\underline{\varepsilon}^{(pl)}$, $\dot{\lambda}$ and σ_F. The parameters $\underset{4}{S}$ and H remain constant.

In a tensile test of an isotropic material, we can assume that the stress state is uniaxial. The stress tensor is thus

$$\underline{\sigma} = \begin{pmatrix} \sigma_{11} & 0 & 0 \\ 0 & 0 & 0 \\ 0 & 0 & 0 \end{pmatrix}.$$

As expected, the equivalent stress is thus simply

$$\sigma_{eq} = \sigma_{11}.$$

Hooke's law in its rate formulation becomes

$$\dot{\underline{\varepsilon}}^{(el)} = \begin{pmatrix} 1/E & 0 & 0 \\ 0 & -\nu/E & 0 \\ 0 & 0 & -\nu/E \end{pmatrix} \dot{\sigma}_{11}. \tag{3.60}$$

The plastic strain rate can be calculated, using the deviatoric stress tensor $\underline{\underline{\sigma}}' = \underline{\underline{\sigma}} - \underline{\underline{1}}\sigma_{\text{hyd}}$,

$$\underline{\underline{\dot{\varepsilon}}}^{(\text{pl})} = \begin{pmatrix} 2/3 & 0 & 0 \\ 0 & -1/3 & 0 \\ 0 & 0 & -1/3 \end{pmatrix} \cdot \begin{cases} 0 & \text{for } \sigma_{11} < \sigma_{\text{F}} \\ \sigma_{11}\dot{\lambda} & \text{for } \sigma_{11} = \sigma_{\text{F}} \end{cases}. \tag{3.61}$$

The equivalent plastic strain rate for the case $\sigma_{11} = \sigma_{\text{F}}$ is thus

$$\dot{\varepsilon}^{(\text{pl})}_{\text{eq}} = \sqrt{\frac{4}{9}\sigma_{11}^2\dot{\lambda}^2} = \frac{2}{3}\left|\sigma_{11}\dot{\lambda}\right|.$$

The evolution of the flow stress can be calculated since we know that no load reversal occurs:

$$\dot{\sigma}_{\text{F}} = H \cdot \begin{cases} 0 & \text{for } \sigma_{11} < \sigma_{\text{F}} \\ \frac{2}{3}\sigma_{11}\dot{\lambda} & \text{for } \sigma_{11} = \sigma_{\text{F}} \end{cases}.$$

Neither the elastic nor the plastic parts of the strain contain shear terms. The 22- and 33-components can be calculated easily, using the equations (3.60) and (3.61). They will not be considered anymore. Thus, we find the following equations:

$$\dot{\varepsilon}_{11} = \dot{\varepsilon}_{11}^{(\text{el})} + \dot{\varepsilon}_{11}^{(\text{pl})}, \tag{3.62}$$

$$\dot{\varepsilon}_{11}^{(\text{el})} = \frac{1}{E}\dot{\sigma}_{11}, \tag{3.63}$$

$$\dot{\varepsilon}_{11}^{(\text{pl})} = \begin{cases} 0 & \text{for } \sigma_{11} < \sigma_{\text{F}}, \\ \frac{2}{3}\sigma_{11}\dot{\lambda} & \text{for } \sigma_{11} = \sigma_{\text{F}}, \end{cases} \tag{3.64}$$

$$\dot{\sigma}_{\text{F}} = H \cdot \begin{cases} 0 & \text{for } \sigma_{11} < \sigma_{\text{F}}, \\ \frac{2}{3}\sigma_{11}\dot{\lambda} & \text{for } \sigma_{11} = \sigma_{\text{F}}. \end{cases} \tag{3.65}$$

These equations, (3.62) to (3.65), are all that is required to solve the exercise.

* Elastic region

The initial conditions are $\sigma_{11} = 0$ and $\varepsilon_{11}^{(\text{pl})} = 0$. During deformation, the total strain rate $\dot{\varepsilon}_{11} = 0.001\,\text{s}^{-1}$ is prescribed. Since the yield criterion is not fulfilled for small stresses, we find $\dot{\varepsilon}_{11} = \dot{\varepsilon}_{11}^{(\text{el})} = 0.001\,\text{s}^{-1}$. Using equation (3.63), the stress rate can be calculated as $\dot{\sigma}_{11} = E\dot{\varepsilon}_{11} = 210\,\text{MPa/s}$. The stress at time t is thus

$$\sigma_{11} = \int_0^t \dot{\sigma}_{11}\mathrm{d}\tilde{t} = \dot{\sigma}_{11}t = 210\,\text{MPa/s} \cdot t. \tag{3.66}$$

* Elastic-plastic region

Yielding starts at a time t_p given by the yield criterion

$$\sigma_{11}(t_p) = \sigma_F(\varepsilon_{eq}^{(pl)} = 0) = R_p,$$
$$210\,\mathrm{MPa/s} \cdot t_p = 210\,\mathrm{MPa}. \tag{3.67}$$

Thus, we find $t_p = 1\,\mathrm{s}$. Starting at this moment, the plasticity terms in the equations have to be taken into account. Since the yield criterion is now fulfilled, we can replace σ_{11} by σ_F.

If we now add the elastic part, equation (3.63), and the plastic part, equation (3.64), to get the total strain rate, equation (3.62), we find

$$\dot{\varepsilon}_{11} = \frac{1}{E}\dot{\sigma}_F + \frac{2}{3}\sigma_F\dot{\lambda}.$$

Replacing $\dot{\lambda}$ by equation (3.65) yields

$$\dot{\sigma}_F = \frac{EH}{E+H}\dot{\varepsilon}_{11}.$$

Using the parameter values provided, we find for $t \geq 1\,\mathrm{s}$ an increase in flow stress of $\dot{\sigma}_F = 9.546\,\mathrm{MPa/s}$. Using the initial value $\sigma_F(t = 1\,\mathrm{s}) = 210\,\mathrm{MPa}$, we find the flow stress

$$\sigma_F = 210\,\mathrm{MPa} + 9.546\,\mathrm{MPa/s} \cdot (t - 1\,\mathrm{s}). \tag{3.68}$$

This directly yields the elastic strains

$$\varepsilon_{11}^{(el)} = \frac{\sigma_F}{E} = 0.001 + 4.546 \times 10^{-5}\,\mathrm{s}^{-1} \cdot (t - 1\,\mathrm{s}). \tag{3.69}$$

The plastic strain is the difference between total and elastic strain:

$$\varepsilon_{11}^{(pl)} = 9.546 \times 10^{-4}\,\mathrm{s}^{-1} \cdot (t - 1\,\mathrm{s}). \tag{3.70}$$

To summarise all results, we find the following stresses and strains:

$$\sigma_{11} = \begin{cases} 210\,\mathrm{MPa/s} \cdot t & \text{for } t < 0\,\mathrm{s}, \\ 210\,\mathrm{MPa} + 9.546\,\mathrm{MPa/s} \cdot (t - 1\,\mathrm{s}) & \text{for } t \geq 0\,\mathrm{s}, \end{cases}$$

$$\varepsilon_{11} = 0.001\,\mathrm{s}^{-1} \cdot t,$$

$$\varepsilon_{11}^{(el)} = \begin{cases} 0.001\,\mathrm{s}^{-1} \cdot t & \text{for } t < 0\,\mathrm{s}, \\ 0.001 + 4.546 \times 10^{-5}\,\mathrm{s}^{-1} \cdot (t - 1\,\mathrm{s}) & \text{for } t \geq 0\,\mathrm{s}, \end{cases}$$

$$\varepsilon_{11}^{(pl)} = \begin{cases} 0 & \text{for } t < 0\,\mathrm{s}, \\ 9.546 \times 10^{-4}\,\mathrm{s}^{-1} \cdot (t - 1\,\mathrm{s}) & \text{for } t \geq 0\,\mathrm{s}, \end{cases}$$

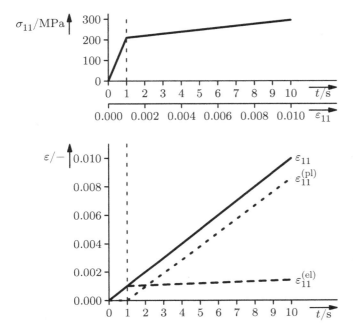

Fig. 3.33. Illustration of various stress and strain measures during a tensile test

Figure 3.33 shows plots of the results.

As can be seen in the figure, the plastic strain $\varepsilon^{(\mathrm{pl})}$ increases markedly stronger than the elastic strain $\varepsilon^{(\mathrm{el})}$ after yielding. The slope of the stress-strain curve is 9546 MPa. It is slightly smaller than the hardening parameter H. This is due to the elastic strains which slightly increase with the stress.

* 3.4 Hardness

Hardness is defined as the resistance of a material to indentation.[28] Since this resistance strongly depends on the shape of the indenter and the force level, there are a large number of different testing methods. These different methods can be classified in three groups, *scratch tests, indentation tests,* and *rebound tests.* In general, it is not possible to calculate a hardness value given a value measured with another method – however, conversion tables for common materials are available.

---

[28] In different branches of engineering, the term 'hardness' is used with slightly different meaning. In tribology, for example, hardness denotes the resistance to wear, in machining, it is used as a measure of machinability. The definition provided here is used in mechanical testing.

Although hardness is not a material parameter that can be easily understood theoretically, hardness tests are of great importance, for they are simple and may even be employed on built-in components. A further advantage is that small test volumes can be investigated, even down to single grains (microhardness testing).

∗ 3.4.1 Scratch tests

Historically, scratch tests are of some importance, for they were the first hardness tests employed. In a scratch test, it is tested whether a material can be scratched using a needle made of another material. Either relative scales that allow to sort materials by their scratchability are used, or the size of the scratch is measured to determine the hardness. Although the method can yield quantitative results, it is not easy to perform precise measurements.

∗ 3.4.2 Indentation tests

Indentation tests are the most common hardness tests, for they are rather easy to perform. A hard indenter with a certain geometry is pressed into the test specimen, and the surface of the indentation or the indentation depth are measured and related to the force required.

One example is the *Brinell* hardness test. In this test, a hardened steel ball with diameter D is pressed into the test surface with a prescribed force, avoiding sudden impact.[29] After unloading, the diameter d of the remaining indentation is measured. The Brinell hardness is defined as the testing force, measured in kp, divided by the total area of the indentation, measured in mm²:

$$HB = \frac{F/\text{kp}}{A/\text{mm}^2} = \frac{0.102 F/\text{N}}{A/\text{mm}^2} \,. \tag{3.71}$$

The surface A is measured from the diameter, using the formula

$$A = \frac{\pi}{2} D(D - \sqrt{D^2 - d^2}) \,. \tag{3.72}$$

The unit of the hardness is that of a pressure. Since the total surface of the indentation was used, the hardness does not correspond to the average pressure between indenter and material. This can be corrected by using only the projected area of the indentation. If such a definition is used, the hardness is almost independent of the testing force, provided the material does not harden and the testing force was sufficient to cause significant plastic deformation. If

---

[29] There are different standards for the size and diameter of the ball. Commonly used values are $D = 10\,\text{mm}$ and a force of $29.43\,\text{kN} = 3000\,\text{kp}$. The choice of parameters depends on the tested material and the thickness of the specimen. If large testing forces are needed, cemented carbide balls can also be used.

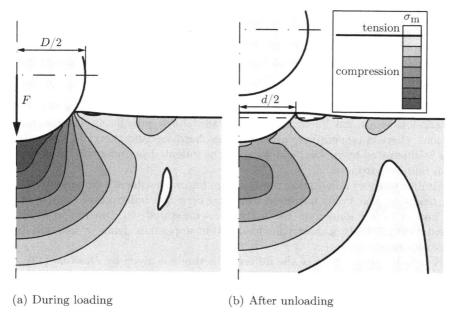

(a) During loading   (b) After unloading

Fig. 3.34. Finite element simulation of the hydrostatic stress in a Brinell hardness measurement using an elastic-plastic material law. The ball is assumed to be rigid and no stresses have been calculated for it. Between ball and tested material, a coefficient of friction of $\mu = 0.3$ has been used. The resulting volume of the indentation is 8% larger than the bulge

the material hardens, the measured hardness increases with the testing force if this definition is used, whereas it decreases at large forces when the Brinell definition is employed.

One disadvantage in theoretically analysing this method is that the geometry of the indentation changes during the test. If the indenter has a pyramidal shape, the shape of the indentation remains unchanged, only its size grows. Such an indenter is used in the *Vickers* hardness test. Again, the hardness is defined as quotient of testing force and total area. In both methods, the stress state beneath the indenter is triaxial, with a large hydrostatic pressure in the material. This is advantageous because it reduces the danger of crack formation in brittle materials. Figure 3.34(a) illustrates the process for a spherical indenter. To understand the indentation process mechanically, a simple model can be used where the material is assumed to be rigid-perfectly plastic. In this case, a relation between the size of the indentation and the yield strength of the material can be derived. The material displaced by the indenter moves and causes a bulge, with a volume that is the same as that of the displaced material because of the constant volume. In reality, the volume of the bulge is usually smaller than that of the indentation, showing that the assumption of a rigid material is incorrect. Figure 3.34(b) illustrates this using a finite element simulation with a spherical indenter. As can be seen, the volume of the indentation is larger than that of the bulge. A more detailed study shows that a plastic zone forms beneath the indenter that elastically compresses the material beneath it, causing residual stresses.

This consideration already shows that hardness is a complex material property because the elastic and plastic properties of the material play a role. In materials that are not linear-elastic and can deform with large elastic deformations, there is no simple relation between hardness and the yield strength. This is illustrated by rubber, which cannot be indented permanently, resulting in an infinite hardness.

Similar to these indentation methods are impact hardness testing methods (for example, the Poldi hardness tester), where the indentation caused by the impact of a hammer on the material is measured. In contrast to other indentation methods, a short-time load is thus applied, causing an increase in the strain rate.

A detailed description of the different methods is given by *Dowling* [43].

* 3.4.3 Rebound tests

In rebound tests, a hammer is used that drops down onto the material, and the rebound height is measured. In a purely elastic impact, the total kinetic energy of the material is transformed to deformation energy and then again to kinetic energy so that the hammer rebounds to its original height. If plastic deformation occurs, energy is dissipated and the rebound height is reduced by the corresponding amount. The advantages of this method are the small size of the indentation and the short testing time. Hardness value obtained with this method can also not be converted directly to other hardness values.

3.5 Material failure

Plastic deformation during service is often considered as a failure criterion. One reason for this is that the deformations are usually intolerably large, another is that the yield strength is usually not small enough compared to the tensile strength so that the safety of the component is not guaranteed. A component, however, may also fail by fracture instead of plastic deformation. There are a large number of possibilities how this fracture can occur which will be discussed only partially in the following. A detailed survey can be found in *Lange* [90]. Fracture of polymers will be discussed in chapter 8; here, we will briefly discuss the failure of metals and ceramics.

Fractures and cracks can be classified in three groups, depending on whether their main cause is *mechanical, thermal,* or *corrosive.*

Mechanical fracture can be due to monotonic increase of the load (*overload fracture* or *forced fracture*) or due to cyclic loads (*fatigue fracture*). Overload

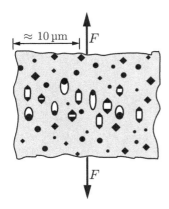

Fig. 3.35. Formation of cavities at particles due to large plastic deformation [90]. The particles (e. g., precipitates) either detach from the matrix (shown as circles) or fracture (shown as squares)

fracture will be discussed in this section; fatigue fracture is the subject of chapter 10. The most important group of thermally caused fractures is the *creep fracture* discussed in chapter 11. One example of corrosive fracture will be discussed in section 3.5.3, see also section 5.2.6.

An *overload fracture* is characterised by a mainly monotonously increasing load that is applied moderately fast or abruptly [90]. These conditions discriminate overload fracture from fatigue fracture (non-monotonous, cyclic load) and creep fracture (long loading times at high temperature).

The fracture can occur as *shear fracture, cleavage fracture,* or a mixture of both. The two characteristic forms will be discussed in the following.

3.5.1 Shear fracture

Shear fracture[30] (microscopically ductile fracture) occurs by plastic deformation with slip in the direction of planes of maximum shear stress (see sections 3.3.2 and 6.2.5). Therefore, it occurs only in ductile materials. In most cases, shear fracture is associated with large macroscopic deformations, as, for example, in a tensile test. However, if this is prevented by the component geometry, the component may fail macroscopically brittle, but still with a shear fracture. This may happen if there are notches or cracks in the material (see chapters 4 and 5).

In very pure metals, large deformations are possible. In a tensile test, the specimen can therefore be drawn to a thin tip (see figure 3.15(a)). Most engineering metals, however, contain particles (e. g., precipitates, see section 6.4.4). During large plastic deformation at high stresses, the particles may fracture or detach from the surrounding matrix, depending on the strength of the particle and the interface (figure 3.35). Particle fracture occurs preferentially in brittle particles and at high tensile stresses (for example, in a triaxial stress

---

[30] Sometimes, the term 'shear fracture' is used for a fracture caused by applying a shear load to a specimen, regardless of the fracture mechanism.

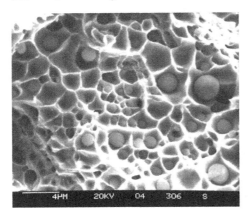

Fig. 3.36. Dimples of a ductile overload failure in a ferritic steel [90]. In some dimples, inclusions can be seen clearly in this scanning electron microscope micrograph. This is, however, not always the case; sometimes, the inclusions fall out of the dimples or cannot be observed although they still are inside the dimple

state). Detachment of the particle from the matrix mainly occurs when the deformation in the matrix is large.

Particle failure induces microcracks in the material. They deform to form ellipsoidal cavities. In between the particles, the matrix is only single-phase and thus has an increased ductility. The cavities thus grow by slipping of the matrix (see sections 6.2.3 and 6.2.5) on planes of maximum shear stress (e. g., in a uniaxial stress state at 45° to the loading direction). The matrix between the cavities is drawn to thin tips or ridges.[31] The finally formed fracture surface is characterised by a large number of *dimples* formed in this way. The size of the dimples is in the range of a few micrometres. Sometimes, this kind of fracture is called *fibrous fracture.*

In most cases, shear fracture is *transcrystalline* (through the grains), but, depending on the material state, *intercrystalline fracture* (fracture along the grain boundaries) may also occur.

In section 3.2.2, we already discussed the failure of a tensile specimen by shear-face fracture or cup-and-cone fracture. This will be elaborated on here. Since the stress level and the plastic deformation are largest in the specimen's centre in the necking region (see figures 3.13 and 3.14), damage by formation and coalescence of cavities starts there. Accordingly, the first cracks also form in this region. They grow along planes of maximum shear stress, at 45° to the loading direction in a tensile test because slip and, thus, damage are concentrated along these planes. During this process, the crack grows slightly beyond the region of the minimum cross section, where the stresses are largest.

---

[31] This is comparable to the drawing of a thin tip in the tensile test of a pure metal.

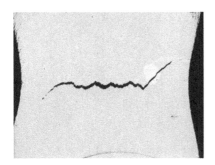

Fig. 3.37. Formation of a crack in a tensile test specimen. The crack initiates at the centre of the specimen and propagates towards the surfaces. Inside the specimen, the crack runs at an angle of 45° only for short distances and switches the orientation to remain in the cross section with the highest stress

How the crack propagates further depends on several parameters, like the hardening behaviour of the material and the strain rate.

> Since the radial stress σ_r and the circumferential stress σ_c (figure 3.13) have the same magnitude, the stress is the same in all direction perpendicular to the loading direction and is equal to them. Thus, all planes at 45° to the loading direction have the same maximum shear stress, and slip can occur on any of them. Within the specimen, there is thus no preferential slip direction, and several different slip planes may be found locally.

If the material softens, for example, at large plastic deformations parallel to the crack, it may be easier to follow the direction of a crack into the less highly stressed region. In this case, the crack extends through the whole specimen at an angle of 45° to the loading direction and a shear-face fracture forms (figure 3.15(b)). *Lange* [90] discusses the conditions for a shear-face fracture in some detail.

In most cases, it is easier for the crack not to depart too far out of the region of smallest cross section. On the one hand, this is due to the smaller stress level in the thicker parts of the specimen. On the other hand, the damage in these regions is less because less inclusions have failed there (due to the smaller stress and plastic deformation). The crack changes its direction and propagates at 45° to the loading direction back to the smallest cross section with its larger stresses and damage. The crack thus zigzags through the interior of the specimen as shown in figure 3.37. As soon as the crack reaches the outer part of the specimen, the stress state becomes two-dimensional.[32] The maximum principal stress is oriented in the loading direction and the

---

[32] In the region of the crack, the specimen is a ring, only. On the outer and inner surface of this ring, there can be no radial stresses, thus the radial stresses within the ring must be small.

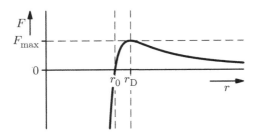

Fig. 3.38. Dependence of the interatomic distance r on the external force F

minimum principal stress in radial direction. The largest shear stress can be found on a cone at 45° to the loading direction. Slip and crack propagation thus occur preferentially on these planes. Since the material cross section is small, slip can occur over larger distances than before without causing a geometrical incompatibility. Furthermore, the specimen is less damaged in its surface region due to the smaller plastic deformation there. For these reasons, the direction of crack propagation is not determined as strongly by cracks in the material as it was before. The crack thus grows at 45° on conical surfaces without changing its propagation direction. Thus, the characteristic cup and cone fracture surface forms on the two halves of the specimen.

Since slip and failure occur simultaneously on the whole circumference of the specimen and since both possible conical surfaces are equivalent, both directions are usually found in a specimen, leading to partial cups and cones on both halves of the specimen.

3.5.2 Cleavage fracture

A *cleavage fracture* (microscopically brittle fracture) occurs (almost) without microscopic deformation perpendicular to the largest tensile stress. Bonds between the atoms break. In face-centred cubic metals, the ductility is so large that cleavage fracture can occur in extreme cases only. In body-centred cubic metals, cleavage fracture can occur at low temperature or high strain rates; in ceramics, cleavage fracture is the standard case.

The binding force, shown in figure 2.6 on page 38, provides a simple model for the cracking of atomic bonds. If we plot the external instead of the internal force for a certain atomic distance r, figure 3.38 results. If the external tensile force F exceeds the maximum, the atoms separate and the bond breaks. The breaking of the bond is due to a tensile force so that bond breaking is caused by the largest tensile stress in the material, the maximum principal stress σ_I.

The component or specimen cross section does not fail simultaneously everywhere. Instead, a crack forms locally by bond breaking. On the one hand, this is due to local stress concentrations in the component, which may be caused by the component geometry, its microstructure, or by previous plastic deformations. This is discussed in detail in *Lange* [90]. On the other

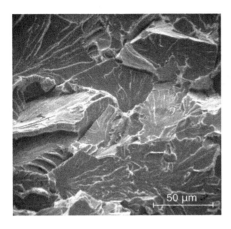

Fig. 3.39. Scanning electron microscope micrograph of a cleavage fracture surface in a journal of a shaft made of 42 CrMo 4

hand, the material may contain microstructurally weak points which may ease crack formation, for example a grain with its cleavage plane (see below) perpendicular to the maximum principal stress. Since the microstructure (e. g., the grain orientation) or the stress level change in the vicinity of the initialised crack, the crack cannot propagate initially. It thus remains stationary [90] and propagates (stably) only upon load increase. At a certain critical crack length or stress level, the crack propagates unstably through the specimen. The stress and crack length required for unstable crack propagation (in the ductile or brittle case) are calculated by *fracture mechanics,* the topic of chapter 5.

Similar to shear fracture, cleavage fracture is usually *transcrystalline,* but may sometimes also be *intercrystalline.* As already mentioned, a transcrystalline cleavage fracture propagates along certain crystallographic planes, the *cleavage planes* (e. g., the {100} planes in body-centred cubic metals). Cleavage fracture surfaces are microscopically smooth, but they may contain steps, for example because of a transition of the crack to a neighbouring grain with slightly different orientation or because of cutting through a screw dislocation (a one-dimensional lattice defect, see sections 6.2 and 6.3.5). The appearance of a cleavage fracture surface may vary [90], one example is shown in figure 3.39.

When grain boundaries are embrittled (for example, by precipitates, see section 6.4.4), cleavage fracture may be intercrystalline. In this case, the grain structure can be clearly seen in a scanning electron microscope picture (see figure 1.10(b)).

As already discussed, the maximum principal stress σ_I determines whether cleavage fracture occurs. If it reaches the *cleavage strength* σ_C (sometimes also called *cohesive strength*), the initially crack-free material fails by cleavage fracture. This stress σ_C is thus sufficient to initiate a crack in a crack-free material and to propagate it. Figure 3.40 illustrates the cleavage strength using Mohr's circle. It is a vertical, straight line at $\sigma = \sigma_C$.

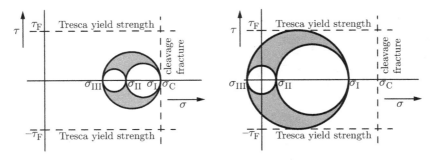

(a) Cleavage fracture. The cleavage strength is reached before the yield strength

(b) Yielding. The yield strength is reached before the cleavage strength

Fig. 3.40. Illustration of the Tresca yield strength and cleavage strength in Mohr's circle

3.5.3 Fracture criteria

Depending on whether the stress state reaches the yield strength or the cleavage strength first, the material will yield or fail by cleavage fracture as sketched in figure 3.40. Cleavage fracture occurs when the cleavage strength is reached first:

$$\sigma_{eq}(\sigma_I, \sigma_{II}, \sigma_{III}) < R_p , \tag{3.73a}$$

$$\sigma_I = \sigma_C , \tag{3.73b}$$

where σ_{eq} is an equivalent stress, for example the Tresca or von Mises equivalent stress. If we use the Tresca yield criterion, equation (3.73a) yields $\tau_{max} < \tau_F$ (figure 3.40(a)).

> Since the flow stress increases due to hardening, failure by cleavage fracture may occur even after plastic yielding. In this case, a (macroscopically) ductile (but microscopically brittle) cleavage fracture develops, a rather seldom case that can occur only in a multiaxial stress state [90].

Cleavage fracture is favoured by the following conditions:

- Loading of the material in a triaxial stress state that keeps Mohr's circle small and shifts it to the right in the direction of the cleavage strength. Such a stress state can be found in the notched bar impact bending test [42]. In components, changes in cross section and notches cause such a stress state (see chapter 4).
- Loading of the material at high strain rates, for example in the notched bar impact bending test, or at low temperature. This is due to the fact that the yield strength always depends on these two parameters (see section 6.3.2), whereas the cleavage strength is almost constant. This is particularly the

case in polymers and body-centred cubic metals. In polymers, the yield strength strongly increases with decreasing temperature if the temperature is near the glass temperature (see chapter 8); in body-centred cubic metals, this happens near the so-called ductile-brittle transition temperature (see section 6.3.3). The increase in the yield strength increases the danger of reaching the cleavage strength before the yield strength.

- Increasing the yield strength of metals e. g., by alloying, heat treatments (like hardening), or cold working (see section 6.4). This implies that high-strength materials have a larger tendency to fail by cleavage fracture than low-strength materials.
- Reduction of the cleavage strength σ_C by weakening of the interatomic bonds. This may happen, for example, when hydrogen or sulfur is dissolved in steel (see below).

The material behaviour is ductile if the yield strength is reached first (figure 3.40(b)):

$$\sigma_{eq}(\sigma_I, \sigma_{II}, \sigma_{III}) = R_p , \tag{3.74a}$$

$$\sigma_I < \sigma_C . \tag{3.74b}$$

This can be achieved by keeping at least one stress component in the compressive region, thus shifting Mohr's circle to the left of the diagram. This method is used in metal forming like forging or rolling.

In many cases, the cleavage strength σ_C cannot be measured experimentally. For example, body-centred cubic metals only fail by cleavage fracture even at temperatures below the ductile-brittle-transition temperature if the stress state is triaxial. For this reason, it is impossible to measure σ_C in a tensile test. Ceramics usually fail because a microcrack, already present in the material, propagates and causes failure at a stress below σ_C (see section 7.3). The tensile strength R_m of a ceramic is thus smaller than its cleavage strength.
    nn

* Hydrogen embrittlement

One important cause for the embrittlement of high-strength metals, especially ferritic steels, is hydrogen dissolved in the material. The hydrogen atoms are situated in the gaps between the atoms in the crystal lattice (*interstitially*, see figure 6.37 on page 204) and weaken the interatomic bonds, thus reducing the cleavage strength. Hydrogen can enter the material in electrochemical reactions in aqueous solutions, for example during corrosion or galvanisation (e. g., electrogalvanising of sheets).

One prerequisite for the accumulation of hydrogen in the material is that it is present in its atomic state because it cannot diffuse into the material otherwise. The electrochemical reaction in aqueous solutions mentioned above is one example:

$$H_3O^+ + e^- \longrightarrow H_2O + H.$$

Welding in humid atmosphere can, due to the dissociation of water molecules, also allow hydrogen to diffuse into the material. The tensile residual stresses produced by the welding process widen the crystal lattice and thus attract hydrogen atoms which then reduce the cleavage strength. This may cause crack propagation by residual stresses alone, without any external stress. Since the diffusion to the most highly stressed regions needs some time, fracture may occur hours or days after the welding process [90]. Therefore, this is often called *desktop effect* (the material fails while lying around somewhere) or delayed fracture. This effect is especially important in high-strength materials, for example in steels, because the residual stresses in these materials can be large without being relieved by plastic deformation.

Dissolved hydrogen can also cause so-called stress corrosion cracking. This will be discussed in section 5.2.6.

4

Notches

Notches are abrupt changes in the geometry of a component. They may be necessary for design reasons, for example in a seat of a rolling bearing, a feather key slot, a drill-hole, or a screw thread for a connection. Notches may also be caused during manufacturing or service. Examples are cavities in casting, tool marks in machining, or wear marks in service. Notches cause local stress concentrations and thus may induce premature failure if not correctly accounted for during component design. In this chapter, we will discuss theories that allow to estimate how notches affect the stresses and thus provide tools for safe design of notched components.

4.1 Stress concentration factor

The stress distribution in a component can be visualised using so-called *stress trajectories.* These trajectories always run in the direction of the maximum principal stress. Their distance is inversely proportional to the stress so that the *stress trajectory density* is a measure of the locally acting stress. Each abrupt change in cross section deflects the stress trajectories which then move closer together. Thus, a local stress concentration arises.

> The term *stress trajectory* is due to the fact that the stress distribution in a component is analogous to the velocity distribution of a laminar, frictionless fluid. The stress concentration at changes in the geometry corresponds to the disturbed flow of the fluid at similar geometries. Different from fluid flows, stress trajectories cannot become turbulent.

Figure 4.1 shows sketches of the stress trajectories near differently shaped notches. If we look at the stress trajectories at a cross section at the notch root, we see that they are not evenly distributed, but become more narrow at the notch root. Thus, there is a local stress concentration, with a maximum stress $\sigma_{\max}$ in the notch root as shown in figure 4.2. The shape and size of the

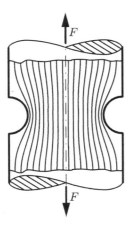

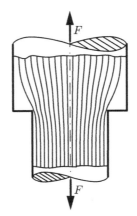

(a) Circumferential notch

(b) Abrupt change of diameter
in a round rod

Fig. 4.1. Stress trajectories in notched components. The stress trajectories are aligned with the maximum principal stress, their density is a measure of the stress level. At the notch root, there is a stress concentration in both geometries

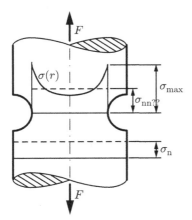

Fig. 4.2. Axial stress distribution in a notched cross section under tensile load

notch determines how strong the stress concentration is. This is quantified by the *stress concentration factor* K_t that is defined for linear-elastic material behaviour as

$$K_t = \frac{\sigma_{max}}{\sigma_{nss}},\tag{4.1}$$

with σ_{max} being the maximum stress in the notch root and σ_{nss} being the *net-section stress*, defined as the force divided by the reduced cross-sectional area at the notch position. The net-section stress is thus larger than the nominal stress σ_n away from the notch.

In the centre of the specimen at the position of the notch, the stress is smaller than the net-section stress σ_{nss}. The reason for this is that the total force transferred through the cross section does not change and that the stress at the notch root is larger than σ_{nss}.

In the context of notches, the stress concentration is always related to the net-section stress σ_{nss}.[1] If the stress concentration at the notch root has to be compared to the nominal stress far away from the notch (σ_n or, more precisely, $\sigma_{n\infty}$, infinitely far away from the notch), the increase in stress due to the reduced cross section and the stress concentration at the notch root have to be multiplied.

To design notched components, knowledge of K_t is required. Therefore, empirical formulae have been determined that can be used to calculate K_t for different geometries and load cases. They are collected in tables e. g., 'Peterson's Stress Concentration Factors' [109] or 'Dubbel' [18]. One example, a shaft with a circumferential notch under tensile load, is shown in figure 4.3. The dimensions in the figure are the outer diameter D, the diameter at the notch root d, the notch depth t (with $2t = D - d$), and the notch radius ϱ.

As an example, consider a shaft with $D = 100\,\mathrm{mm}$ and a notch with radius $\varrho = 5\,\mathrm{mm}$ and depth $t = 5\,\mathrm{mm}$ (a semi-circular notch). The diameter at the notch position is thus $d = D - 2t = 90\,\mathrm{mm}$. Using $d/D = 0.9$ and $\varrho/t = 1.0$, we can read off the stress concentration factor from the diagram (see figure 4.3): $K_t \approx 2.7$. The exact value is $K_t = 2.734$.

Imagine the shaft to be loaded in tension with a force of 1200 kN. Far away from the notch, the stress is thus 152.8 MPa. Due to the smaller cross section at the notch position, the net-section stress is $\sigma_{nss} = 188.6\,\mathrm{MPa}$. If the material is linear elastic, as we assumed so far in this section, and use $K_t \approx 2.7$, the maximum stress is $\sigma_{max} = K_t\,\sigma_{nss} = 516\,\mathrm{MPa}$.

If the available materials to construct the shaft are a ceramic with $R_m = 400\,\mathrm{MPa}$ or the aluminium alloy AlSi 1 MgMn with $R_{p0.2} = 202\,\mathrm{MPa}$ and $R_m = 237\,\mathrm{MPa}$, we can expect the ceramic to fail because the stress at the notch root is much larger than the tensile strength. For the aluminium alloy, the tensile strength is also exceeded, and we thus might expect its failure as well. However, the calculation is not valid in the case of a ductile material, for equation (4.1) is valid only for a linear-elastic material, whereas the alloy AlSi 1 MgMn yields at $R_{p0.2} = 202\,\mathrm{MPa}$. This increases the strain at the notch root and reduces the stress concentration. The actual stress at the notch root cannot be calculated with the tools introduced so far. In the next section, we will discuss Neuber's rule that allows to estimate the stresses.

---

[1] This is different from fracture mechanics (chapter 5) where the nominal stress is always calculated by using the total cross section (σ_n).

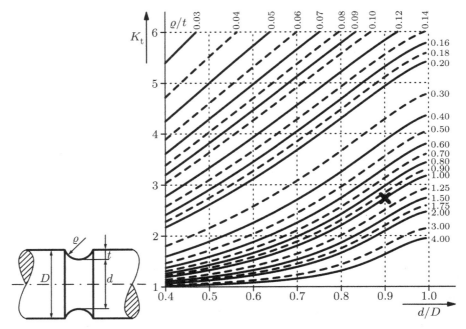

Fig. 4.3. Diagram of the stress concentration factor K_t of a shaft with a circumferential notch under tensile loading (after [18]). From the lines in the diagram, the line with the appropriate ratio ϱ/t has to be selected. Next, the intersection with the vertical line at the correct ratio d/D is determined. The value of K_t can be read off at the ordinate. The cross marks the example point discussed in the text

4.2 Neuber's rule

In the previous section, we defined the stress concentration factor K_t (equation (4.1)) for linear-elastic materials. As the example at the end of the previous section shows, it cannot be used directly for the case of ductile materials, for yielding at the notch root reduces the stresses. In this section, we discuss how the influence of a notch can be taken into account even in ductile materials.

Because the stress concentration factor is calculated using stresses, it is now denoted as $K_{t,\sigma}$, resulting in

$$K_{t,\sigma} = \frac{\sigma_{\max}}{\sigma_{\text{nss}}}.$$ (4.2)

A similar formula can be used for the strains:

$$K_{t,\varepsilon} = \frac{\varepsilon_{\max}}{\varepsilon_{\text{nss}}},$$ (4.3)

with ε_{nss} being the strain at a stress σ_{nss}. The maximal strain in the notched cross section is $\varepsilon_{\max}$. In a linear-elastic material, Hooke's law ensures $K_{t,\sigma} =$

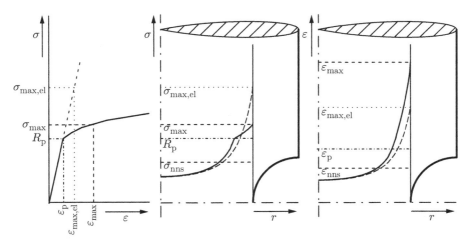

Fig. 4.4. Axial stress and strain in the notched cross section for two different materials: Linear-elastic (dashed line) and elastic-plastic (solid line). The figure shows the nominal stresses and strains (σ_{nss}, ε_{nss}) and the maximum values in the linear-elastic ($\sigma_{\text{max,el}}$, $\varepsilon_{\text{max,el}}$) and elastic-plastic case (σ_{max}, ε_{max})

$K_{\text{t},\varepsilon}$. If σ_{max} exceeds the yield strength of the material,[2] the material yields at the notch root and Hooke's law is no longer valid. As shown in figure 4.4, this increases ε_{max} compared to the linear-elastic case. The maximum stress σ_{max}, on the other hand, is reduced due to local unloading. Therefore, $K_{\text{t},\sigma} < K_{\text{t},\varepsilon}$ holds. The numerical values of $K_{\text{t},\sigma}$ and $K_{\text{t},\varepsilon}$ are still unknown, though.

> Strictly speaking, equivalent stresses (for example, the von Mises equivalent stress) should be used to calculate stresses and strains due to the multiaxial stress state. Furthermore, the equation $K_{\text{t},\sigma} = K_{\text{t},\varepsilon}$ is only approximately valid in the elastic region because of the transversal contraction caused by the radial and circumferential stresses. For engineering purposes, a uniaxial calculation is sufficient, especially so if we consider the scatter in the material parameters. The multiaxiality of the stress state at the notch root is discussed in section 4.3.

Neuber [106] suggested that the geometric mean of $K_{\text{t},\varepsilon}$ and $K_{\text{t},\sigma}$ remains unchanged even if the material yields:

$$K_{\text{t},\varepsilon} \cdot K_{\text{t},\sigma} = K_{\text{t}}^2 \,. \tag{4.4}$$

Figure 4.5 shows a qualitative plot of $K_{\text{t},\varepsilon}$ and $K_{\text{t},\sigma}$ with increasing load. Until the material yields, both quantities are equal; after yielding, $K_{\text{t},\varepsilon}$ increases and $K_{\text{t},\sigma}$ decreases. If we insert equations (4.2) and (4.3) into equation (4.4), we find

---

[2] Here we assume that σ_{nss} is smaller than the yield strength.

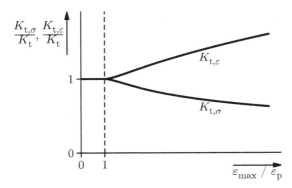

Fig. 4.5. Qualitative dependence of the stress concentration factors on the load. ε_p is the strain at yielding (stress R_p) in the notch root

$$\frac{\varepsilon_\mathrm{max}}{\varepsilon_\mathrm{nss}} \cdot \frac{\sigma_\mathrm{max}}{\sigma_\mathrm{nss}} = K_\mathrm{t}^2 \;.$$

If we assume that the net-section stress σ_nss is smaller than the yield strength, we can use Hooke's law $\varepsilon_\mathrm{nss} = \sigma_\mathrm{nss}/E$ to derive *Neuber's rule*

$$\varepsilon_\mathrm{max} \cdot \sigma_\mathrm{max} = \frac{\sigma_\mathrm{nss}^2}{E} K_\mathrm{t}^2 \;. \tag{4.5}$$

According to Neuber, this equation holds in the notch root for a given load and geometry.

> Neuber's rule can be derived as an approximation. In the case of a parabolic notch and under the assumption of a certain simple stress-strain curve, a formula can be derived that simplifies to Neuber's rule in the case of a very sharp notch. For large notch radii, Neuber's rule is a conservative approximation i.e., it overestimates stresses and strains [106]. This conservative property of the rule is valid for most other notch geometries as well [59].

If we approximate the stress state at the notch root as uniaxial,[3] the material state must lie on the stress-strain curve measured in tensile tests. This provides another relation between σ_max and ε_max, which are therefore uniquely determined. Graphically, equation (4.5) corresponds to a hyperbola in the σ-ε space of the stress-strain diagram, since the right side is constant for a given load case. The stresses and strains at the notch root can be found as the intersection of the hyperbola and the stress-strain curve as shown in figure 4.6.

---

[3] In reality, the stress state is biaxial at the notch root (the radial stress at the surface is zero), so that there is no difference to the uniaxial case if the Tresca yield criterion is used. If the von Mises yield criterion is used, there is a slight difference which is neglected here.

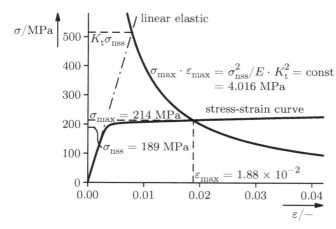

Fig. 4.6. Determination of σ_{max} and ε_{max} using the Neuber's rule. The specified numbers refer to the example in the text (AlSi 1 MgMn)

We now resume the example from page 121, using the stress-strain curve of AlSi 1 MgMn from figure 4.6. Taking the value of $\sigma_{nss} = 188.6$ MPa and Young's modulus of $E = 66\,200$ MPa, we find from equation (4.5)

$$\sigma_{max} \times \varepsilon_{max} = \frac{(188.6\,\text{MPa})^2}{66\,200\,\text{MPa}} \times 2.734^2 \approx 4.016\,\text{MPa}\,,$$

the Neuber's hyperbola shown in the figure. Reading off the intersection of the two curves yields $\sigma_{max} = 214$ MPa and $\varepsilon_{max} = 1.88 \times 10^{-2}$. Since σ_{max} is significantly smaller then the tensile strength $R_m = 237$ MPa and since the plastic strain of about 1.88×10^{-2} is small compared to the fracture strain ($A = 0.17$), the material can bear the load. Thus, the metal that is weaker without a notch can bear larger loads than the ceramic in the notched component.

These considerations show that the component does not fracture although it yields at the notch root. However, plastic deformation is confined to a small volume. The global deformation is thus small so that limited plastic flow in the notch root is also acceptable from this point of view.

To summarise, it can be stated that in notched components the strongest material is not always the best choice since weaker materials have a larger ductility. This is even more important under cyclic loads, a fact to be discussed in chapter 10.

* 4.3 Tensile testing of notched specimens

In this section, we discuss the influence of a notch on a tensile test specimen. We compare an un-notched and a notched tensile specimen as shown in fig-

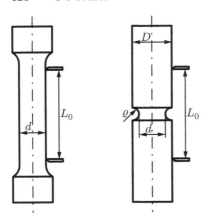

Fig. 4.7. Un-notched and notched tensile specimen

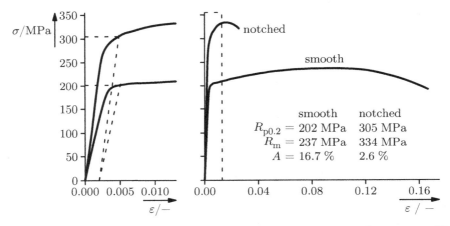

	smooth	notched
$R_{p0.2}$ =	202 MPa	305 MPa
R_m =	237 MPa	334 MPa
A =	16.7 %	2.6 %

Fig. 4.8. Stress-strain curve of an un-notched and a notched tensile specimen. The left part is a detailed view at small strains

ure 4.7, made of the aluminium alloy AlSi 1 MgMn. The specimen dimensions are as follows: The diameter $d = 8\,\text{mm}$ of the un-notched specimen is the same as that of the notched specimen at the notch position, d. The outer diameter of the notched specimen is $D = 1.5d = 12\,\text{mm}$, the notch radius is $\varrho = 0.25d = 2\,\text{mm}$. Thus, the notch depth is $t = \varrho$. From diagram 4.3, we can read off a stress concentration factor of $K_t = 1.70$. The original gauge length is $L_0 = 5d = 40\,\text{mm}$.

Figure 4.8 shows the stress-strain diagrams measured for the specimens. In the diagram, the nominal strain $\varepsilon = \Delta L/L_0$ according to figure 4.7 and the nominal stress $\sigma = F/(\pi d^2/4)$ are plotted. It has to be noted that the

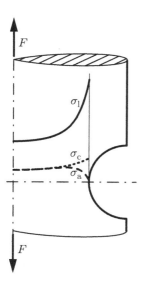

Fig. 4.9. Stress distribution in a purely elastic stress state

diameter at the narrowest cross section is used in both specimens. Therefore, the numbers in the diagram are not the true stresses and strains in the notched specimen. Nevertheless, they can be used to determine a relative stiffness and strength.

We now discuss the differences between the two stress-strain curves:

It is apparent that the elastic stiffness of the notched specimen is larger than that of the un-notched one. The reason for this is that the cross section of the un-notched specimen is constant throughout the specimen: $S_{\text{smooth}} = \pi d^2/4$. In the notched specimen, the largest part of the gauge length has the larger cross-sectional area of $S_{\text{notched}} = \pi D^2/4 = 2.25 S_{\text{smooth}}$ and thus a higher specimen stiffness.

Significant yielding occurs at a larger nominal stress in the notched than in the un-notched specimen. To understand this, we need to take a closer look at the stress distribution. In the un-notched specimen, the stress state is uniaxial, and the specimen starts to yield when the longitudinal stress is $\sigma_l = R_{\text{p}}$. In the notched specimen, the stress state at the notch position is more complicated. The longitudinal stress σ_l is distributed according to figure 4.2. Furthermore, there are stresses in radial and circumferential direction (σ_r and σ_c). Figure 4.9 shows the three stress components for a purely elastic deformation without yielding at the notch root. The longitudinal stress σ_l is maximal at the notch root. The radial stress σ_r, however, has to be zero since no stress can be transmitted at the specimen surface. The stress state is thus biaxial. This only slightly impedes yielding which therefore still starts

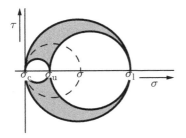
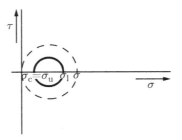

(a) At the notch root. Mohr's circle is large and thus causes early yielding

(b) In the centre of the specimen. Due to the triaxial stress state, Mohr's circle is small, and yielding is severely hampered

Fig. 4.10. Mohr's circle of the notched (solid line) and un-notched (dashed line) tensile specimen for purely elastic behaviour at the same external load

at $\sigma_1 \approx R_\mathrm{p}$.[4] Because of the stress concentration at the notch root, yielding begins here at an early stage as we saw in section 4.2 (see figure 4.10(a)). The interior of the specimen is in a state of triaxial tensile stress. Therefore, a higher external load is required to make the material yield (see figure 4.10(b)).

The maximum external force during the test is larger in the notched than in the un-notched case. This is again due to the triaxial stress state and can be explained in the same way as for the onset of yielding above. It has to be kept in mind that the diameter of the un-notched specimen is the same as that of the notched specimen at the notch position. Thus, it should not be assumed that a specimen can be made stronger by notching it.

The nominal strain at fracture is much larger in the un-notched specimen than in the notched one. This is due to the large gauge length used for measuring the strains. In the un-notched specimen, plastic deformation occurs throughout the entire gauge length (until necking starts), and the measured strain is the same as the strain within the material. In the notched specimen, yielding occurs locally within the notched cross section. Large plastic deformations occur in this region, whereas the rest of the specimen remains elastic and its strains are only small. If the strain was measured locally, the difference would be smaller, as we already saw in section 3.2.2.

---

[4] If we use the Tresca yield criterion, yielding occurs exactly at $\sigma_1 = R_\mathrm{p}$. With the von Mises yield criterion, the result is $\sqrt{1/2 \cdot [(\sigma_1 - \sigma_\mathrm{c})^2 + \sigma_1^2 + \sigma_\mathrm{c}^2]} = R_\mathrm{p}$. Depending on the value of the circumferential stress, the axial stress at which yielding starts may be up to 15.5% larger than with the Tresca criterion (see section 3.3.1).

5

Fracture mechanics

5.1 Introduction to fracture mechanics

Frequently, components contain cracks, microcracks, or crack-like defects. Possible causes are machining flaws, manufacturing defects (e. g., casting pores in metals or sintering pores in ceramics), or crack formation during service (e. g., by cyclic loads – see chapter 10 – or corrosive attack).[1] Some examples are shown in figure 5.1. Particles, for example precipitates, can initiate cracks as well: On the one hand, they often possess an unfavourable geometry (e. g., being plate-like or sharp-edged), on the other hand, they can fail easily by detaching from the matrix or by breaking. In all these cases, it is not sufficient to design components against the yield strength because failure by crack propagation can occur at much smaller loads. *Fracture mechanics* deals with such problems. Its main objective is to predict at what loads cracks may grow, in order to enable a safe design. Cracks can be considered as extremely (or even infinitely) sharp notches. As the theoretically calculated stress field at the crack tip becomes infinite in this case, the methods used for assessing notches are not applicable. The tools of fracture mechanics are capable to deal with these infinities.

In this chapter, we assume a monotonically increasing or static load. The application of fracture mechanics to the case of cyclic loads is the topic of section 10.6.1.

5.1.1 Definitions

As already stated, fracture mechanics deals with the growth of cracks, also called *crack propagation, crack growth,* or *crack extension.* A non-propagating crack is called *stationary* and does not cause component failure.

---

[1] Pores are not cracks, but are rather notch-like. However, their radius of curvature is frequently small enough to justify their treatment as cracks.

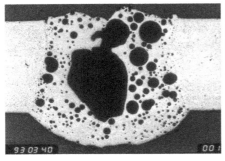
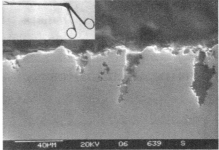

(a) Welding line in pressure die cast alu- (b) Corrosion attack on a rongeur for-
minium. Courtesy of Institut für Füge- ceps made of a martensitic stainless steel
und Schweißtechnik, Technische Univer-
sität Braunschweig, Germany

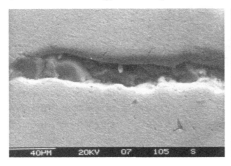
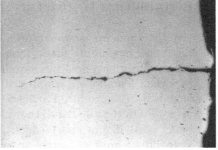

(c) Slag inclusion in a forged casing   (d) Crack in a connection rod

Fig. 5.1. Examples of initial cracks and other damage phenomena that might initiate cracks

Fracture mechanics is based on continuum mechanics and is therefore only applicable when all relevant length scales are large compared to the length scale of the microstructure (for example, the grain size).

Three characteristic load cases are distinguished in fracture mechanics, which differ in the orientation of the stress field to the crack. They are called *mode I* to *III*. In mode I, the largest principal stress σ_I is oriented perpendicularly to the crack surface as shown in figure 5.2(a). Tensile stresses open the crack and thus separate the surfaces. Compressive stresses close the crack so that forces can be transmitted almost identically to a case without a crack. In modes II and III, the crack surfaces are loaded in shear (see figures 5.2(b) and 5.2(c)). These modes do not open the crack. When the load is applied, the crack surfaces slide with friction and thus dissipate part of the external work. Mixed-mode loads can also occur. As we will see later, crack propagation is determined by an energy balance. Because the energy dissipated in modes II or III is not available for crack propagation, a crack propagates at smaller

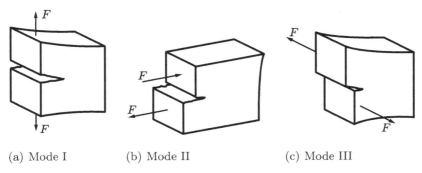

(a) Mode I   (b) Mode II   (c) Mode III

Fig. 5.2. Load cases in fracture mechanics

loads in mode I. Independent of its initial orientation, a growing crack thus changes its orientation to be perpendicular to the maximum principal stress i. e., to be loaded in mode I, if stress field and material are homogeneous. This load case is therefore the most important one and will be the only one considered in the following. The maximum principal stress σ_I therefore determines the material behaviour in crack propagation.

When the crack propagates, the crack surface can be formed by either shear or cleavage fracture, or a mixture of both, leading to fracture surfaces as discussed in section 3.5. If fracture occurs by crack propagation at stresses below the yield strength, the global plastic deformation of the component is usually small because plastic deformation is localised at the crack tip.

5.2 Linear-elastic fracture mechanics

As the name suggests, linear-elastic material behaviour is the precondition to allow applying the theory of *linear-elastic fracture mechanics* (LEFM), discussed in this section. Strictly speaking, this precondition is fulfilled only in brittle materials like ceramics. In good approximation, it can also be used in ductile materials if the region of plastic deformation is restricted to the vicinity of the crack tip. Therefore, it can in many cases also be used to analyse metals.

We start by considering the stress field near the crack tip and the energy release during crack propagation. Next, we will discuss how to design components against failure by crack propagation and how to determine relevant material parameters.

5.2.1 The stress field near a crack tip

A crack can be considered as an infinitely sharp notch. If we estimate the maximal stress by setting the notch radius to zero, the stress concentration

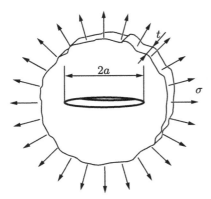

Fig. 5.3. Infinite plate containing an internal crack that is opened by a load

factor K_t and thus the stress at the crack tip becomes infinite.[2] Therefore, the stress field is singular near the crack tip.[3] Because the stress field in the vicinity of the crack tip is similar, independently on the geometry, one geometry is investigated first; the stress distributions for other geometries can be derived from it. We consider an infinitely extended plate of thickness t containing an internal crack of length $2a$. The plate is loaded on all sides with a stress σ, see figure 5.3. We assume that the stress field can be described as plane stress[4] and that the material is linear elastic, homogeneous, and isotropic. This configuration is called a *Griffith crack* and is loaded in mode I because there are no shear stresses in the crack plane. The stress state within the plate can then be calculated analytically. If terms that are small near the crack tip are neglected, the following near-field approximation can be calculated for the stress field [75, 147]:

$$\tilde{\sigma}_{11}(r,\varphi) = \frac{K_\mathrm{I}}{\sqrt{2\pi r}} \cos\frac{\varphi}{2} \left[1 - \sin\frac{\varphi}{2} \sin\frac{3\varphi}{2}\right],$$

$$\tilde{\sigma}_{22}(r,\varphi) = \frac{K_\mathrm{I}}{\sqrt{2\pi r}} \cos\frac{\varphi}{2} \left[1 + \sin\frac{\varphi}{2} \sin\frac{3\varphi}{2}\right], \qquad (5.1)$$

$$\tilde{\tau}_{12}(r,\varphi) = \frac{K_\mathrm{I}}{\sqrt{2\pi r}} \cos\frac{\varphi}{2} \sin\frac{\varphi}{2} \cos\frac{3\varphi}{2}.$$

r and φ are the coordinates of a polar coordinate system centred at the crack tip (figure 5.4), and the tilde on the stress marks it as an approximation. The

---

[2] For very small notch radii, the calculation of the stress concentration factor K_t is problematic and the methods of fracture mechanics are more precise. Nevertheless, the fact that there is a singularity at the crack tip is reflected correctly by K_t.

[3] This does not mean that the force on each atomic bond becomes infinite as well, see exercise 13.

[4] This assumption is correct if the thickness is small compared to all other dimensions i. e., $t \ll a$.

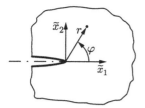

Fig. 5.4. Coordinate system near the crack tip

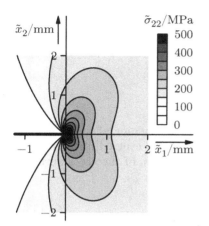

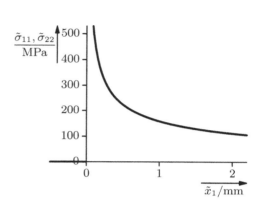

(a) Stress distribution

(b) Stress distribution along the $\tilde{x}_1$ axis ($\varphi = 0$).
On the $\tilde{x}_1$ axis, $\tilde{\sigma}_{11}$ and $\tilde{\sigma}_{22}$ are identical

Fig. 5.5. σ_{22} stress field near the crack tip. The crack of length $2a = 10\,\text{mm}$ lies along the negative $\tilde{x}_1$ axis and ends in the centre of the coordinate system shown in figure 5.4. The material is loaded with a stress of $\sigma = 100\,\text{MPa}$

stress intensity factor K_I (SIF) is defined using the external stress σ:

$$K_\text{I} = \sigma\sqrt{\pi a}\,Y\,. \tag{5.2}$$

The index 'I' denotes the mode I load case. Y is a *geometry factor* that takes different geometries into account and is 1 for the case considered here. It will be discussed in greater detail below (see page 139). All other terms in the equations depend only on the spatial position so that only the stress intensity factor determines the stress level near the crack tip. The stress functions are singular because the distance from the crack tip r enters the denominator. Figures 5.5(a) and 5.5(b) show examples of the stress fields. The physical unit of the stress intensity factor K is $\text{MPa}\sqrt{\text{m}}$ according to equation (5.1) because the stress near the crack tip is proportional to the quotient of the stress intensity factor and the square root of the distance to the crack tip $\sqrt{r}$: $\sigma \propto K/\sqrt{r}$.

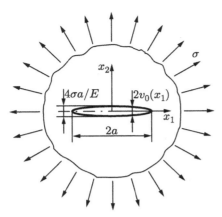

Fig. 5.6. Elliptical crack opening

The stress field at the crack causes an ellipsoidal opening of the crack due to Hooke's law (figure 5.6). Using a coordinate system centred in the middle of the crack as shown in figure 5.6, the displacement v_0 of the crack surfaces is given by

$$v_0(x_1) = \frac{2\sigma}{E}\sqrt{a^2 - x_1^2}\,. \tag{5.3}$$

The ellipsoidally opened crack has the semiaxes a and $2\sigma a/E$.

As we have neglected small terms in the derivation of equations (5.1), they are valid only near the crack tip. This can immediately be seen by the fact that the stresses in equations (5.1) approach zero for large r, whereas the true far-field stress is σ. Directly at the crack tip, the equations are also not valid because they are based on continuum mechanics which is not applicable on the atomic scale (see also exercise 13).

If we increase the load and thus the stress intensity factor K_I, the crack will propagate when K_I reaches a certain critical value, the *fracture toughness* K_Ic (again, 'I' denotes loading in 'mode I'). The fracture toughness is a material property. Consequently, the failure criterion for loading in mode one is given by

$$K_\mathrm{Ic} = \sigma_\mathrm{c}\sqrt{\pi a}\,Y\,, \tag{5.4}$$

where σ_c is the critical stress which will lead to crack propagation for a crack of initial length $2a$.

5.2.2 The energy balance of crack propagation

The propagation of a crack can be understood by calculating an energy balance. As the stress state of the material changes when the crack propagates, the elastic energy changes as well. In addition, energy is needed to break the atomic bonds. Finally, external work can be performed on the system, for instance by a shift in the loading point. The crack will only propagate if the overall energy balance calculated from these contributions is positive i. e., if energy is released during crack propagation. To define criteria for crack propagation, these energy contributions have to be considered.

As before, we study the case of a infinite plate of thickness t that contains an internal crack of length $2a$ and is loaded with a homogeneous stress σ (see figure 5.3). In the following, we will only consider one half of the crack ($x_1 \geq 0$, see figure 5.6); the other half is identical due to symmetry. All energy contributions for the whole crack are thus twice the values given here.

Depending on the load case, the loading points of a load F may shift by distance δ when the crack propagates, so an external work of

$$W = \frac{1}{2} \int_0^\delta F \mathrm{d}\delta \tag{5.5}$$

is done.[5] The factor $1/2$ stems from the fact that only half of the plate is considered. A special case occurs if the loading points do not shift at all. This may happen if the force is a constraining force necessary to enforce a displacement boundary condition, a so-called *dead load*. It can also occur if the crack is very small compared to the component, for instance – as in the case here – if the component is assumed to be infinite.[6] In both cases, the external work is $W = 0$.

During crack propagation, changes in the stress and strain fields cause a change in the stored elastic strain energy

$$U^{(\mathrm{el})} = \iiint_V w^{(\mathrm{el})} \, \mathrm{d}V \tag{5.6}$$

with an *elastic energy density* (see section 2.4.1)

$$w^{(\mathrm{el})} = \int \sigma_{ij} \, \mathrm{d}\varepsilon_{ij} \,. \tag{5.7}$$

When the crack propagates by an amount $\mathrm{d}a$, new surfaces are formed on both sides of the crack with an area of $t \, \mathrm{d}a$.[7] To create new surface in a

---

[5] As we consider an infinitely extended plate, the force F is, strictly speaking, infinite as well. For a mathematically exact description it would be necessary to normalise the energy and the force on a unit length in x_1 direction. As we will consider only stresses and energy densities in the following, this mathematical nicety is irrelevant here.

[6] This can also be explained by assuming that the forces at infinity are dead loads. In this case, they will change only by an infinitesimal amount when the crack propagates. If fixed loading points do not cause a change in the stress, a constant stress can also not cause a change in the loading points.

[7] Crack propagation by $\mathrm{d}a$ means that both crack tips advance by $\mathrm{d}a$, so the total crack length increases by $2\mathrm{d}a$.

material, energy is needed because atomic bonds have to be split. A measure of this energy is the *specific surface energy* γ_0. If the crack propagates by da, the total surface energy Γ_0 changes by

$$d\Gamma_0 = 2\gamma_0\, t\, da\,. \tag{5.8}$$

The energy is multiplied by 2 because of the two crack surfaces. In many cases, additional work has to be done to create new surface by crack propagation, for example when plastic deformation near the crack tip occurs in a metal This energy contribution has also to be taken into account in calculating the energy balance. This will be discussed in more detail below.

For the crack to propagate, the external work dW must at least equal the change in the elastic energy and the surface energy.[8] The condition for crack propagation is therefore

$$dW = dU^{(el)} + d\Gamma_0\,.$$

If we relate this energy differential to the crack propagation da and the thickness t, the following equation results:

$$\frac{1}{t}\left(\frac{dW}{da} - \frac{dU^{(el)}}{da}\right) = \frac{1}{t}\frac{d\Gamma_0}{da} = 2\gamma_0\,. \tag{5.9}$$

The energy available to create a new surface

$$\mathcal{G}_I = \frac{1}{t}\left(\frac{dW}{da} - \frac{dU^{(el)}}{da}\right) \tag{5.10}$$

is called the *energy release rate* or, occasionally, the *crack-extension force*. $\mathcal{G}_I$ can be considered as the energy per newly created crack surface (units J/m^2) or as a force acting on the crack and being normalised to unit thickness (units N/m).

As already said, the crack can propagate if equation (5.9) is valid so that the energy release rate reaches a critical value

$$\mathcal{G}_{Ic} = 2\gamma_0\,. \tag{5.11}$$

Accordingly, $\mathcal{G}_{Ic}$ is called the *critical energy release rate* for loading in mode I. It is a material parameter that depends only on γ_0 but not on the load or the geometry.

We now want to estimate the change $dU^{(el)}$ in the *stored elastic strain energy* during crack propagation. To do this, we again consider the case of vanishing external work and assume a constant stress σ and a state of plane stress.[9] Furthermore, we define a state 1 in which a stress σ_R is applied to the crack surfaces (see figure 5.7). At $\sigma_R = \sigma$, the crack is completely closed

---

[8] In the case considered here, the external work is zero. The calculation is universally valid if dW is kept in the equations.

[9] This is a valid assumption if the plate is thin compared to the crack length. Due to the stress concentration near the crack tip, a state of plane strain will occur when the plate is thick (see section 5.2.7).

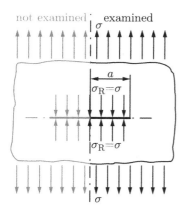
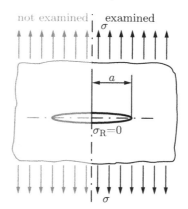

(a) Stage 1: Crack closed   (b) Stage 2: Crack opened

Fig. 5.7. Opening of a crack by reducing a stress σ_R on the crack surface from σ to 0

because the stress is then equal to the stress that would be present in a crack-free material. In state 2 ($\sigma_R = 0$) the crack is opened with an ellipsoidal shape according to equation (5.3).

The transition from state 1 to state 2 can be performed by reducing the additional stress σ_R from σ to 0 while shifting the loading points with the crack surfaces. To calculate the work ΔW, we first consider an infinitesimal part of the crack with a length $\mathrm{d}x_1$ that is shifted from state 1 with a crack opening of $v = 0$ to state 2 with a crack opening of $v = v_0(\sigma, x_1)$. At each point in time, a force $\sigma_R t\,\mathrm{d}x_1$ acts on this crack element because its area is $t\,\mathrm{d}x_1$. The work needed to reduce the stress σ_R from σ to 0 is thus

$$\mathrm{d}(\Delta W) = \int_0^{v_0(\sigma,x_1)} (\sigma_R(v)\, t\, \mathrm{d}x_1)\, \mathrm{d}v = t\, \mathrm{d}x_1 \int_0^{v_0(\sigma,x_1)} \sigma_R(v)\, \mathrm{d}v\,.$$

σ_R is a function of the momentary crack opening v that is given by equation (5.3) when σ is replaced by $\sigma - \sigma_R$.

As there is a linear relation between σ_R and v, the integral $\int \sigma_R \, \mathrm{d}v$ can be changed to $\int v \, \mathrm{d}\sigma_R$ when the integration limits are changed accordingly:

$$\mathrm{d}(\Delta W) = t\, \mathrm{d}x_1 \int_\sigma^0 v_0(\sigma - \sigma_R, x_1)\, \mathrm{d}\sigma_R\,.$$

To calculate the work done on half of the crack length i.e., from $0 \leq x_1 \leq a$, we have to integrate over this distance:

$$\Delta W = 2t \int_0^a \int_\sigma^0 v_0(\sigma - \sigma_R, x_1)\, \mathrm{d}\sigma_R \, \mathrm{d}x_1\,, \tag{5.12}$$

The factor 2 is necessary because work is done on both sides of the crack. This process does not create any additional surface; therefore ΔW is equal to the change in the stored elastic strain energy $\Delta U^{(\mathrm{el})}$. Using equation (5.3) yields

$$
\begin{aligned}
\Delta U^{(\mathrm{el})} &= 2t \int_0^a \int_\sigma^0 \frac{2(\sigma - \sigma_{\mathrm R})}{E} \sqrt{a^2 - x_1^2}\,\mathrm d\sigma_{\mathrm R}\,\mathrm dx_1 \\
&= -2t \frac{\sigma^2}{E} \int_0^a \sqrt{a^2 - x_1^2}\,\mathrm dx_1 \\
&= -t \frac{\sigma^2}{E}\left[x_1\sqrt{a^2 - x_1^2} + a^2 \arcsin \frac{x_1}{a} \right]_0^a = -\frac{t}{2}\frac{\pi a^2 \sigma^2}{E}\,.
\end{aligned}
\tag{5.13}
$$

If the crack length increases by an amount $\mathrm da$ when a critical stress $\sigma_{\mathrm c}$ is applied, the stored elastic strain energy changes by

$$
\mathrm dU^{(\mathrm{el})} = \frac{\partial \Delta U^{(\mathrm{el})}}{\partial a}\,\mathrm da = -t\frac{\pi a \sigma_{\mathrm c}^2}{E}\,\mathrm da\,.
\tag{5.14}
$$

If we insert this into equation (5.9) and use the condition $\mathrm dW/\mathrm da = 0$, we finally get

$$
\begin{aligned}
2\gamma_0 &= \frac{\sigma_{\mathrm c}^2 \pi a}{E}\,, \\
\sqrt{2\gamma_0 E} &= \sigma_{\mathrm c}\sqrt{\pi a}\,, \\
\sqrt{\mathcal G_{\mathrm{Ic}} E} &= \sigma_{\mathrm c}\sqrt{\pi a}\,.
\end{aligned}
\tag{5.15}
$$

The left-hand side of the equation is again a material parameter. For any fixed a, it determines the load at which the component will fail by crack propagation. The right-hand side is (for the case $Y = 1$) equal to that of equation (5.4). Fracture toughness and critical energy release rate are therefore related by the following equation:

$$
K_{\mathrm{Ic}}^2 = \mathcal G_{\mathrm{Ic}} E\,.
\tag{5.16}
$$

For arbitrary stresses in a state of plane stress, $K_{\mathrm I}$ and $\mathcal G_{\mathrm I}$ are also related by

$$
K_{\mathrm I}^2 = \mathcal G_{\mathrm I} E\,.
\tag{5.17}
$$

So far, we have considered a state of plane stress. If the component is in plane strain e. g., because it is so thick that transversal contraction is constrained, the relation between $K_{\mathrm I}$ and $\mathcal G_{\mathrm I}$ changes to – stated here without derivation –

$$
K_{\mathrm I}^2 = \mathcal G_{\mathrm I} \frac{E}{1 - \nu^2}\,.
\tag{5.18}
$$

The relation (5.11) between the critical energy release rate and the specific surface energy ($\mathcal G_{\mathrm{Ic}} = 2\gamma_0$) is only valid for a completely elastic material. In

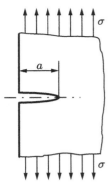

Fig. 5.8. Plate of thickness t with a surface crack

metals, the high stresses near the crack tip always cause plastic deformations that dissipate energy. In this case, crack propagation can occur only if the energy is large enough to create new surface and to deform the material plastically. The dissipated plastic energy is included in the critical energy release rate $\mathcal{G}_{Ic}$, making it and also K_{Ic} larger in metals and polymers than expected by the surface energy alone. In ceramics, the formation of secondary cracks may also cause an increase in $\mathcal{G}_{Ic}$.

We now consider another geometry, a semi-infinite plate with a surface crack of length a (figure 5.8). As this is just half of the geometry we used so far, it might be assumed that there will be no change at all in the energy balance. This, however, is not true because for the semi-infinite plate the left boundary must be free of normal forces and can be displaced horizontally. If the state is plane stress, the stress at the left boundary is uniaxial. This is not true in the case of the infinite plane. Therefore, part of the stored elastic strain energy will be released if the infinite plate is cut in two. As the elastic energy density of a semi-infinite plate without a crack is the same as that of the infinite plate, the semi-infinite plate can release more energy when the crack advances. Consequently, the energy release rate $\mathcal{G}_{Ic}$ is larger and the crack propagates at smaller stresses. This influence of the geometry is taken into account by the geometry factor Y in equation (5.2). For the case considered here it is given by $Y = 1.1215$.

If the crack length a is not small compared to the extension of the component, the geometry factor can depend on the crack length: $Y = Y(a)$. Table 5.1 lists some geometry factors for common crack configurations. It has to be kept in mind that the length of surface cracks is denoted by a, that of internal cracks by $2a$.

In using equation (5.2), it is important to always use the nominal stress σ that does not take the reduction of the cross-sectional area by the crack into account. This reduction is already accounted for by the crack-length depen-

Table 5.1. Geometry factors for some geometries [58]

geometry	geometry factor
	$Y = 1$
	$Y = 1.1215$
	$Y = \sqrt{\dfrac{2b}{\pi a} \tan \dfrac{\pi a}{2b}}$
	$Y = \dfrac{1 - 0.025(a/b)^2 + 0.06(a/b)^4}{\sqrt{\cos(\pi a/2b)}}$
(plate-shaped crack)	$Y = \dfrac{2}{\pi}$

dence of Y.[10] This is an important difference to the case of notches, where the reduced cross section in the notch root is used to calculate the stresses (see chapter 4).

So far, we only considered the case of vanishing external work. We already stated that this is possible without restricting the generality of the results for $\mathcal{G}_{\mathrm{Ic}}$. This will now be shown by comparing two examples.

* Loading by constraints (dead loads)

A plate of width w, length l, and thickness t with a surface crack is elongated by a displacement $\delta = \delta_1$ and then clamped between two walls so that δ remains fixed (figure 5.9(a)). The force needed for this elongation is $F(a)$ and we assume that the crack length a does not change during the elongation.

---

[10] For the infinite geometries we used so far, the geometry factor is constant because a finite crack length does not cause a reduction of the infinite cross section.

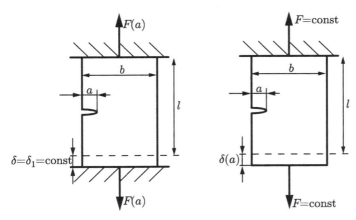

(a) Plate with constraints   (b) Plate with free loads

Fig. 5.9. Plate with surface crack and different boundary conditions

If λ is the *compliance*, the inverse of the stiffness,

$$\lambda = \frac{\delta}{F}\,, \tag{5.19}$$

displacement by δ_1 stores an elastic energy of

$$U^{(\mathrm{el})} = \int_0^{\delta_1} F\mathrm{d}\delta = \frac{1}{2}\frac{\delta_1^2}{\lambda}\,. \tag{5.20}$$

If we now propagate the crack by $\mathrm{d}a$ at a constant displacement δ_1, the change in elastic energy is

$$\frac{\mathrm{d}U^{(\mathrm{el})}}{\mathrm{d}a} = -\frac{1}{2}\frac{\delta_1^2}{\lambda^2}\frac{\mathrm{d}\lambda}{\mathrm{d}a}\,. \tag{5.21}$$

As the compliance increases during crack propagation, the stored energy decreases.

Because no external work is done (the loading points are fixed during crack propagation), the energy release rate for crack propagation is (according to equation (5.10)) given by

$$\mathcal{G}_\mathrm{I} = -\frac{1}{t}\frac{\mathrm{d}U^{(\mathrm{el})}}{\mathrm{d}a}\,.$$

Using equation (5.21), we obtain

$$\mathcal{G}_\mathrm{I} = \frac{1}{2t}\frac{\delta^2}{\lambda^2}\frac{\mathrm{d}\lambda}{\mathrm{d}a} = \frac{F^2}{2t}\frac{\mathrm{d}\lambda}{\mathrm{d}a}\,. \tag{5.22}$$

∗ Loading by constant force (free loads)

Now we load an identical plate until a force F is reached and afterwards keep the force constant (figure 5.9(b)). Again, the crack length is a. According to equation (5.20), the stored elastic strain energy is

$$U^{(\mathrm{el})} = \frac{1}{2}F^2\lambda\,. \tag{5.23}$$

If the crack propagates by $\mathrm{d}a$, the change in energy is

$$\frac{\mathrm{d}U^{(\mathrm{el})}}{\mathrm{d}a} = \frac{1}{2}F^2\frac{\mathrm{d}\lambda}{\mathrm{d}a}\,. \tag{5.24}$$

Because of $\mathrm{d}\lambda/\mathrm{d}a > 0$, the stored elastic strain energy increases in this load case. During crack propagation by $\mathrm{d}a$, an external work $\mathrm{d}W = F\mathrm{d}\delta = F^2\mathrm{d}\lambda$ is done. The energy release rate can be calculated using equation (5.10):

$$\mathcal{G}_{\mathrm{I}} = \frac{1}{t}\left(\frac{\mathrm{d}W}{\mathrm{d}a} - \frac{\mathrm{d}U^{(\mathrm{el})}}{\mathrm{d}a}\right) = \frac{1}{t}\left(F^2\frac{\mathrm{d}\lambda}{\mathrm{d}a} - \frac{1}{2}F^2\frac{\mathrm{d}\lambda}{\mathrm{d}a}\right) = \frac{F^2}{2t}\frac{\mathrm{d}\lambda}{\mathrm{d}a}\,, \tag{5.25}$$

proving that both load cases yield the same result for the energy release rate. As long as the stress fields are the same, the energy release rate is independent of the load case. This assumption from section 5.2.2 is thus vindicated.

5.2.3 Dimensioning pre-cracked components under static loads

In sections 3.3.1 to 3.3.3 and 3.5.2, failure criteria for plastic deformation and cleavage fracture were introduced. If a component contains a crack, it has to be designed to avoid crack propagation as well. A component with a known crack length a can resist a given stress $\underline{\sigma}$ only if

- the yield strength is not reached: $\sigma_{\mathrm{eq}} < R_{\mathrm{p}}$, with an equivalent stress σ_{eq} defined according to Tresca or von Mises,
- the cleavage strength is not reached: $\sigma_{\mathrm{I}} < \sigma_{\mathrm{C}}$, and if
- the stress intensity factor is smaller than the fracture toughness: $K_{\mathrm{I}} < K_{\mathrm{Ic}}$. Thus,

$$\sigma_{\mathrm{I}} < \frac{K_{\mathrm{Ic}}}{\sqrt{\pi a}\,Y} \tag{5.26}$$

is required according to (5.2). Here we assumed, for reasons discussed in section 5.1.1, that the largest principal stress σ_{I} is perpendicular to the largest crack so that mode I loading is relevant.

If the component is loaded cyclically or at elevated temperatures, further failure criteria have to be considered that will be discussed in chapters 10 and 11.

Depending on the crack length, a ductile, pre-cracked material will fail by plastic collapse or crack propagation. The crack length at which the material will fail by crack propagation, and not by plastic collapse, depends on the yield strength R_p and the fracture toughness K_{Ic}. The transition between the two can be estimated by calculating the *critical crack length* a_c for which both criteria are met simultaneously:

$$K_{Ic} = R_p\sqrt{\pi a_c}\,Y\,. \tag{5.27}$$

If the crack length is larger than the critical crack length, the material can be expected to fail by crack propagation. The critical crack length is thus given by

$$a_c = \frac{K_{Ic}^2}{\pi R_p^2 Y^2}\,. \tag{5.28}$$

For brittle materials, the limiting stress for failure in the absence of a crack (denoted σ_{limit} in the following) is given by the cleavage strength σ_C which is approximately equal to the compressive strength R_{cm}, the maximal nominal failure stress under compression (analogous to the tensile strength R_m).

> Although the critical crack length is an indicator of the crack sensitivity of a material, it is not a material parameter. This can be seen directly from equation (5.28) which contains the geometry factor Y. For this reason, analytically calculating a_c is frequently impossible.
>
> Even if a material starts to flow plastically without crack propagation, it is not guaranteed that crack propagation does not start later. This is due to the hardening of the material that causes an increase in the stress and thus in the stress intensity factor. For this reason, another quantity is often used as limiting stress σ_{limit} to calculate the critical crack length in ductile materials, for instance $(R_p + R_m)/2$.

If the crack length is significantly smaller than the critical value, the material fails by *plastic collapse.* Large plastic deformations occur near the crack tip, rounding the crack tip without significant propagation. The specimen behaves similar to a notched specimen without a crack. This can be visualised by forming a 'tensile specimen' of plasticine and 'pre-cracking' it by cutting it with a pair of scissors. If one pulls on the plasticine specimen, the crack vanishes, leaving a notch.

If, on the other hand, $a \geq a_c$, the material fails by crack propagation. This can also be visualised easily by cutting into a piece of paper and pulling on the two halves.

In reality, the transition between plastic collapse and crack propagation is not abrupt, but gradual. If we draw both failure limits into a diagram with the stress σ and the stress intensity factor K_I as axes, we obtain a *failure-assessment diagram* (FAD, figure 5.10). The dashed lines depict the idealised behaviour with an abrupt transition between crack propagation and plastic collapse; the solid line shows a more realistic behaviour.

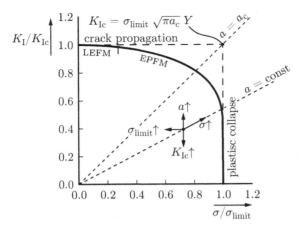

Fig. 5.10. Failure-assessment diagram (after [21]) in standardised form for $Y =$ const. The arrows show how the state of the loaded component changes when the corresponding parameter is raised. Thus, increasing the limiting stress σ_{limit} or the initial crack length a increases the tendency for failure by crack propagation, whereas increasing the fracture toughness K_{Ic} decreases it. Within the region marked with EPFM, elastic-plastic fracture mechanics has to be used because the plastic zone near the crack tip is not small. This theory is the topic of section 5.3

> It has to be noted that even below $\sigma/\sigma_{\text{limit}} = 1$, strong plastic deformation occurs near the crack tip so that linear-elastic fracture mechanics is not applicable here. Therefore, it is not surprising that failure by crack propagation will occur exactly at K_{Ic} only if the relative stress $\sigma/\sigma_{\text{limit}}$ is small.

Safety-critical components are frequently inspected by ultrasonic or X-ray testing. If no indication of a crack is found (otherwise the component is scrapped), the detection limit of the testing method is assumed as the size of the largest crack. The component is dimensioned to ensure that even with a crack of this size the safety margin to the failure limit in figure 5.10 is sufficient.

5.2.4 Fracture parameters of different materials

Table 5.2 shows the fracture toughness and the critical crack length of some materials. For the calculation of the critical crack length with equation (5.28), the yield strength R_{p} was used for metals and polymers and the compressive strength R_{cm} for ceramics. The geometry factor Y was taken to be 1. It is rather obvious that the fracture toughness K_{Ic} of metals is about one or two orders of magnitude larger than that of ceramics. Accordingly, the critical crack length in metals is much larger as well. This is due to the fact that metals, with a yield strength that is comparable to the compressive strength of

Table 5.2. Fracture toughness and crack sensitivity of some materials. The latter is quantified by the critical crack length a_c for $Y = 1$ [8, 11, 21, 27]. Here, R_p is used for metals and polymers and R_{cm} for ceramics. Typical values were taken from the literature to calculate the critical crack length

material	$K_{Ic}/\text{MPa}\sqrt{\text{m}}$	$R_p, R_{cm}/\text{MPa}$	a_c/mm
40 CrMo 4	60	480	5.0
40 NiCrMo 6	60	1 550	0.5
30 CrMoV 21-14	124	1 080	4.2
chrome-nickel steel	50	1 640	0.3
	90	1 420	1.3
Ti Al6 V4	55	900	1.2
	100	860	4.3
AlCu alloy	25	455	1.0
	35	325	3.7
Al_2O_3	4.0	3 000	5.7×10^{-4}
Si_3N_4	5.0	1 200	5.5×10^{-3}
ZrO_2	10.0	2 000	8.0×10^{-3}
porcelain	1.0	350	2.6×10^{-3}
polymethylmethacrylate (PMMA)	1.6	64	0.2
polycarbonate (PC)	3.3	56	1.1
polyethylene (HDPE)	3.5	30	4.3

ceramics, deform plastically near the crack tip even if the material behaviour is macroscopically linear-elastic. Thus, energy is needed not only to create fresh surface, but also to deform the material. In ductile materials, this latter contribution to the energy balance is much larger than the former.

The low fracture toughness of polymers does not mean that they are more sensitive to cracks than metals. Due to their lower strength, they are loaded less heavily than are metals so that their crack sensitivity is similar as can be seen from the critical crack length values given.

We saw in section 5.2.2 that a crack propagates when the energy release rate $\mathcal{G}_I$ reaches the critical value $\mathcal{G}_{Ic}$. Within each group of materials (for example, low-alloy steels), the fracture toughness usually decreases with increasing strength. This is plausible because the size of the plastic zone at a certain stress level decreases with increasing R_p, thus reducing the energy needed for plastic deformation, whereas the specific surface energy γ_0 remains approximately constant.

Figure 5.11 shows this fact graphically in form of a toughness-strength diagram. In this diagram, the crucial parameter for failure by crack propagation, the fracture toughness K_{Ic}, is plotted on the ordinate. Depending on the material, the parameter on the abscissa is a quantity relevant for failure by plastic deformation or cleavage fracture. Materials situated in the upper left of the diagram show large critical crack lengths and tend to fail by plastic

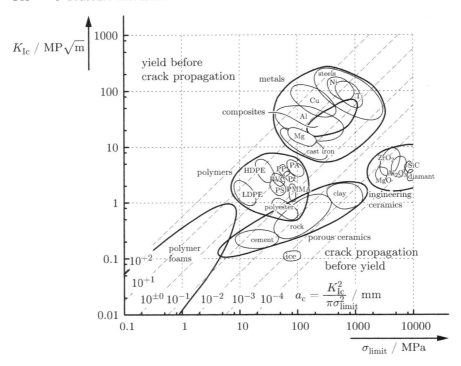

Fig. 5.11. Toughness-strength diagram after *Ashby* [7]. The failure stress σ_{limit} of metals and polymers is the yield strength R_{p}; for ceramics, the compressive strength R_{cm} is used which is approximately the same as the cleavage strength σ_{C}. For composites, the tensile strength R_{m} has been used. 'Yield' is to be understood as failure by plastic collapse instead of crack propagation. The diagonals depict pairs of K_{Ic} and σ_{limit} with constant critical crack length a_{c} in units of mm

collapse; for materials in the lower right part, the critical crack length is small and the tendency to fail by crack propagation is large.

5.2.5 Material behaviour during crack propagation

According to equation (5.4), the critical stress σ_{c} of a pre-cracked specimen is given by

$$K_{\text{Ic}} = \sigma_{\text{c}}\sqrt{\pi a}\, Y \,.$$

If we load the specimen with a stress $\sigma = \sigma_{\text{c}}$, we should expect that the crack propagates without stopping because σ_{c} decreases with increasing crack length. This can be illustrated by plotting the actual stress intensity factor K_{I} against the crack length a and comparing it to the fracture toughness K_{Ic} (figure 5.12(a)). If a_1 is the initial crack length, crack propagation starts at the critical stress σ_c. Afterwards, the calculated stress intensity factor would be larger than the fracture toughness, $K_I > K_{Ic}$ for all crack lengths. Accordingly, the crack propagates fast and the specimen fails. In reality, the stress intensity factor K_I can never exceed the fracture toughness K_{Ic}. We found a corresponding situation for other failure criteria (yield condition, cleavage fracture). Therefore, it is impossible to keep the load constant during crack propagation. Instead, the load decreases as shown in figure 5.12(a). Crack propagation occurring under this condition of decreasing load is called *unstable*.[11]

So far, we have assumed that the fracture toughness is constant and does not change during crack propagation. This is not necessarily the case. In many materials, the resistance against crack propagation initially increases during crack propagation. This can happen when the propagating crack forms a so-called *process zone* near the crack tip where energy is dissipated. In ductile metallic materials, this is connected to the hardening in the plastic region near the crack tip. In single-phase ceramics, the formation of secondary cracks can dissipate additional energy [21]. This will be discussed further in section 7.2. Increasing the crack-growth resistance by energy dissipation is also one major aim in reinforcing materials with particles or fibres and will be treated in chapters 7 and 9.

As the fracture toughness can change during crack propagation, another material parameter is needed that describes the current resistance against further crack growth, called *crack-growth resistance* or *crack-extension resistance* K_{IR}.[12] The crack-growth resistance initially increases in materials forming a process zone, a behaviour frequently denoted *R curve behaviour*. This can be illustrated with a *crack-growth resistance curve* (*R curve*), where K_{IR} is plotted against the crack length a or the increase in crack length Δa.

How crack propagation begins depends on the shape of the crack-growth resistance curve and the initial crack length. Figure 5.12(b) shows the example of a material with an increasing crack-growth resistance and a relatively large initial crack length a_2. Crack propagation starts at a stress σ_{c2}, but it can only proceed if the stress is increased further. This case is called *stable crack propagation*. The crack becomes *unstable* when the stress does not increase on increasing the crack length i. e., at a stress $\sigma = \sigma_2^*$.[13] The associated crack-growth resistance is denoted by K_2^*.

---

[11] There are differing definitions of stable and unstable crack propagation in the literature, based on the external load, as done here and in *Gross* [58], or on the energy balance [21].

[12] The situation is analogous to that in plastic deformation: The yield strength R_p characterises the stress at the onset of plastic deformation, the actual flow stress σ_F denotes the current stress during yielding.

[13] Sometimes, this transition between stable and unstable crack propagation at K_I^* is called K_{Ic}, for example in the standard ASTM E 561. In this book, K_{Ic} always denotes the value of the stress intensity factor at which crack propagation starts.

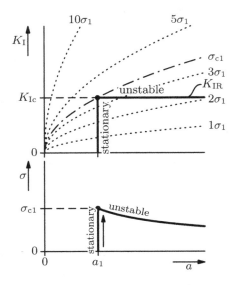

(a) During crack propagation, the stress decreases and crack propagation is unstable from the start at σ_c

Fig. 5.12. Plot of the crack-growth resistance K_{IR} and the stress σ versus the crack length a for the case of constant K_{IR} during crack propagation (1/2)

In contrast, figure 5.12(c) shows a configuration with the same crack-growth resistance curve, but a smaller initial crack length a_3. In this case, the stress needed for the crack to propagate decreases from the start, although the crack-growth resistance increases. Hence, crack propagation is unstable as soon as K_{Ic} is reached. The critical stress σ_{c3} is nevertheless larger than that for the longer crack (σ_{c2}).

As the figures 5.12(b) and 5.12(c) illustrate, an increasing crack-growth resistance K_{IR} does not guarantee stable crack propagation because the stress can still decrease.

If the load propagating the crack is displacement-controlled, a decreasing load does not necessarily cause failure of the component. As the compliance of the component increases with increasing crack length, the stress may become small enough to stabilise the crack, a phenomenon called *crack arresting*.

We now want to estimate the stress intensity factor K_I that marks the onset of unstable crack propagation. This happens when the required external stress σ decreases or at least stops to increase on crack propagation by a distance da. During crack propagation, the current stress intensity factor K_I, which depends on the stress σ and the crack length a, must equal the crack-growth resistance K_{IR}:

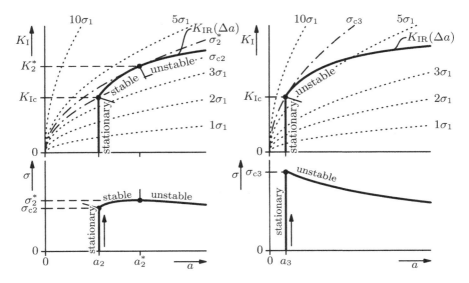

(b) For a crack of length a_2, stable crack propagation begins at a stress σ_c. At a stress of σ^*, the crack becomes unstable

(c) For a short crack of length a_3, crack propagation is unstable from the start at σ_c

Fig. 5.12. Plot of the crack-growth resistance K_{IR} and the stress σ versus the crack length a for the case of increasing K_{IR} during crack propagation. In both figures, the same crack-growth resistance curve $K_{\mathrm{IR}}(\Delta a)$ is plotted (after [104]) (2/2)

$$K_{\mathrm{I}}(\sigma, a) = K_{\mathrm{IR}}(a).$$

In order to meet this condition at any arbitrary time, the derivatives must also be equal:

$$\frac{\mathrm{d}K_{\mathrm{I}}(\sigma, a)}{\mathrm{d}a} = \frac{\mathrm{d}K_{\mathrm{IR}}(a)}{\mathrm{d}a}.$$

If we separate the left-hand side into its constituent terms, we get

$$\frac{\partial K_{\mathrm{I}}(\sigma, a)}{\partial \sigma}\frac{\partial \sigma}{\partial a} + \frac{\partial K_{\mathrm{I}}(\sigma, a)}{\partial a} = \frac{\mathrm{d}K_{\mathrm{IR}}(a)}{\mathrm{d}a}.$$

As long as the load increases during crack propagation, $\partial \sigma/\partial a > 0$ holds, and the crack propagation is stable. If the external load does not increase anymore, $\partial \sigma/\partial a \leq 0$ holds. The condition for the transition between stable and unstable crack propagation is thus $\partial \sigma/\partial a = 0$. K_{I}^* can be calculated by

$$\left.\frac{\partial K_{\mathrm{I}}(\sigma, a)}{\partial a}\right|_{K_{\mathrm{I}} = K_{\mathrm{I}}^*} \geq \left.\frac{\mathrm{d}K_{\mathrm{IR}}(a)}{\mathrm{d}a}\right|_{K_{\mathrm{I}} = K_{\mathrm{I}}^*}. \tag{5.29}$$

If the crack-growth resistance curve $K_{\mathrm{IR}}(a)$ is known, this criterion together with the condition $K_{\mathrm{I}}(\sigma, a) = K_{\mathrm{IR}}(a)$ allows to calculate the beginning of unstable crack propagation.

* 5.2.6 Subcritical crack propagation

So far, we have assumed that a crack is stationary if the stress intensity factor is smaller than the fracture toughness K_{Ic}. This, however, is not always the case. If, for instance, a corrosive medium penetrates the material along the crack surface, it might weaken the material near the crack tip. This decreases the fracture toughness locally and the crack can propagate, but only until it reaches undamaged material. If this happens, a crack can slowly propagate at loads below the critical load, a phenomenon called *subcritical crack growth*. The crack propagates subcritically until the stress intensity factor K_{I} equals the fracture toughness K_{Ic} and the crack propagation becomes unstable.[14] As the subcritical crack growth is time-dependent, it can be described by the *crack-growth rate* $\mathrm{d}a/\mathrm{d}t$, specifying the increment $\mathrm{d}a$ by which the crack propagates during an infinitesimal time $\mathrm{d}t$.

Generally, subcritical crack growth occurs when the material near the crack tip is weakened by time-dependent processes. Different physical phenomena may be responsible for this.

In many metals, corrosive media like electrolytes can cause *stress corrosion cracking*. Two different kinds of stress corrosion cracking can be distinguished.

In anodic stress corrosion cracking, loading accelerates corrosion at the crack tip, resulting in crack propagation. This may happen in metals whose surface is usually protected by a passivating layer. The crack tip can be activated, for instance by local plastic deformation, and the crack can propagate. If a new passivating layer forms immediately on the freshly formed surface, corrosion can only proceed at the crack tip, causing the crack to remain sharp-edged. Whether anodic stress corrosion cracking can occur depends on the surrounding media, the material and its state, the temperature, and the mechanical stress. In unfavourable circumstances, the required stress level may be very small and even residual stresses may be sufficient to cause failure by a delayed fracture (see section 3.5.3). Anodic stress corrosion cracking can, for example, occur in the presence of chloride ions (e. g., in saline air) in aluminium alloys and austenitic chrome nickel steels.[15]

---

[14] In section 5.2.5, we saw that even above K_{Ic} a crack may still be stable if the crack-growth resistance increases. If subcritical crack growth occurs, this effect is usually negligible because the crack grows fast at stress intensity values close to K_{Ic}.

[15] This latter case is especially problematic because these steels are usually called 'stainless', implying a high level of corrosion resistance. However, in saline atmosphere, the corrosion resistance of these steels is reduced.

A different case is hydrogen-induced stress corrosion cracking. It occurs in materials showing hydrogen embrittlement (see section 3.5.3, page 117). The stress concentration near the crack tip dilates the crystal lattice. Hydrogen, generated during the corrosion process, is therefore preferentially stored in this region near the crack tip, reducing the fracture toughness in the most stressed region. The crack propagates through the weakened material, and, subsequently, hydrogen diffuses to the new crack tip. Hydrogen-induced stress corrosion cracking is observed particularly in high-strength steels because here elastic strains near the crack tip can be large and large amounts of hydrogen can thus be stored.

Polymers show a similar effect in the presence of solvents. Solvents preferredly enter the material near the crack tip because the distance between the molecules is increased there by the large tensile stresses. If, for instance, a rod made of polymethylmethacrylate (*Plexiglas*) is bent and the tensile side is wetted with acetone or alcohol, brittle fracture can occur after a short exposure time. In this case, the cleavage strength is reduced because the dipole bonds between the molecules are replaced by bonds formed with the solvent (see also section 8.8).

Ceramics can also fail by subcritical crack growth due to stresses and localised chemical reactions. In glasses, for instance, water can enter surface defects and can attack the bonds between the silicon and the oxygen atoms if these are strained by an external stress [9]:

$$Si-O-Si + H_2O \rightarrow Si-O-H + H-O-Si.$$

This reaction can cause a seemingly sudden failure of glasses. Subcritical crack growth also occurs in crystalline ceramics, especially if they have a glassy phase (see section 7.1) on the grain boundaries. Among the most sensitive ceramics are silicate ceramics like porcelain or mullite, for they usually contain a large amount of more than 20 % glassy phases [19, 142]. Subcritical crack growth can also be present in engineering ceramics, for instance in aluminium oxide (Al_2O_3) in humid atmosphere or saline solution [104].

At elevated temperatures, metals and ceramics exhibit time-dependent plastic deformation, called creep, a phenomenon to be discussed in detail in chapter 11. If a pre-cracked material is loaded at high temperatures, the crack can grow. In metals and ceramics, pores are frequently responsible for this because they form and coalesce in the highly-stressed region in front of the crack tip, often on grain boundaries [119] (see section 11.3). The crack thus frequently propagates between the grains (intercrystalline fracture). This process is called *creep crack growth* (CCG). In polymers, time-dependent plastic deformation occurs already at ambient temperature (see section 8), and they are thus also susceptible to creep crack growth.

Subcritical crack growth is determined by the temperature, the material, and, for the low-temperature processes, the environment. In a given system, the crack-growth rate frequently depends on the stress intensity factor K_I only. Below a certain, temperature-dependent limiting value K_{I0}, crack growth

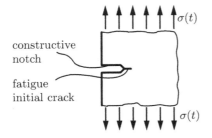
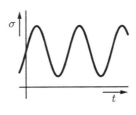

constructive
notch

fatigue
initial crack

Fig. 5.13. Creating an initial crack by cyclic loading

vanishes completely or almost completely [104]. If K_I exceeds K_I0, the crack-growth rate rapidly increases with increasing stress intensity factor. In many cases (for example, in stress corrosion cracking), a plateau region follows i.e., the crack-growth rate is nearly constant for a certain range of K_I-values. When the stress intensity factor approaches K_Ic, the crack-growth rate increases rapidly again. Examples for crack-growth rate curves and a mathematical description of the crack-growth rate are discussed in section 7.2.6.

*5.2.7 Measuring fracture parameters

In the previous sections, we introduced several important material parameters: The fracture toughness K_Ic, the critical energy release rate $\mathcal{G}_\mathrm{Ic}$, and the crack-growth resistance curve. We now want to see how these quantities are measured.

A common feature of all experiments is that the test specimens are pre-cracked. To achieve this, a notched sample is used and a crack is propagated from this notch by cyclic loading (see chapter 10) as shown in figure 5.13. Creating the initial crack in this way is necessary because the notch tip is usually not sharp enough to behave like a true crack. Cyclic loading allows to produce the initial crack at a load that is much smaller than that needed for static experiments (see chapter 10).

Several specimen geometries are standardised, and the corresponding geometry factors are given in tabular form or by approximation functions [21,133, 138]. Among the most common specimen geometries are the *three-point bending specimen* (figure 5.14(a)) and the *compact tension specimen*, or *CT specimen* for short (figure 5.14(b)).

For the compact tension specimen, the stress intensity factor can be calculated from the external load F by

$$K_\mathrm{I} = \frac{F}{B\sqrt{W}} f(a/w) \tag{5.30}$$

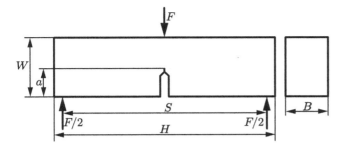

(a) Three-point bending specimen

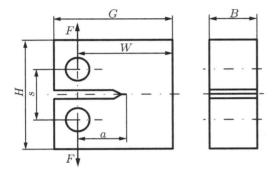

(b) CT specimen

Fig. 5.14. Standardised specimen geometries for fracture mechanical experiments

where

$$f(a/w) = \frac{2 + \frac{a}{W}}{\left(1 - \frac{a}{W}\right)^{3/2}} \times \tag{5.31}$$

$$\left[0.886 + 4.64\frac{a}{W} - 13.32\left(\frac{a}{W}\right)^2 + 14.72\left(\frac{a}{W}\right)^3 - 5.6\left(\frac{a}{W}\right)^4\right]$$

is the geometry factor. In calculating K_{I}, it is important to note that the crack length a is measured from the point of loading, not from the beginning of the initial crack. This is easily understood because it is completely irrelevant for the stress state near the crack tip how 'wide' the crack is some distance away. It is only important that the initial crack is sharp-edged with a small radius of curvature, and this is ensured by creating it through cyclic loading. The quantities G, H, B, W, and s from figure 5.14(b) must, according to the standard ASTM E 399, be related in a certain way: As an example, $B = W/2$, $s = 0.55W$, $H = 1.2W$, and $G = 1.25W$ are used for a standard CT specimen. The initial crack length should be limited by $0.45W \leq a \leq 0.55W$.

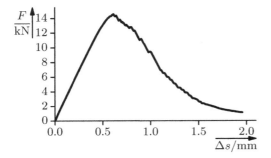

Fig. 5.15. Load-displacement curve of an AlCuMg 2 CT specimen

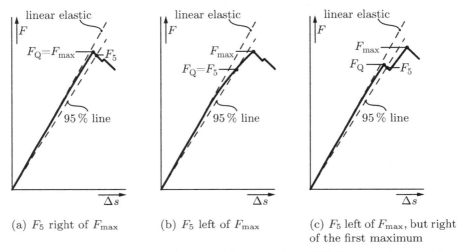

(a) F_5 right of F_{max} (b) F_5 left of F_{max} (c) F_5 left of F_{max}, but right of the first maximum

Fig. 5.16. Determination of the critical force F_Q. The 95% lines are drawn with a slope of 90% only

* Measuring the fracture toughness

The procedure to measure K_{Ic} or $\mathcal{G}_{Ic}$ is independent of the specimen geometry. The loading points are displaced with constant speed and the required force is measured. If the force is plotted against the displacement of the loading points, a *load-displacement curve* results as shown in figure 5.15. The onset of unstable crack propagation can be seen from this curve because the force reaches a maximum and drops, resulting in a larger compliance of the specimen. As the experiment is displacement-controlled, the crack usually stabilises again due to the unloading and propagates only when the crack is opened further. If the size of the plastic zone near the crack tip is small compared to the volume of the specimen, crack propagation starts without any noticeable deviation of the loading curve from linear-elastic behaviour (see figure 5.16(a)).

In contrast, the material behaviour shown in figure 5.16(b) indicates significant plastic deformation. The crack propagation is stable at first and becomes unstable at a force F_{max}. This shape of the curve is typical for ductile materials. In this case, it has to be ensured that linear-elastic fracture mechanics is still valid. Furthermore, it is not possible to determine from the curve alone at which force crack propagation has started because stable crack propagation and plastic deformation both reduce the slope of the curve. To determine the facture toughness K_{Ic}, a pragmatic approach is taken, similar to the definition of the yield strength $R_{\text{p0.2}}$. This will be described below.

A special case is shown in figure 5.16(c). On reaching a load F_{Q}, the crack propagates unstably for a certain distance and then becomes arrested. This is called *pop-in*.

To determine the fracture toughness K_{Ic}, the following procedure has been agreed upon (see e. g., standards ASTM E 399 and ISO 12737):

We start by drawing a line with a slope of 95 % of that of the elastic line from the experiment (figure 5.16). The intersection of this line with the load-displacement curve determines the force F_5.[16] Two cases can be distinguished:

- If F_5 lies to the right of the force value at which the first reduction in the load occurs, the force F_{Q} is determined by this maximum ($F_{\text{Q}} = F_{\text{max}}$ in figure 5.16(a), F_{Q} in figure 5.16(c)).
- If F_5 is left to this maximum, $F_{\text{Q}} = F_5$ is used as a critical value (figure 5.16(b)). In this case, it has to be ensured that plastic deformation was small enough to allow using linear-elastic fracture mechanics. A necessary condition for this is

$$\frac{F_{\text{max}}}{F_{\text{Q}}} \leq 1.1 \,. \tag{5.32}$$

If this condition does not hold, the experiment has to be evaluated according to the rules of elastic-plastic fracture mechanics, discussed in section 5.3.

For a CT specimen, the critical stress intensity factor K_{Q} is calculated from F_{Q} by using equation (5.30),

$$K_{\text{Q}} = \frac{F_{\text{Q}}}{B\sqrt{W}} f(a/w) \,.$$

For this, it is necessary to know the initial crack length. This can be measured optically after the specimen has been fractured because the fracture surface of the initial crack produced by cyclic loading can easily be discerned from the statically produced crack surface (figure 5.17).

The value of K_{Q} determined in this way does not depend on the material only. Instead, the stress state near the crack tip, that in itself depends on

---

[16] The subscript '5' denotes the reduction of the slope by 5 % used in constructing the line.

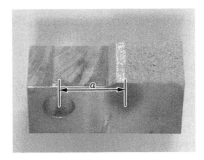

Fig. 5.17. Measuring the initial crack length

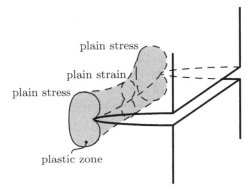

Fig. 5.18. Size and shape of the plastic zone near the crack tip (dog bone, after [58])

the specimen geometry, may influence this value. This can be illustrated by inspecting the plastic zone near the crack tip (see figure 5.18).

At the surface, the specimen is in a state of plane stress because no normal forces can be transmitted here. The smallest principal stress is thus zero. Within the specimen, the stress state is a state of nearly plane strain because the transversal contraction near the crack tip is constrained by the surrounding material. The stress state is thus a state of triaxial tension. Therefore, the equivalent stresses and thus the plastic deformations are larger near the surface than within the specimen. As plastic deformation dissipates energy, the crack-growth resistance increases with decreasing thickness of the specimen (see figure 5.19). For sufficiently thick specimens, the influence of the surface zone with its state of plane stress can be neglected and the crack-growth resistance K_Q approaches a constant value, the fracture toughness K_{Ic}.

To ensure independence of the geometry and to determine the fracture toughness as lower (and therefore safe) limiting value for the crack-growth resistance of a material, a state of plane strain is required. Only if this can be guaranteed, the measured value K_Q is called fracture toughness K_{Ic}. According to the standards ASTM E 399 and ISO 12737, this requires

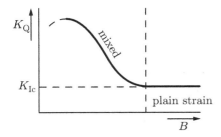

Fig. 5.19. Dependence of the measured K_Q-values on the specimen thickness B. For very thin specimens, the state of plane stress is dominant, for thick specimens the behaviour is determined by the state of plane strain within the specimen. In the region denoted as 'mixed', both states contribute

$$\left\{ \begin{array}{c} B \\ a \\ W - a \end{array} \right\} \geq 2.5 \left(\frac{K_Q}{R_p} \right)^2 . \tag{5.33}$$

This is a reasonable requirement as can be seen by estimating the size r of the plastic zone from equation (5.1). If we insert $K_I = K_Q$ in this equation, we can estimate this size by setting $\tilde{\sigma}_{22}(\tilde{x}_2 = 0) = R_p$ and solving for $r = \tilde{x}_1$:

$$r \approx \frac{1}{2\pi} \left(\frac{K_Q}{R_p} \right)^2 . \tag{5.34}$$

Thus, equation (5.33) ensures that the plastic zone is small compared to the specimen size.

* Measuring the crack-growth resistance curve

As explained in section 5.2.5, the crack-growth resistance curve is a plot of the stress intensity factor versus the crack length a. Experiments are usually displacement-controlled to enable measurement of the load-displacement curve after the maximum force has been exceeded.

To measure the crack-growth resistance curve according to ASTM E 561, the load is applied step-wise, and the crack length is measured for each load value after the crack has stabilised. For a complete curve, 10 to 15 measurement values are needed. To avoid using several specimens for a single curve, the crack length is measured during the experiment. This can be done in several ways [21, 43].

> Optical methods measure the crack length directly on the polished surface of the specimen. This method is rather simple, but its main disadvantage is that the crack is only measured on the surface; the crack length within the specimen is unknown.

The *compliance method* measures the compliance of the specimen by unloading it during the experiment. Comparing the measured value with a calibration curve determined on specimens with known crack length, the crack length can be determined.

The *electrical potential drop method* uses the electrical resistance of the specimen to measure the crack length. A constant electrical current is applied between two points of the specimen far away from the crack and the potential drop in the vicinity of the crack is measured. Comparison with a calibration curve allows calculation of the crack length. Obviously, the specimen has to be electrically isolated from the testing machine and the displacement transducer.

As before, it is necessary that the deformation of the specimen is mainly elastic to allow use of linear-elastic fracture mechanics. For the crack-growth resistance curve measurement, this can be ensured according to ASTM E 561 by the following condition:

$$W - a \geq \frac{4}{\pi} \left(\frac{K_{\max}}{R_\mathrm{p}} \right)^2 .$$

From the measured values for the force F_i and the crack length increment Δa_i, the crack-growth resistance curve is calculated using equation (5.30).

* 5.3 Elastic-plastic fracture mechanics

In the previous sections, it was frequently stressed that linear-elastic fracture mechanics can only be used if the plastic zone near the crack tip is sufficiently small. If this is not the case, we enter the domain of *elastic-plastic fracture mechanics* (EPFM) which can deal with a large plastic zone. The method, however, cannot be used for arbitrarily large plastic zones – plastic behaviour must still be restricted to the region around the crack tip and must be mainly determined by the surrounding elastic stress field.

Two alternative methods are commonly used to describe the state near the crack tip: The crack tip opening displacement and the J integral. Both methods can be shown to be mathematically equivalent.

* 5.3.1 Crack tip opening displacement (CTOD)

In the *crack tip opening displacement* method (or *CTOD-method* for short), it is assumed that crack propagation is not determined by the stress intensity factor, but by the amount of plastic deformation near the crack tip. This can be measured by the opening δ_t of the crack tip. If this reaches a critical value δ_c, the crack propagates.

The crack tip opening displacement can be defined in different ways. They all have in common that it is assumed that the crack tip is blunted by the

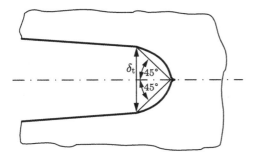

Fig. 5.20. One possible definition of the crack tip opening displacement δ_t

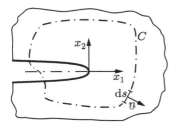

Fig. 5.21. Coordinate system and integration contour for the J integral

plastic deformation and that both crack surfaces are almost parallel, see figure 5.20. One possibility to determine δ_t is to draw two lines at an angle of $45°$ to the crack line and measure the distance between their intersections with the crack surface.

∗ 5.3.2 J integral

In section 5.2.2, we calculated the energy balance of a propagating plane crack and predicted crack growth if the energy release rate reaches a critical value $\mathcal{G}_\mathrm{Ic}$. The definition of $\mathcal{G}_\mathrm{I}$ in equation (5.10) was independent of the material behaviour. Linear-elastic behaviour was only assumed to calculate the terms in equation (5.12). The so-called J *integral* can quantify the energy release rate without this assumption. For a crack geometry as sketched in figure 5.21, J is defined as

$$ J = \int_C \left[w\,\mathrm{d}x_2 - \left(\underline{\underline{\sigma}} \cdot \frac{\partial \underline{u}}{\partial x_1} \right) \cdot \underline{n}\,\mathrm{d}s \right] . \tag{5.35} $$

Here C is a closed curve encircling the crack tip, $w = \int \sigma_{ij}\,\mathrm{d}\varepsilon_{ij}$ is the energy density, $\underline{u}$ the displacement vector and $\underline{n}$ an outwardly pointing normal vector on C as shown in figure 5.21. The derivation of the J integral and detailed explanations can be found in appendix D. If no crack is present, equation (5.35) yields $J = 0$. Strictly speaking, equation (5.35) is applicable only for elastic (but not necessarily linear-elastic) material behaviour. As long as the loading is monotonous and no unloading occurs, it can also be used for plastic deformations.[17]

The choice of the path C is arbitrary as long as it encloses the crack tip. In practice – for instance when doing finite element simulations – it is usually wise to put it not too close to the crack tip so that it runs only through elastically deformed regions.

Equations (5.10) and (5.35) are equivalent. Therefore, for linear-elastic materials and a state of plane stress, the equation

$$\mathcal{G}_{\mathrm{I}} = J = \frac{K_{\mathrm{I}}^2}{E} \tag{5.36}$$

holds. As for $\mathcal{G}$, there is a critical value J_{c}, denoting the onset of crack propagation. This will be discussed below in section 5.3.3.

> We now want to show the equivalence of the J integral and the energy release rate (equation (5.10)) by calculating the energy balance. This is elaborated further in appendix D.6.
>
> We start by looking at the second term of equation (5.10), $\mathrm{d}U^{(\mathrm{el})}/\mathrm{d}a$. Using the elastic energy density $w^{(\mathrm{el})}$, we get
>
> $$\frac{\mathrm{d}U^{(\mathrm{el})}}{\mathrm{d}a} = \iiint_V \frac{\mathrm{d}w^{(\mathrm{el})}}{\mathrm{d}a} \mathrm{d}V$$
>
> or, for a plane geometry with specimen thickness t and a region A containing the crack tip,
>
> $$\frac{\mathrm{d}U^{(\mathrm{el})}}{\mathrm{d}a} = t \iint_A \frac{\mathrm{d}w^{(\mathrm{el})}}{\mathrm{d}a} \mathrm{d}x_1 \mathrm{d}x_2 . \tag{5.37}$$
>
> For an infinite plate, crack propagation by $\mathrm{d}a$ in positive x_1 direction can also be considered as a shift of the integration region A in negative x_1 direction, resulting in
>
> $$\frac{\mathrm{d}U^{(\mathrm{el})}}{\mathrm{d}a} = -t \iint_A \frac{\mathrm{d}w^{(\mathrm{el})}}{\mathrm{d}x_1} \mathrm{d}x_1 \mathrm{d}x_2 .$$
>
> *Gauss' theorem* [24] allows to convert an area integral over an area A into a line integral along its boundary C.[18] For the integral in question, we thus get
>
> $$-\frac{1}{t} \frac{\mathrm{d}U^{(\mathrm{el})}}{\mathrm{d}a} = \int_C w^{(\mathrm{el})} \mathrm{d}x_2 ,$$
>
> which corresponds to the first term in equation (5.35).

---

[17] This is the reason why the energy density in equation (5.35) is called w, not $w^{(\mathrm{el})}$.

[18] Here we use a simplified version of Gauss' theorem for two dimensions (see appendix D.1).

We now look at the first term in equation (5.10), $\mathrm{d}W/\mathrm{d}a$. To calculate the work $\mathrm{d}W$ done during an infinitesimal propagation of the crack, a closed surface can be put around the crack tip. The work $\mathrm{d}(\mathrm{d}W)$ done during this infinitesimal crack growth by $\mathrm{d}a$ on an area element $\mathrm{d}S$ is given by the product of the force $\mathrm{d}\underline{F}$ acting on the area element and the change in the displacement field $\mathrm{d}\underline{u}$:

$$\mathrm{d}(\mathrm{d}W) = \mathrm{d}\underline{F} \cdot \mathrm{d}\underline{u} \,. \tag{5.38}$$

The force $\mathrm{d}\underline{F}$ acting on the surface element $\mathrm{d}S$ can be calculated by multiplying ΔS with the normal stress on the surface, $\underline{\underline{\sigma}}\,\underline{n}$, where $\underline{n}$ is the normal vector of the surface:

$$\mathrm{d}\underline{F} = \left(\underline{\underline{\sigma}}\,\underline{n}\right) \cdot \mathrm{d}S \,.$$

For infinitesimal crack propagation by $\mathrm{d}a$, the displacement of the material is given by

$$\mathrm{d}\underline{u} = \frac{\partial \underline{u}}{\partial a}\,\mathrm{d}a \,.$$

If we collect all these relations, put them into equation (5.38), and normalise by $\mathrm{d}a$, we get

$$\frac{\mathrm{d}(\mathrm{d}W)}{\mathrm{d}a} = \left(\underline{\underline{\sigma}}\,\underline{n}\right) \cdot \frac{\partial \underline{u}}{\partial a}\,\mathrm{d}S \,.$$

Integrating over the area S yields

$$\frac{\mathrm{d}W}{\mathrm{d}a} = \iint_S \left(\underline{\underline{\sigma}}\,\underline{n}\right) \cdot \frac{\partial \underline{u}}{\partial a}\,\mathrm{d}S$$

or, for a plane geometry with specimen thickness t,

$$\frac{\mathrm{d}W}{\mathrm{d}a} = t \int_C \left(\underline{\underline{\sigma}}\,\underline{n}\right) \cdot \frac{\partial \underline{u}}{\partial a}\,\mathrm{d}s \,.$$

As $\underline{\underline{\sigma}}$ is symmetric, the vectors $\underline{n}$ and $\partial \underline{u}/\partial a$ can be exchanged. If we again use $\mathrm{d}x_1 = -\mathrm{d}a$, we finally arrive at

$$\frac{1}{t}\frac{\mathrm{d}W}{\mathrm{d}a} = -\int_C \left(\underline{\underline{\sigma}}\,\frac{\partial \underline{u}}{\partial x_1}\right) \cdot \underline{n}\,\mathrm{d}s \,. \tag{5.39}$$

This is the second term from equation (5.35). So, finally, we have shown the equivalence between equation (5.10) and (5.35) as shown in equation (5.36).

* 5.3.3 Material behaviour during crack propagation

We saw in section 5.2.5 that the crack-growth resistance K_{IR} in linear-elastic fracture mechanics depends on the crack length increment Δa. Similarly, the

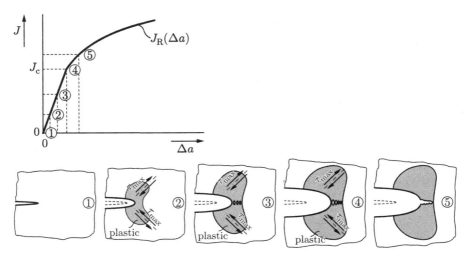

Fig. 5.22. J_R crack-growth resistance curve in a ductile material up to the point of crack initiation (after [21]). The different states of crack propagation are described in the text

value of the J integral also changes during crack propagation. Analogous to K_{IR}, the value of the J integral during crack propagation is called J_R *crack-growth resistance*. As for K_{IR}, a crack-growth resistance curve can be drawn by plotting J_R against the crack length increment Δa (figure 5.22).

Crack propagation in ductile materials is different from that in brittle ones. With increasing load, the material deforms plastically near the crack tip and blunts it (subfigure ② in figure 5.22). This blunted region is called the *stretch zone*.

Subsequently, due to the high stresses and large plastic deformations, cavities form in front of the crack tip (cf. section 3.5.1), depicted in subfigures ③ and ④. During formation of the stretch zone and of cavities (subfigures ① to ④), the relation between Δa and J_R is almost linear.

With increasing load, the cavities coalesce with each other and the crack, causing a 'true' growth of the crack (subfigure ⑤). This is called *crack initiation* and occurs when J reaches J_c [58].[19] The slope of the crack-growth resistance curve now decreases (see figure 5.22). The formation and coalescence of cavities or pores during crack propagation is characteristic for shear fracture (section 3.5.1), producing the typical dimple fracture surface. The state ⑤ is different from the initial configuration ① for two reasons: The material near the crack tip is now plastically deformed and has thus hardened, and the crack surface is dimpled, resulting in a blunted crack tip.

---

[19] Sometimes, this value of the J integral is denoted as J_i (where 'i' stands for 'initiation') [21]. In this case the value at the transition between stable and unstable behaviour (called J^* here) is called J_c.

Contrary to brittle materials, in ductile materials even small loads can cause (very small) crack growth. If the load is not raised further, this will not cause any trouble, but if the load is cyclic, each repetition of the load will cause a small crack propagation and finally cause a sufficiently large crack to destroy the component. This so-called fatigue fracture is the topic of chapter 10.

The smaller the plastic region near the crack tip is, the steeper is the line characterising the formation of the stretch zone. In the limiting case of linear-elastic material behaviour, the line is vertical and the curve is similar to that in figure 5.12. The smaller the unloading of the crack tip due to plastic deformation is, the higher are the stresses near the crack tip. As the stress state is triaxial, the danger of crack propagation by cleavage fracture (see section 3.5.2) grows. The transition between dimple surface fracture for ductile materials and cleavage fracture in brittle ones is not clear-cut, and mixtures of both cases can occur.

The value of the J integral that marks the beginning of unstable crack propagation can be calculated analogously to the stress intensity factor K_{I} in section 5.2.5. The critical J value J^* is *not* the crack initiation value J_{c}. At J_{c}, stable crack growth by coalescence of cavities begins; at J^*, the crack starts to become unstable.

As unstable crack propagation causes an unloading of the material, the J integral must not be used during this stage because, according to section 5.3.2, the equations are not valid in this case, even when the experiment is stabilised by displacement control.

* 5.3.4 Measuring elastic-plastic fracture mechanics parameters

As in linear-elastic fracture mechanics, specimens in elastic-plastic fracture mechanics are also standardised. An initial crack is produced in the same way by cyclic loading. The values of the J integral for crack propagation, J_{c} and J^*, are read off the J_{R} crack-growth resistance curve, so there is no need to perform additional experiments.

We will now discuss the procedure using the example of a CT specimen (figure 5.14(b)). The specimen is loaded using displacement-control, and a load-displacement curve is measured. The resulting graph will look like the one in figure 5.23. The area beneath the curve corresponds to the work done

$$W_F = \int_0^{\Delta s_1} F(\Delta s, a) \, \mathrm{d}(\Delta s) \,, \tag{5.40}$$

with Δs denoting the displacement of the loading points. From this, the J integral for a CT specimen is given by [21]

$$J = \frac{W_F}{B(W - a)} \cdot \eta \tag{5.41}$$

with

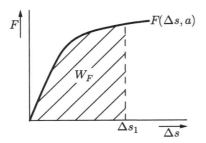

Fig. 5.23. Load-displacement curve to determine the J_R crack-growth resistance

$$\eta = 2 + 0.522 \left(1 - \frac{a}{W}\right) \, .$$

As can be expected because of the definition of the J integral as an energy release rate, the value of the J integral is directly connected to the external work W_F. The derivation of equation (5.41) can be found in *Gross / Seelig* [58]

To measure the J_R crack-growth resistance curve, a direct or indirect measurement of the crack length during the experiment is needed. This is done using one of the methods described in section 5.2.7. It is not possible to measure the development of the crack length during the test after the experiment on the fractured specimen.

6

Mechanical behaviour of metals

As already mentioned in section 1.2, metals are characterised by their excellent plastic deformability that is of great technical importance. On the one hand, it allows to produce complex metallic components by forming processes like forging or drawing. On the other hand, it causes metals to deform plastically when the yield strength is reached, instead of failing catastrophically by fracturing. This improves safety because overloading can often be detected before disaster happens.

In this chapter, we will explain the mechanisms behind the plastic deformation of metals. Afterwards, we will discuss how the stress required for plastic deformation can be increased, thus strengthening the material.

6.1 Theoretical strength

Plastic deformation is irreversible. Therefore, the configuration of the atoms must be changed during plastic deformation, for otherwise they would return to their original position on unloading. If we consider shearing a single crystal as an example, it can be deformed plastically by sliding whole layers of atoms against each other as shown in figure 6.1.[1] For this sliding to happen, the bonds between the atoms have to be stretched elastically until they can switch to the next atom. The stress required for this process can be estimated (see exercise 16) and is of the order of one fifth of the shear modulus of the crystal. The yield strength predicted this way for metallic single crystals is thus between 1 GPa and 25 GPa.

If we measure the strength of single crystals of pure metals, the values found are several orders of magnitudes below this theoretical value and even lie below that of engineering alloys. Typical values are in the range of a few

---

[1] For simplicity, we usually use a simple cubic lattice in the sketches of crystals shown in this chapter, although this is not a Bravais lattice found in any technically important metal.

Fig. 6.1. Sliding of atomic planes in a perfect crystal

megapascal. As single crystals always contain lattice defects, one possible explanation could be that these are responsible for the reduced strength. If, however, the number of defects is reduced further, for instance by a heat treatment, the yield strength becomes even smaller. Only an absolutely perfect single crystal without any defects would possess a yield strength agreeing with the theoretical prediction. This can only be nearly realised in so-called whiskers (see section 6.2.8) which, however, are extremely small.

The reason for this spectacular failure of the theoretical prediction is that plastic deformation does not occur by sliding of complete layers of atoms. Instead, it proceeds by a mechanism that is based on a special type of lattice defect, the *dislocations*. To understand plastic deformation of metals thus requires an understanding of dislocations.

6.2 Dislocations

6.2.1 Types of dislocations

Dislocations are one-dimensional (line-shaped) lattice defects. Figure 6.2(a) shows an *edge dislocation,* one of the two basic types. Its spatial structure can most easily be visualised by imagining that an additional half-plane of atoms is put into the crystal. In the vicinity of the line where this half plane ends, the crystal is distorted, further away from it, it still is perfect.

An edge dislocation can be described by two vectors. The first is the *line vector $\underline{t}$*, the vector pointing in the direction of the *dislocation line.* The second vector is the *Burgers vector $\underline{b}$* that can be determined in the following way: We draw a so-called *Burgers circuit* around the dislocation line that takes the same amount of steps from one atom to the next in each direction as visualised in figure 6.2. If the crystal were perfect, the circuit would be closed, but, due to the lattice defect, it is not. An additional step is required to get back to the starting point. The vector describing this step is the Burgers vector $\underline{b}$. As long as the Burgers circuit encloses the dislocation line, the Burgers vector defined in this way is independent of the size and the shape of the circuit. As shown in figure 6.2(a), the Burgers vector and the line vector of an edge dislocation are perpendicular.

In our definition, we did not specify the direction of the circuit, but it has to be chosen consistently. One simple way of doing this is to use a 'right-hand rule' oriented on the line vector $\underline{t}$ of the dislocation line as shown in figure 6.3.

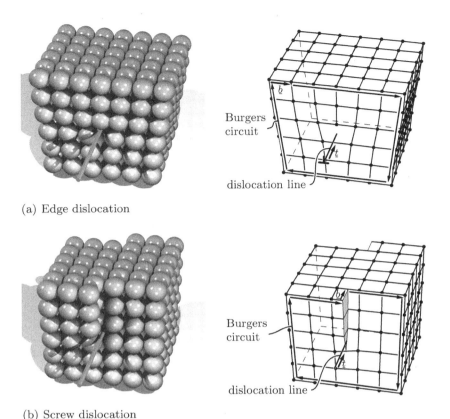

(a) Edge dislocation

(b) Screw dislocation

Fig. 6.2. Types of dislocations. In the lattice models, a possible Burgers circuit is marked

The choice for the direction of the line vector is arbitrary as well. If it is reversed, the orientation of the Burgers vector reverses as well.

The second basic type of a dislocation, the *screw dislocation* is shown in figure 6.2(b). It can be visualised by imagining that the crystal has slipped by one atomic distance on a half plane ending at the dislocation line. The screw dislocation can also be characterised by its line vector and Burgers vector. The figure shows that both are parallel. If we move along a crystal plane around the dislocation, the resulting path is helical and thus looks like a screw, which explains the name of this dislocation type.

Dislocation lines are always either closed or end at the surface of the crystal, but they can never end within the crystal. Why this is so can be seen from figure 6.4. Imagine that a dislocation would end somewhere within the crystal. The crystal is distorted in the vicinity of the dislocation line, but is perfect at a sufficient distance away from the dislocation. We now walk on a

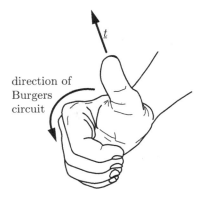

direction of
Burgers
circuit

Fig. 6.3. Right-hand rule for determining the direction of a Burgers circuit

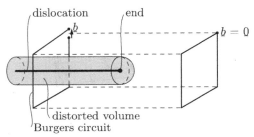

Fig. 6.4. It is impossible for a dislocation line to end within the crystal because it has to be possible to transfer one of the drawn circuits into the other in a continuous way. This is not the case here because only one of the circuits has a non-zero Burgers vector

Burgers circuit around the dislocation line and find a non-vanishing Burgers vector. If we take an identical circuit somewhere far away from the dislocation line, the Burgers vector would vanish. As both paths lie completely within the undistorted perfect region of the crystal, it should be possible to create one from the other by a parallel shift, but then, both should have the same Burgers vector. Therefore, getting from one path to the other is only possible if we intersect the dislocation line somewhere.

A dislocation line is usually not straight, but undulates through the crystal on an intricately curved path with constantly changing line vector. The Burgers vector, on the other hand, always remains constant. Therefore, a dislocation can be edge-like in some region, screw-like in another and can have a mixed character in between. The angle between a Burgers vector and a line vector can vary between 0° und 90° as visualised in figure 6.5.

6.2.2 The stress field of a dislocation

Because dislocations distort the crystal lattice, an elastic stress field forms around the dislocation line. This will now be shown using the example of an

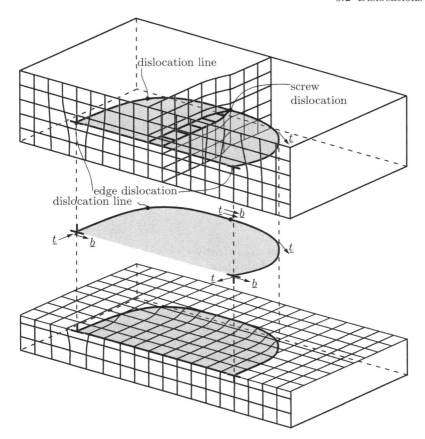

Fig. 6.5. Dislocation line in a crystal. The dislocation has the same constant Burgers vector $\underline{b}$ over its whole length. The positions with edge dislocation character differ by opposite line vectors $\underline{t}$. Thus, the crystal half-planes are added from the top in the one case and from the bottom in the other case

edge dislocation with a line vector along the x_3 axis. Due to the insertion of the half plane, we can expect compressive stresses on this side of the dislocation and tensile stresses on the opposite side. The following equations for the components of the stress tensor hold, given here without derivation [40]:

$$
\begin{aligned}
\sigma_{11} &= -\frac{Gb}{2\pi(1-\nu)} \cdot \frac{x_2(3x_1^2 + x_2^2)}{(x_1^2 + x_2^2)^2}, \\
\sigma_{22} &= \frac{Gb}{2\pi(1-\nu)} \cdot \frac{x_2(x_1^2 - x_2^2)}{(x_1^2 + x_2^2)^2}, \\
\tau_{12} &= \frac{Gb}{2\pi(1-\nu)} \cdot \frac{x_1(3x_1^2 - x_2^2)}{(x_1^2 + x_2^2)^2}.
\end{aligned}
\tag{6.1}
$$

As the local strain at the core of the dislocation in the direction of its line vector, $\varepsilon_{33}(x_1 = x_2 = 0)$, must be the same as everywhere else in the material, the stress state is a state of plane strain with $\varepsilon_{33} = 0$. Using Hooke's law, we thus find

$$\sigma_{33} = \nu(\sigma_{11} + \sigma_{22}).$$

These equations have been derived using continuum mechanics. They are thus only valid a few atomic distances away from the core of the dislocation; too close to it the assumptions of continuum mechanics do not hold.

Figure 6.6 shows the components of the stress field and the hydrostatic stress

$$\sigma_{\mathrm{hyd}} = \frac{1}{3}(\sigma_{11} + \sigma_{22} + \sigma_{33}) \tag{6.2}$$

around the dislocation. The plot of σ_{hyd} clearly shows the regions of compression on the side of the additional half-plane and the tensile stresses on the other side. In figure 6.7, the stress distribution from figure 6.6 is illustrated qualitatively.

Because the dislocation elastically deforms the lattice in its vicinity, elastic energy is stored here. The more dislocations there are in a crystal, the higher its stored elastic energy. If we try to elongate a dislocation line, for instance by bending it, energy is needed. The stress near the dislocation line is, according to equation (6.1), proportional to the product of the shear modulus G and the Burgers vector b, while the displacement is proportional to the Burgers vector. Thus, the stored energy T per unit length of dislocation is approximately[2]

$$T \approx \frac{Gb^2}{2}. \tag{6.3}$$

The energy per unit length T has the unit of a force. Analogous to a taut string, T can be considered as a force 'stretching' the dislocation line. Therefore, T is often called the *line tension* of the dislocation. Frequently, its value is of the order of $10^{-9}\,\mathrm{N}$.

6.2.3 Dislocation movement

If a sufficiently large shear stress acts on a dislocation, the dislocation moves through the crystal. How this happens is shown in figure 6.8 for an edge dislocation: Near the dislocation line, the atoms are displaced from their equilibrium positions, stretching and compressing the atomic bonds. If an external shear stress is applied, trying to shift the upper crystal plane relative to the lower,

---

[2] The exact value of T depends on the type of dislocation i. e., on the orientation of $\underline{b}$ and $\underline{t}$, and on the curvature of the dislocation line. For the considerations that follow, the estimate given here is sufficient.

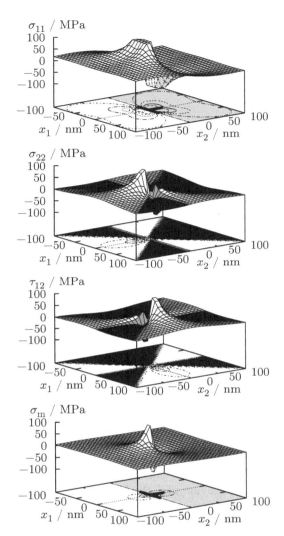

Fig. 6.6. Stress distribution near an edge dislocation oriented in the x_3 direction in aluminium. Values are cut at $\pm 100\,\mathrm{MPa}$. Compressive stresses are printed in dark colour

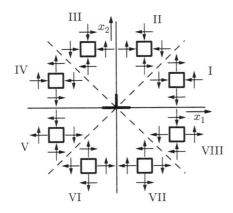

Fig. 6.7. Qualitative illustration of the stress distribution around an edge dislocation using infinitesimal material elements

one bond of the atom in the dislocation line will be stretched, but the next will be compressed. Thus the atom can exchange its partner by flipping the bond from one atom to the next. This process is repeated until the dislocation has travelled to the surface of the crystal and the upper half of the crystal has been displaced relative to the lower one. In the case of an edge dislocation, the slip of the crystal is on a plane in the direction of the dislocation movement. The dislocation eases plastic deformation of the crystal in two ways: The atomic bonds are already stretched and compressed, and the slip of the crystal does not have to occur simultaneously for all atoms on the slip plane. The deformation is irreversible, hence plastic, because the dislocation will not move back to its starting point when the crystal is unloaded. The volume of the crystal does not change in this process, thus finally explaining the statement from chapter 3 that plastic deformation keeps the volume constant and is independent of the hydrostatic stress.

Figure 6.9 shows the slipping of the crystal for a screw dislocation. Again, atomic bonds are switched during dislocation movement. The slip plane for a screw dislocation contains the dislocation line and the Burgers vector. In contrast to the edge dislocation, the slip plane is not uniquely determined as shown in figure 6.9. The screw dislocation moves perpendicular to the applied shear stress and to the slip.

The slip of a mixed dislocation follows from the cases already discussed. If we consider the example of a dislocation loop (figure 6.10), the loop increases or decreases its diameter when a shear stress is applied because the edge dislocation moves in the direction of the shear stress and the screw dislocation moves perpendicular to it. A dislocation loop changes its shape uniformly if both types of dislocation have the same mobility.

If, on the other hand, one type of dislocation moves less easily than the other, dislocation movement is at first dominated by the more mobile type as sketched in figure 6.11. This increases the length of the less mobile type. In

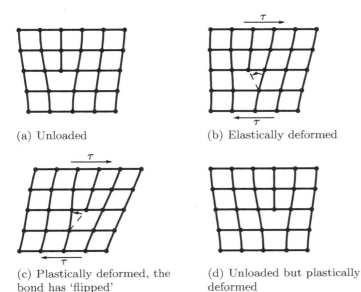

(a) Unloaded   (b) Elastically deformed

(c) Plastically deformed, the   (d) Unloaded but plastically
bond has 'flipped'   deformed

Fig. 6.8. Schematic illustration of the slip of an edge dislocation. The applied shear stress strains bonds in the vicinity of the dislocation, causing the bond to flip from one atom to the other. The dislocation moves by one lattice position. The distortions are strongly exaggerated

the end, this type will therefore strongly influence the plastic behaviour and the required stresses.

6.2.4 Slip systems

During dislocation movement, parts of the crystal slip relative to each other. The *slip direction* and the amount of slip are determined by the Burgers vector $\underline{b}$. For an edge dislocation, the slip direction is also the direction of the dislocation movement, for a screw dislocation, these directions are perpendicular. The plane separating the two slipped crystal parts is called the *slip plane*, and the combination of slip direction and slip plane is called a *slip system*.

Figure 6.12 illustrates slipping of a crystal with the slip direction and the slip plane. As slipping occurs by dislocation movement along the slip plane, the dislocation line $\underline{t}$, the Burgers vector $\underline{b}$ and the direction of movement $\underline{v}$ must all lie within it. The normal vector on the slip plane must thus fulfil the conditions

$$\underline{n} \perp \underline{b} \quad \wedge \quad \underline{n} \perp \underline{t} \quad \wedge \quad \underline{n} \perp \underline{v}. \tag{6.4}$$

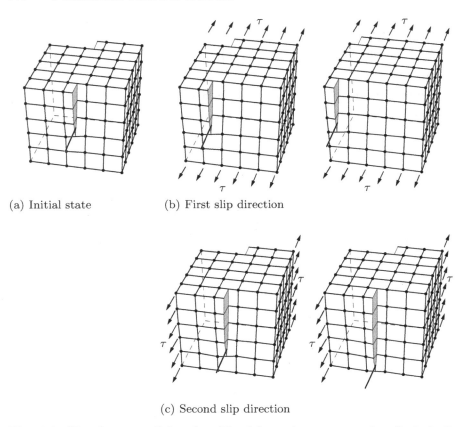

(a) Initial state   (b) First slip direction

(c) Second slip direction

Fig. 6.9. Slip of a screw dislocation. The deformation can occur by slip in both shown directions (cf. section 6.2.4)

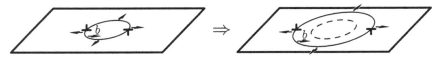

Fig. 6.10. Slip of a dislocation loop

For an edge or a mixed dislocation, $\underline{b}$ and $\underline{t}$ are not parallel and the slip plane is uniquely determined by the dislocation itself:

$$\underline{n} = \frac{\underline{b} \times \underline{t}}{|\underline{b} \times \underline{t}|}.$$

For a screw dislocation, this is different because $\underline{b}$ and $\underline{t}$ are parallel. The direction of movement is thus not determined uniquely. The dislocation can move on different planes and can thus overcome obstacles by *cross slip* (cf. figure 6.9). During cross slip, it changes from the slip plane with the largest

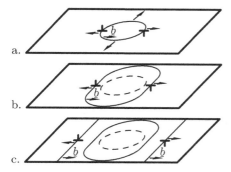

a.

b.

c.

Fig. 6.11. Non-uniform slip of a dislocation loop

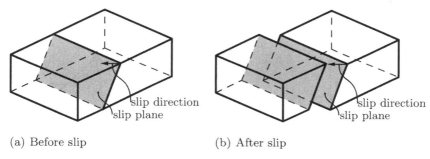

(a) Before slip   (b) After slip

Fig. 6.12. Slip plane and slip direction in a crystal

shear stress to another one with smaller resultant stress (see section 6.2.5). Edge dislocations can use a similar mechanism only at elevated temperatures by so-called *climb*, explained in sections 6.3.4 and 11.2.2.

Movement of a dislocation is the easier, the closer packed the slip plane and the slip direction are. Therefore, slip planes and directions are preferredly close-packed. If we assume that atoms are simple spheres and that slipping occurs simultaneously on the whole plane, this fact can be visualised easily (figure 6.13). To slip one atomic plane against the other, the atoms have to be lifted only by a small amount in the case of closest packing (figure 6.13(a)). If the packing is less dense, the upper plane has to be lifted more (figure 6.13(b)), and slipping is more difficult. Any crystal lattice can only form certain distinct Burgers vectors, and, therefore, the crystal can slip only along certain planes in certain directions. In the following sections, we discuss the slip systems of the most important lattices.

Face-centred cubic crystal

The face-centred cubic crystal is close-packed (see section 1.2.2). Planes of type $\{111\}$ and directions of the type $\langle 1\bar{1}0 \rangle$ are close-packed and thus form the slip systems (figure 6.14). If we consider planes as identical differing only

(a) Close-packed plane or direction. Slip is relatively easy, as the upper array of spheres has to be lifted only slightly

(b) Non-close-packed plane or direction. Slip is hampered, as the upper array of spheres has to be lifted further

Fig. 6.13. Illustration of slip on the basis of spheres

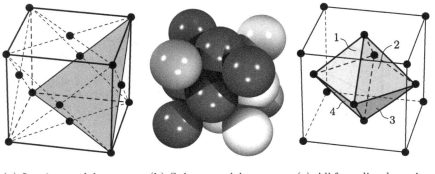

(a) Lattice model   (b) Sphere model   (c) All four slip planes in one figure

Fig. 6.14. Slip systems in face-centred cubic metals. The slip planes are the body diagonals; the slip directions lie on the plane diagonals or on the edges of the octahedron in figure (c), respectively

Table 6.1. Slip systems in face-centred cubic metals

slip plane	slip direction	number of planes	directions per plane	number (total)
{111}		$\langle 1\bar{1}0 \rangle$	4	3	12

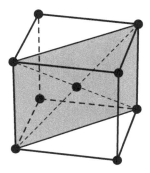

(a) Illustration of a slip plane in the unit cell. Both slip directions are plotted as dashed lines

(b) Sphere model. The stacking of the type of slip planes shown in (a) is marked by colours

Fig. 6.15. $\{110\}\langle\bar{1}11\rangle$ slip systems in body-centred cubic metals

Table 6.2. Slip systems in body-centred cubic metals

slip plane		slip direction	count of planes	directions per plane	count (total)
$\{110\}$	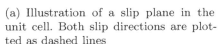	$\langle\bar{1}11\rangle$	6	2	12
$\{112\}$		$\langle11\bar{1}\rangle$	12	1	12
$\{123\}$		$\langle11\bar{1}\rangle$	24	1	24

in the orientation, but not in the direction of the normal vector, there are four independent slip planes as visualised in figure 6.14(c). Each of these planes has three independent slip directions. Altogether, we thus have $4 \times 3 = 12$ independent slip systems in this Bravais lattice. Table 6.1 lists the slip systems in face-centred cubic crystals.

Body-centred cubic crystal

The body-centred cubic crystal is not close-packed. The slip systems with the closest packed directions and planes in this lattice are of the type $\{110\}\langle\bar{1}11\rangle$ (figure 6.15). With two slip directions per plane and six different slip planes, twelve slip systems result. As summarised in table 6.2, slip is also possible on other crystallographic planes that are only slightly more difficult to activate [55].

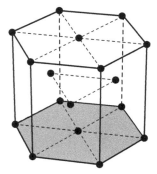
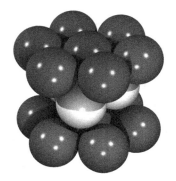

(a) Illustration of one slip plane in the unit cell. The dashed diagonals are the three slip directions

(b) Sphere model. The stacking of close-packed slip planes is marked by colours

Fig. 6.16. Basal slip systems in hexagonal metals

Hexagonal close-packed crystal

As the name implies, the hexagonal close-packed crystal has the highest possible packing density. Its stacking sequence (cf. figure 1.9) differs from that of the face-centred cubic lattice. Only the $\{0001\}$-basal planes are close-packed. They contain the three $\langle 11\bar{2}0 \rangle$ close-packed directions, resulting in only three independent slip systems (figure 6.16).

Three independent slip systems are not sufficient for arbitrary deformations. For the hexagonal crystal, this is easily understood because shear deformation out of the common slip plane of the three systems is impossible. Therefore, other, more difficult, slip systems must be activated. Because real metals never show the ideal hexagonal structure, but possess either a stretched or a compressed unit cell (varying ratio c/a), it depends on the chemical element which other systems are activated. Table 6.3 gives a synopsis of the most important slip systems. The slip systems with the horizontal slip plane are called basal slip systems. If the slip planes are on the prism faces of the unit cell, they are called prismatic slip systems. The other slip systems are called pyramidal slip systems.

6.2.5 The critical resolved shear stress

It was stated in section 6.2.3 that a dislocation will start to move if a sufficiently large shear stress acts on the slip system. This stress value is called critical resolved shear stress τ_{crit}. It is not equal to the yield strength τ_{F} of an isotropic material under shear loading because in the latter case different slip systems have to be activated that are usually not parallel to the shear stress. For a single crystal, the yield criterion (cf. section 3.3.1) is

Table 6.3. Overview over slip systems in hexagonal metals

slip plane count		slip direction count/plane	count (total)	examples
		Burgers vector in basal plane		
$\{0001\}$ 1		$\langle 11\bar{2}0\rangle$ 3	3	Cd, Zn, Mg, Ti, Zr
$\{01\bar{1}0\}$ 3		$\langle \bar{2}110\rangle$ 1	3	Ti, Zr
$\{01\bar{1}1\}$ 6		$\langle \bar{2}110\rangle$ 1	6	Ti, Mg, Zr
$\{01\bar{1}2\}$ 6		$\langle \bar{2}110\rangle$ 1	6	Zn
$\{11\bar{2}2\}$ 6		$\langle 1\bar{1}00\rangle$ 1	6	Ti
		Burgers vector out of basal plane		
$\{01\bar{1}0\}$ 3		$\langle \bar{2}113\rangle$ 2	6	Zn
$\{01\bar{1}1\}$ 6		$\langle \bar{2}113\rangle$ 2	12	Zr
$\{11\bar{2}1\}$ 6		$\langle \bar{2}113\rangle$ 2	12	Zn, Zr
$\{11\bar{2}2\}$ 6		$\langle \bar{2}113\rangle$ 2	12	Zn

$$\tau^{(\mathrm{ss})} = \tau_{\mathrm{crit}}, \tag{6.5}$$

where $\tau^{(\mathrm{ss})}$ is the shear stress in the slip system.

Only in special cases will the external shear stress be exactly parallel to the slip system. Usually, it is thus necessary to calculate the *resolved shear stress* i. e., the stress component acting as shear stress on the considered slip system in the slip direction. If we restrict ourselves to the case of uniaxial loading as in a tensile test, the calculation of this component is not too difficult.[3] We

---

[3] The more general derivation for arbitrary stress states will be discussed at the end of the section.

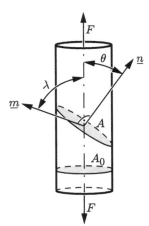

Fig. 6.17. Orientation of a slip system in a tensile test specimen

consider a certain slip system in the uniaxially loaded tensile test specimen sketched in figure 6.17. Let $\underline{n}$ be the normal vector of the slip plane, $\underline{m}$ the slip direction, and A_0 the cross section of the specimen. The area of the inclined slip plane is A. To calculate the component of the force in the slip direction, we have to project the external force F onto the direction $\underline{m}$:

$$F_{\underline{m}} = F \cos \lambda .$$

If we now relate both forces to the area they are acting upon and use $\tau^{(ss)} = F_{\underline{m}}/A$ and $\sigma = F/A_0$, we arrive at

$$\tau^{(ss)} A = \sigma A_0 \cos \lambda . \tag{6.6}$$

The area of the slip system in the specimen is

$$A = \frac{A_0}{\cos \theta} .$$

Putting this into equation (6.6) results in the *resolved shear stress* (or *Schmid stress*)

$$\tau^{(ss)} = \sigma \cos \lambda \cos \theta . \tag{6.7}$$

This equation determines the shear stress in a slip system resulting from the external stress and the orientation of the system. The factor $\cos \lambda \cos \theta$ is known as the *Schmid factor*. If the resolved shear stress reaches the critical value τ_{crit}, the material yields. The yield criterion for uniaxial loading is thus

$$\sigma \cos \lambda \cos \theta = \tau_{\text{crit}} . \tag{6.8}$$

As $\underline{n}$ and $\underline{m}$ are always perpendicular, the Schmid factor has a maximum value of 0.5 when $\lambda = \theta = 45°$. In this case, the yield strength measured in the tensile test is twice the shear stress needed to activate the slip system. If the slip system is oriented in a non-optimal way, the yield strength is even larger. Note that this calculation is purely continuum mechanical. Dislocation theory is only needed to ascertain that it is the shear stresses that determine the plastic deformation and that slip can occur only on certain planes.

That the largest shear stress component in a uniaxially loaded specimen is half of the normal stress can also be seen from Mohr's circle [58] (see section 2.2.1). The principal stresses are $\sigma_{\mathrm{I}} = \sigma$ and $\sigma_{\mathrm{II}} = \sigma_{\mathrm{III}} = 0$, resulting in a Mohr's circle with radius $\sigma/2$, leading to a maximum shear stress of $\tau_{\mathrm{max}} = 0.5\,\sigma$.

The resolved shear stress can also be calculated for arbitrary stress states $\underline{\sigma}$. To calculate, we first calculate the traction vector $\underline{t}$ on the slip plane with normal vector $\underline{n}$

$$\underline{t} = \underline{\sigma} \cdot \underline{n} \,.$$

The resolved shear stress in the slip direction $\underline{m}$ can be calculated from this by projecting $\underline{t}$ onto $\underline{m}$:

$$\tau^{(\mathrm{ss})} = \underline{t} \cdot \underline{m} \,.$$

If we insert one equation into the other, we obtain

$$\tau^{(\mathrm{ss})} = \left(\underline{\sigma} \cdot \underline{n}\right) \cdot \underline{m} \,. \tag{6.9}$$

This results in the yield criterion

$$\left(\underline{\sigma} \cdot \underline{n}\right) \cdot \underline{m} = \tau_{\mathrm{crit}} \,. \tag{6.10}$$

To compare the general equation (6.9) with the uniaxial case, we chose the x_1 axis parallel to the external load. The stress tensor is then

$$\underline{\underline{\sigma}} = \begin{pmatrix} \sigma_{11} & 0 & 0 \\ 0 & 0 & 0 \\ 0 & 0 & 0 \end{pmatrix} \,.$$

Thus, only the x_1 components of $\underline{n}$ und $\underline{m}$ are important; they are $n_1 = \cos\theta$ and $m_1 = \cos\lambda$. Inserting this into equation (6.9), we get

$$\tau^{(\mathrm{ss})} = \sigma_{11} \cos\theta \cos\lambda \,.$$

This is identical to equation (6.7) derived previously for the uniaxial case.

Since crystals have several different slip systems, as explained in section 6.2.4, the resolved shear stress on all possible slip planes has to be calculated to determine whether the material will yield. The slip system that is oriented most favourably towards the external load and thus possesses the largest Schmid factor will be activated first.

6.2.6 Taylor factor

In the previous section, we saw how the resolved shear stress required to activate a slip system in a single crystal can be calculated from the external stress. If τ_{crit} is the critical stress needed to activate slip, the external stress is connected to τ_{crit} by the Schmid factor $\cos\lambda\cos\theta$ for a single crystal.

Most engineering alloys are polycrystalline. To calculate the yield strength from the critical resolved shear stress in an isotropic, polycrystalline material, we have to take into account that the grains are oriented in an arbitrary manner. We thus have to take the average of all possible crystal orientations.

Furthermore, the deformation of neighbouring grains has to be compatible. For instance, it is not possible to deform a certain grain with favourable orientation without also deforming its neighbours, because the grains would overlap or gaps would open between them. A plausible assumption is that all grains deform similarly. For this, at least five slip systems must be activated in each grain.

That the number of slip systems to be activated is five can be explained as follows: An arbitrary deformation has six independent components of the strain tensor (see section 2.4.2). Because plastic deformation does not change the volume, each one of these components is dependent on the others, and five independent components remain, corresponding to the required five slip systems.

This relation between the components of the strain tensor at constant volume can be derived by considering the deformation of a cuboid with edges l_1, l_2, l_3 that is deformed until the edge lengths are $l_1 + \Delta l_1$, $l_2 + \Delta l_2$, $l_3 + \Delta l_3$. Since the volume is constant, the relation

$$(l_1 + \Delta l_1)(l_2 + \Delta l_2)(l_3 + \Delta l_3) = l_1 l_2 l_3$$

holds, resulting in

$$(1 + \varepsilon_{11})(1 + \varepsilon_{22})(1 + \varepsilon_{33}) = 1\,.$$

For small strains, products of strains can be neglected. This yields the equation

$$\varepsilon_{11} + \varepsilon_{22} + \varepsilon_{33} = 0\,. \tag{6.11}$$

Equation (6.11) relates the three components ε_{11}, ε_{22}, and ε_{33}, so only two of them are independent. If we add the three off-diagonal components ε_{23}, ε_{13}, and ε_{12} that do not change the volume, the number of independent components is five.

Because the assumption of small strains is not always valid in plastic deformation, equation (6.11) becomes more and more imprecise with increasing strains. To circumvent this problem, we can consider only strain increments:

$$d\varepsilon_{11} + d\varepsilon_{22} + d\varepsilon_{33} = 0 \,,$$

or, using time derivatives,

$$\dot{\varepsilon}_{11} + \dot{\varepsilon}_{22} + \dot{\varepsilon}_{33} = 0$$

at any arbitrary time.

If we take these effects into account, the Schmid factor has to be replaced in a polycrystalline material by another number, the *Taylor factor M*. For a face-centred cubic material, M takes a value of 3.1 [34]. The relation between the critical resolved shear stress τ_{crit} and the yield strength measured in uniaxial tension σ_{F} thus is

$$\sigma_{\text{F}} = M\,\tau_{\text{crit}} \,. \tag{6.12}$$

This value of the Taylor factor has also been confirmed experimentally. Throughout section 6.4, we will use the Taylor factor to calculate the influence of strengthening mechanisms, which affect the critical shear stress, on the uniaxially measured yield strength.

The derivation of the Taylor factor is rather involved. Here, we only want to sketch the main ideas for the example of a face-centred cubic lattice. A detailed discussion can be found in *Cottrell* [34].

To calculate the average over all possible grain orientations is not too difficult. It is necessary to calculate the probability that an arbitrarily oriented grain has a certain value of the Schmid factor $\cos\lambda\cos\theta$. Then, the average of the resulting probability distribution has to be taken. If this is done, an incorrect value of 2.2 results for the Taylor factor.

For a correct calculation, it has to be taken into account that five different slip systems must be activated in each grain to enable an arbitrary deformation. Some of these are oriented less favourably, thus increasing the value of the Taylor factor. To precisely determine its value, the five best-oriented slip systems have to be determined for each crystal orientation. As there are $\binom{12}{5} = 792$ possibilities to choose five systems out of twelve, this calculation is involved. In addition, normal stresses on the grain boundaries have to be continuous, and this has also to be accounted for. Finally, the average over all possible grain orientations is taken to arrive at the Taylor factor.

In this calculation of the yield strength, the average over all possible grain orientations is taken. Therefore, we can expect that plastic deformation will occur at smaller values of stress in favourably oriented

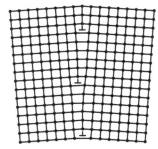

Fig. 6.18. Formation of a low-angle grain boundary by an arrangement of dislocations

grains. This is indeed the case, but the amount of deformation is small because of the compatibility problem discussed above. Larger plastic deformations can therefore occur only when dislocation movement has started in a sufficient number of grains. This consideration shows that the transition between elastic and plastic behaviour of a metal is not clear-cut, but gradual. This is reflected in the typical flow curve of a metal and is the reason why the yield strength is usually defined using the 0.2% proof stress $R_{p0.2}$ (cf. section 3.2).

6.2.7 Dislocation interaction

The stress field around a dislocation (cf. figures 6.6 and 6.7) can interact with the fields of other dislocations. If, for instance, two edge dislocations with the same line vector $\underline{t}$ and Burgers vector $\underline{b}$ are situated on the same slip plane, their tensile and compressive stresses are added, resulting in an increase in the stored elastic energy. This can easily be seen by the following experiment of thought: If we unite the two dislocations to a single one with Burgers vector $2\underline{b}$, the stored line energy is $T = G(2b)^2/2 = 2Gb^2$. If the two dislocations are far apart, the sum of their line energies is only $T = 2Gb^2/2 = Gb^2$. The stored energy is reduced if the two dislocations move apart, resulting in a repulsion.

If two identically oriented edge dislocations are parallel and lie almost on top of each other, the tensile stress field of one overlaps with the compressive stress field of the other. This is energetically favourable, so the dislocations attract and, in the ideal case, finally stop if one is exactly on top of the other. If several dislocations are arranged in this way, the crystal regions on both sides of the dislocation lines are tilted (figure 6.18). This is called a *low-angle grain boundary*.

It can be seen from figure 6.7 that the edge dislocation shown will repel another, identically oriented, dislocation when it is positioned in regions I, IV, V, or VIII. In the other regions, the dislocation is attracted. In both cases, dislocation movement may be hampered, depending on the direction of movement. If the dislocations repel each other, this is obvious because energy is needed to overcome the barrier. If they attract, the released energy will dissipate as heat. To dislodge the dislocations from their energetically favourable position, energy is needed.

For oppositely oriented dislocations, the attractive and repulsive regions are exactly reversed. Similar considerations are also valid for screw dislocations.

6.2.8 Generation, multiplication and annihilation of dislocations

The line energy of dislocations is rather large. To create a dislocation by thermal activation (see appendix C.1) is therefore a highly improbable process (calculated in exercise 18), and the dislocation density in a metal in thermal equilibrium would be vanishingly small. Real metals usually show *dislocation densities*[4] between $10^{12}\,\mathrm{m}^{-2}$ and $10^{16}\,\mathrm{m}^{-2}$. Even in extremely pure single crystals, for instance made of 99.999 999 9 % germanium, the dislocation density is about $10^7\,\mathrm{m}^{-2}$ [98]. This discrepancy between realistic values of the dislocation density and those expected from thermal equilibrium is due to the fact that dislocations are created when the crystal solidifies from the melt.

If one tries to calculate the speed of crystallisation, it is found that the calculated value for a perfect crystal is much smaller than the value measured for real crystals. The energy gain to attach a single atom to a smooth surface is rather small because the number of bonds is small. The probability of such a process is thus low because the attached atom can easily be removed by thermal activation. Therefore, crystals grow preferredly at lattice defects, such as screw dislocations. If such a screw ends at the crystal surface, atoms can attach more easily because they can develop more atomic bonds than on a smooth surface. It is thus almost impossible to create an absolutely perfect crystal without a single dislocation. This mechanism of crystal growth is exploited to produce so-called *whiskers*: Whiskers are long, thin fibres grown around a single screw dislocation in their centre. In the ideal case, they contain only this single dislocation and can thus possess a strength that is close to the theoretically expected value for a single crystal. Whiskers are used as fibres in fibre-reinforced materials (see chapter 9).

Furthermore, dislocations are generated during plastic deformation of metals, thus providing another cause for the high dislocation densities observed. As dislocation lines cannot begin or end within the crystal, new dislocations can be created either at (inner) surfaces of the crystal, especially at grain boundaries, or in special configurations. One such configuration is the *Frank-Read source* (Fig 6.19). It consists of a dislocation that is pinned at two points in the crystal, where the dislocation leaves the slip plane. With increasing stress, the pinned segment bows out, until it becomes unstable when a semi-circular geometry is attained (see also section 6.3.1). Further bow-out of

---

[4] Because dislocations are one-dimensional, the dislocation density can be measured as length per volume or as the number of penetration points in a plane within the crystal. It unit is thus $1/\mathrm{length}^2$.

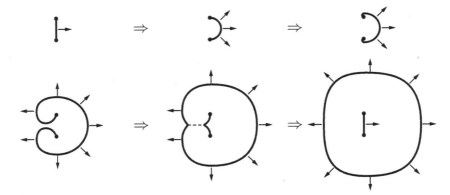

Fig. 6.19. Formation of dislocations by a Frank-Read source

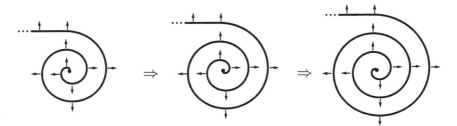

Fig. 6.20. Spiral propagation of a dislocation

the dislocation needs no more increase in the stress. The two arms of the dislocation approach and, as discussed in section 6.2.7, attract each other. They can then annihilate, as will be explained below, resulting in a dislocation loop enclosing the original dislocation, so the process can repeat itself.

Another possibility to increase the dislocation density is a spiral dislocation, pinned at the centre. Similar to a dislocation loop, the spiral dislocation extends when shear stresses are acting on it. As its centre is pinned, the length of the spiral grows (figure 6.20). This process does not increase the number of dislocations, but their density.

As mentioned above in the context of the Frank-Read source, dislocations can also be annihilated. A simple example are two opposite edge dislocations moving on the same slip plane. As explained in section 6.2.7, the superposition of their stress fields causes an attraction. If they approach, the additional upper and lower half planes can unite and form a complete plane, causing the dislocations to vanish.

Generally, annihilation of dislocations can only happen if they meet exactly on the same slip plane. In addition, they have to be oriented so that the newly generated dislocation segments have the same Burgers vector and a continuous

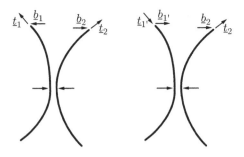

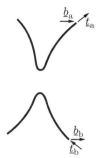

(a) Before the annihilation. In the left figure, both Burgers vectors have the opposite orientation; in the right figure, the line vectors have the oposite orientation. Both notations of the vectors $(\underline{b}_1, \underline{t}_1)$ and $(\underline{b}_{1'}, \underline{t}_{1'})$ are equivalent. The adjacent dislocation segments attract and annihilate

(b) After the annihilation

Fig. 6.21. Annihilation of dislocation segments

line vector. Thus before annihilation, both dislocations must either have the same line vector, but opposite Burgers vectors $(\underline{t}_1 = \underline{t}_2, \underline{b}_1 = -\underline{b}_2)$, or opposite line vectors and the same Burgers vector $(\underline{t}_1 = -\underline{t}_2, \underline{b}_1 = \underline{b}_2)$. This is illustrated in figure 6.21.

That all these conditions are fulfilled simultaneously seems rather improbable. At the onset of plastic deformation, this is indeed true, and more dislocations are generated than annihilated. If plastic strains are large, the number of dislocations becomes larger, and there are more annihilations, until a stationary value of the dislocation density is reached.

If we raise the temperature in a metal, edge dislocations can change their slip plane by a process called climb, increasing the probability of annihilation. The dislocation density thus reduces. In addition, the remaining dislocations arrange in an energetically more favourable configuration. This process is called *recovery*.

6.2.9 Forces acting on dislocations

Force by external shear stress

We first want to calculate the force exerted on a dislocation by an external shear stress. Consider a straight dislocation line of length l_1 that is moved over a distance l_2 by the external stress τ (figure 6.22). For simplicity, we assume that the shear stress is perpendicular to the dislocation line and parallel to the slip plane. The external force F_{ext} is $F_{\text{ext}} = \tau l_1 l_2$ because the stress τ acts

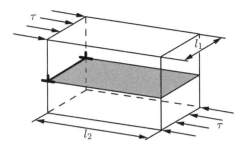

Fig. 6.22. Straight dislocation line subjected to a shear stress

on an area $l_1 l_2$. When the dislocation has moved by l_2, the upper half of the crystal has slipped by one Burgers vector $\underline{b}$. This requires a work of

$$E = F_{\text{ext}} \cdot b = \tau l_1 l_2 b. \tag{6.13}$$

The dislocation has moved by a distance l_2. The work needed can also be calculated by $E = F_{\text{d}} l_2$, when F_{d} denotes the force on the dislocation. As both energies are equal, the force on the dislocation is

$$F_{\text{d}} = \tau l_1 b. \tag{6.14}$$

Here we used the fact that the force is perpendicular on the dislocation line. If the orientation between the stress tensor $\underline{\underline{\sigma}}$, the dislocation line $\underline{l}_1$ and the Burgers vector $\underline{b}$ is arbitrary, the *Peach-Koehler equation*

$$\underline{F}_{\text{d}} = (\underline{\underline{\sigma}} \cdot \underline{b}) \times \underline{l}_1 \tag{6.15}$$

holds.

Equation (6.15) can be derived in a similar way to equation (6.14) by calculating the energy. If the dislocation line is displaced by $\underline{l}_2$, the crystal above the covered area has slipped. The normal vector in this area is given by the cross product $\underline{l}_1 \times \underline{l}_2 / |\underline{l}_1 \times \underline{l}_2|$. The stress in this area is $\underline{\underline{\sigma}} \cdot (\underline{l}_1 \times \underline{l}_2) / |\underline{l}_1 \times \underline{l}_2|$, resulting in a force of $\underline{\underline{\sigma}} \cdot (\underline{l}_1 \times \underline{l}_2)$. Because the crystal has slipped by a Burgers vector, the work is

$$E = \big(\underline{\underline{\sigma}} \cdot (\underline{l}_1 \times \underline{l}_2)\big) \cdot \underline{b}.$$

Using rules for scalar and vector products, this equation can be rewritten due to symmetry of the stress tensor:

$$E = (\underline{\underline{\sigma}} \cdot \underline{b}) \cdot (\underline{l}_1 \times \underline{l}_2)$$
$$= \big((\underline{\underline{\sigma}} \cdot \underline{b}) \times \underline{l}_1\big) \cdot \underline{l}_2.$$

This energy equals the force on the dislocation, multiplied by $\underline{l}_2$:

$$E = \underline{F}_{\text{d}} \cdot \underline{l}_2.$$

As the energies must agree for arbitrary $\underline{l}_2$, the Peach-Koehler equation results:

$$\underline{F}_d = (\underline{\underline{\sigma}} \cdot \underline{b}) \times \underline{l}_1 \, .$$

Peierls force

We already saw in section 6.2.3 that atomic bonds have to flip for a dislocation to move. This requires stretching of the bonds and therefore needs energy. The resulting force fixes the dislocation at its momentary position and has to be overcome to move it. Thus, if the applied stress is too small, no dislocation movement is possible and the crystal cannot deform plastically. Figure 6.13 above illustrates this using the sphere model of atoms. This retaining force is called *Peierls force* (or *Peierls-Nabarro force*). It determines the yield strength (or critical resolved shear stress, see section 6.2.5) of single crystals if their impurity content is small. In face-centred cubic or hexagonal close-packed metals, the Peierls stress is about $10^{-5}G$ (where G is the shear modulus) and can therefore not explain the strength of engineering alloys. In these, other obstacles for the movement of dislocations play a role, to be discussed in section 6.3. In body-centred cubic metals, the Peierls force is larger than in the close-packed structures, especially at low temperatures, and influences the yield strength significantly. This will be explained in section 6.3.2.

After the dislocation has moved by half a Burgers vector, the Peierls force pushes it forwards and moves it to the position of the next energy minimum. The stored energy is usually dissipated as heat (i. e., as random crystal vibration) in the crystal. The Peierls force thus acts as a kind of frictional force and reduces the *effective stress* that can be used to drive the dislocation to overcome other obstacles.

Other inner stresses, caused for example by other obstacles (see the next section), can counteract the external stress τ in a similar way to the Peierls stress. They can thus also be considered as inner frictional forces or stresses. The stress τ^* that is effectively available to move the dislocation is thus $\tau^* = \tau - \tau_i$. If a certain kind of obstacle is investigated, it is often useful to combine the contributions of all other obstacles to a single frictional stress τ_i and to assume that the dislocation is driven by the effective stress τ^*.

6.3 Overcoming obstacles

Dislocations can be retarded by different kinds of obstacles. We already know one of these, the Peierls force. Other types, such as precipitates of a second phase, grain boundaries, or impurity atoms, will be discussed below in section 6.4 when we look at strengthening mechanisms. Here we want to understand in what ways a dislocation can overcome an obstacle. As we will see,

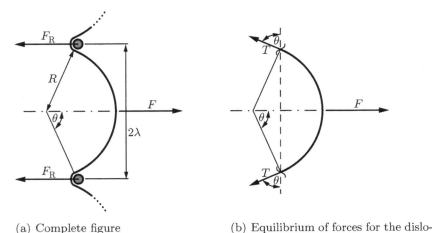

(a) Complete figure

(b) Equilibrium of forces for the dislocation line

Fig. 6.23. Deflection of a dislocation line pinned by obstacles (only part of the dislocation between two obstacles is shown)

it is important whether the dislocation movement is aided by the temperature or not. The first case is called a thermally activated process, the second an athermal one.

6.3.1 Athermal processes

Let there be several obstacles in our material with a distance of 2λ between them (figure 6.23). Consider a dislocation pinned on these obstacles. When the external stress τ acts on the dislocation, it tries to move on and bows out. Its shape is a segment of a circle because this covers the greatest area with the least-most energy to create new length of dislocation line.

The component of the force in the direction of movement is, according to equation (6.14), $F = 2\lambda b\tau$. Therefore, each obstacle exerts a retaining force F_R with opposite orientation and identical magnitude, for each obstacle takes half of the force F from two dislocation segments. If T is the line tension of the dislocation (see equation (6.3)), this force is

$$F_R = 2T \sin\theta = 2T\frac{\lambda}{R} = Gb^2\frac{\lambda}{R},$$

where G is the shear modulus, b the Burgers vector, and R the radius of the dislocation segment. Equalling F and F_R yields

$$\tau = \frac{Gb}{2\lambda}\sin\theta. \qquad (6.16)$$

Here it is crucial that the obstacle cannot bear arbitrarily large forces. If F_{max} is the maximum force the obstacle can bear, the dislocation can detach

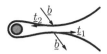

(a) Before the annihilation. The dislocation segments with opposite orientation attract

(b) After the annihilation. Since both dislocation segments lie on the same lattice plane they can annihilate, resulting in a dislocation loop around the obstacle and a free dislocation

Fig. 6.24. Annihilation of dislocation segments in the Orowan mechanism

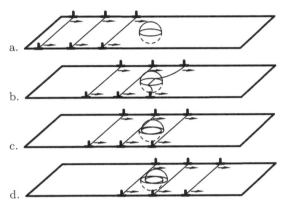

Fig. 6.25. Overcoming an obstacle by cutting of dislocations (after [74])

from the obstacle if F_R exceeds F_{max}. Therefore, there is a critical value $\sin\theta = F_{max}/Gb^2$, and $\sin\theta$ can be considered as dimensionless measure of the obstacle strength. It may seem contradictory that $\sin\theta$ takes only a limited range of values but F_{max} does not. This is resolved by realising that at $\sin\theta = 1$, the dislocation will have bowed out so far that it becomes a semi-circle, resulting in an annihilation of neighbouring dislocation segments (figure 6.24). The dislocation can move on, regardless of the strength of the obstacle. During this process, small dislocation loops remain around the obstacles. The region they enclose did thus not slip by a Burgers vector. This process of overcoming an obstacle is called *Orowan mechanism*, and the required *Orowan stress* is

$$\tau = \frac{Gb}{2\lambda}. \tag{6.17}$$

If $F_{max}/Gb^2 < 1$, the strength of the obstacle is not sufficient to retain the dislocation until the Orowan mechanism starts. In this case, the dislocation passes through the obstacle, thus *cutting* it and shearing one part of the obstacle against the other as shown in figure 6.25. This can only happen if the obstacle can slip in the same slip system as the surrounding material. This is always the case when the obstacle is another dislocation. If the obstacle is a

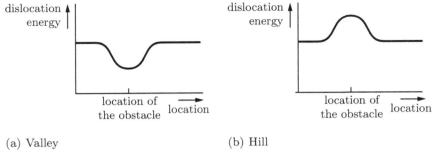

(a) Valley   (b) Hill

Fig. 6.26. Obstacles with different energy

particle (for instance, a precipitate), it can slip in the same slip system if it is coherent. A semi-coherent particle can also slip in the same slip system as the surrounding material, but the dislocation may have to move to another slip system by climb or cross slip because the matrix material and the particle have some, but not all, of the slip systems in common. Incoherent particles cannot be cut by a dislocation.

In summary, the following parameters are the main factors in determining the stress needed to overcome obstacles by cutting or the Orowan mechanism: The strength of the obstacle, the distance between the obstacles, and the elastic stiffness of the material. If we use aluminium as an example ($G = 25.4\,\text{GPa}$, $b = 2.86 \times 10^{-10}\,\text{m}$), we immediately see that the obstacles can only be effective when their distance is significantly smaller than a micrometre (cf. exercise 21). Obstacles must thus be distributed finely to increase the stress needed to move a dislocation appreciably. We can also see from this consideration that materials with a small shear modulus, like magnesium or aluminium, can never be as strong as materials with high modulus. For instance, precipitation-hardened[5] aluminium alloys ($G = 26\,500\,\text{MPa}$) have a yield strength R_p of 600 MPa at most. If the same strengthening method is used in nickel-base alloys ($G = 74\,500\,\text{MPa}$) the yield strength can be as high as 1400 MPa, in good agreement with the value expected from the shear moduli.

It does not matter for the efficiency of an obstacle whether the energy of the dislocation is increased or decreased within it (see figure 6.26). In the first case, energy is needed for the dislocation to penetrate the obstacle i.e., the dislocation is stopped in front of the obstacle. In the second case, the dislocation easily enters the obstacle – releasing some energy as heat –, but additional energy is required to detach it again.

Screw dislocations can use another mechanism, cross slip, to overcome obstacles (see section 6.2.4). As their slip plane is not fixed, they can evade to another plane not blocked by the obstacle as illustrated in figure 6.27.

---

[5] In precipitation hardening, finely distributed particles of a second phase are created by a special heat treatment. This method will be explained in section 6.4.4.

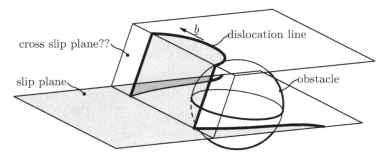

Fig. 6.27. Overcoming obstacles by cross slip of a screw dislocation. The resulting edge dislocations in the cross slipped segment cannot continue to slip in the original direction because they are oriented inappropriately

Although the external shear stress on this so-called *secondary slip plane* or *cross slip plane* is smaller than on the primary one, moving along this path can be easier than trying to overcome the obstacle by cutting or the Orowan mechanism. This is the case if the effective shear stress τ^* (see section 6.2.9) on the secondary slip plane is larger than on the primary one due to the absence of the obstacle force. Because screw dislocations can use this additional mechanism, they are frequently able to overcome obstacles more easily than edge dislocations.

In general, it is important to notice that it is not simply the more mobile type of dislocation that determines plastic deformation: If we consider a dislocation loop with different mobility of the segments as an example, we see that the more mobile type will at first cover a greater distance, but only a small amount of slip is caused by this movement. During the process, the dislocation line reorients itself, increasing the amount of the less mobile type. Thus, the importance of the more mobile one is reduced (see section 6.2.3 and figure 6.11). Because the majority of the slip has now to be performed by the less mobile type, it considerably affects the resistance against plastic deformation.

6.3.2 Thermally activated processes

Aided by thermal energy, dislocations may overcome obstacles even when the external stress is not sufficient to exert a force that exceeds the strength of the obstacle. This is called a thermally activated process (appendix C.1 provides a general introduction to this concept).

Consider a dislocation trying to move through an arrangement of obstacles as sketched in figure 6.28. We assume that the energy of the dislocation is larger within the obstacle than far away from it.[6] The stress needed to move

---

[6] As explained above, the obstacles are still obstacles if they attract the dislocation because energy is needed to leave the obstacle. All arguments made here can easily be converted to this case.

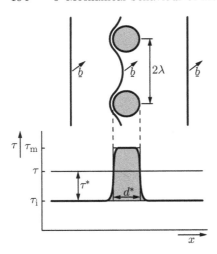

Fig. 6.28. Critical shear stress for overcoming obstacles by a dislocation. Further explanations in the text

the dislocation through the obstacle is plotted in the figure. The position of the dislocation is characterised by a single coordinate x because it moves from left to right in the figure.

Far away from the obstacle, a frictional stress τ_i is required to move the dislocation (see section 6.2.9). In the region of the obstacle, the required stress increases and then decreases again behind it. If we assume that the effect of the obstacle is restricted to its vicinity, the required stress increases steeply. To simplify the calculations, we approximate the resulting stress curve by a rectangular one with the appropriate height and width. The width d^* of the rectangle is then a measure of the width of the obstacle.

A stress of τ_m has to be exerted to move the dislocation through the obstacle. The work Q done by this stress can be calculated, using equation (6.14) for the force on a dislocation:

$$Q = (\tau_m - \tau_i) \cdot 2b\lambda d^* \,. \tag{6.18}$$

We subtracted the frictional stress τ_i because it does not describe the effect of the obstacle, but of the material without it. Q is the *obstacle energy,* the energy barrier the dislocation has to overcome.

If the effective stress τ^* is larger than $\tau_m - \tau_i$, the dislocation can overcome the obstacle. If it is smaller, a certain amount of energy is missing, given by $\Delta E = Q - 2\lambda bd^*\tau^*$. This can be provided by thermal activation. The probability P for this is, according to appendix C.1, given by

$$P \propto \exp\left(-\frac{\Delta E}{kT}\right) = \exp\left(-\frac{Q - 2\lambda bd^*\tau^*}{kT}\right) \,. \tag{6.19}$$

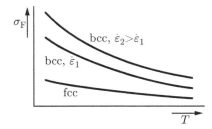

Fig. 6.29. Schematic illustration of the temperature dependence of the yield strength of face-centred and body-centred cubic metals

Here k is *Boltzmann's constant* and T the absolute temperature. The quantity $b \cdot 2\lambda d^*$ has the unit of a volume und is thus frequently called *activation volume* V^*.

Equation (6.19) states that overcoming obstacles becomes easier, the higher the temperature is, and the smaller the energy barrier. It is valid for any kind of obstacle. If kT is larger than the obstacle energy Q, the effect of the obstacle is negligible.

If we consider the Peierls force from section 6.2.9 as obstacle, it can also be overcome by thermal activation. This is especially relevant if the Peierls force is large i. e., when slip is along planes that are not close-packed, for example in body-centred cubic lattices. For this reason, the yield strength of body-centred cubic lattices is strongly dependent on the temperature, different from face-centred cubic metals (figure 6.29). The Peierls stress can reach values of up to several hundred megapascal.

> It may seem contradictory that the Peierls stress is on the one hand able to determine the yield strength of a metal and can nevertheless be overcome by thermal activation already at room temperature. The reason for this is that its activation volume is rather small. The stress τ_m needed to athermally overcome the barrier is large, but due to the small size of the activation volume, the obstacle energy Q is still small enough to be provided by thermal activation at room temperature.

The stronger dependence of the flow stress on the in body-centred cubic metals can also be explained by equation (6.19). To see this, we have to take a closer look at the meaning of the equation. So far, we talked only about the probability of the dislocation overcoming the obstacle, but not about the time needed to do so. Intuitively, it is rather obvious that the probability has to increase with time, but it is not so obvious how this can be seen from equation (6.19). The equation has to be interpreted as stating the probability to overcome the obstacle in a single 'trial'. Thermal fluctuations cause the dislocation to vibrate with a characteristic frequency. Each vibration can be considered as one trial to overcome the obstacle. This explains that with increasing strain rate $\dot{\varepsilon}$, the available number of trials becomes smaller. The yield strength must therefore increase with increasing $\dot{\varepsilon}$; this is more pronounced in body-centred cubic metals. This agrees with experimental observation (see figure 6.29).

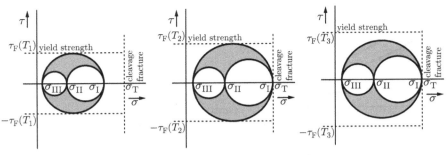

(a) Ductile behaviour at   (b) Transition temperature (c) Brittle behaviour at
temperature T_1 $T_2 < T_1$ $T_3 < T_2$

Fig. 6.30. Illustration of the brittle-ductile transition using Mohr's circle

6.3.3 Ductile-brittle transition

As we saw above, thermal activation is needed to overcome the Peierls barrier in body-centred cubic metals. This does not only cause strong hardening with decreasing temperature, it can also lead to a transition between ductile and brittle behaviour in a rather narrow temperature range. Figure 6.30 shows the transition between ductile and brittle fracture, using Mohr's circle. At elevated temperatures, the material flows plastically before the maximum tensile stress has reached the cleavage strength. At low temperatures, the yield strength has increased, but the cleavage strength is almost unchanged, so the material fractures before plastic flow starts. There is a transition regime between these two regions, the so-called *ductile-brittle transition*. It is not a material parameter because it depends on the stress state and the strain rate. As the equivalent stress, governing the onset of plastic flow (see section 3.3.1), is independent of the hydrostatic stress state, while brittle fracture depends on the maximal principal stress, brittle fracture is especially easy if the state is one of triaxial tension.

6.3.4 Climb

So far, we assumed that the dislocation segment considered stays in its slip plane. This is not always true as we already saw for the case of a cross-slipping screw dislocation. We also saw that an edge dislocation cannot by-pass an obstacle in this way. However, they can leave their slip plane by another mechanism, the thermally activated *climb* process. During climb, the dislocation either incorporates vacancies or emits them, see figure 6.31. The dislocation thus moves perpendicularly to its slip plane. For this process to be relevant, the vacancy density and mobility within the crystal must be large. As explained in appendix C.1, the vacancy density and mobility increase exponentially with the temperature. Therefore, significant climb can occur only at high temper-

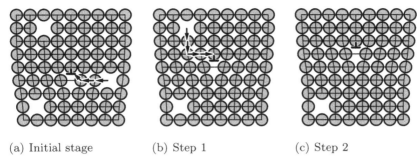

(a) Initial stage   (b) Step 1   (c) Step 2

Fig. 6.31. Climb process of an edge dislocation. By incorporating or emitting vacancies, the dislocation line can leave its original slip plane

atures (approximately above 40% of the melting temperature). This process will be discussed in more detail in section 11.2.2.

6.3.5 Intersection of dislocations

Dislocations are a particularly important type of obstacles for the movement of other dislocations.

Dislocations oriented in parallel interact and exert forces on each other as we already learned in section 6.2.7. Repulsive forces hinder the approach of the dislocations, attractive forces hinder their separation. Both forces impede dislocation movement.

If the dislocations are not parallel, their movement can nevertheless be influenced. If one dislocation by-passes the other, they create, depending on their Burgers vectors, kinks or jogs in the other dislocation [40, 61]. The difference between kinks and jogs is that kinks are within the slip plane while jogs leave it. Figure 6.32 shows the effect of a vertically drawn dislocation on a passing, horizontally drawn one for different configurations, illustrating the generation of a kink or a jog. It has to be noted that the passing dislocation will also create a kink or jog in the vertical dislocation, but for clarity this has not been included in the figure. Kinks and jogs create edge-like segments in screw dislocations and vice versa. The length of the dislocation grows in many configurations by one Burgers vector of the other dislocation. Due to the energy stored in a dislocation, energy has to be provided by the passing dislocation so that the dislocation is an energy barrier. Additional energy is needed because of the interaction of the stress fields. Dislocations that are not parallel to the moving dislocation and act as obstacles are descriptively called *forest dislocation.*

The additional edge segments created in a screw dislocation have another consequence: A jog in a screw dislocation (figure 6.33) can move only in the original slip plane by incorporating or emitting vacancies, thus reducing the mobility. This is the reason why screw dislocations are slower than edge dislocations at low temperatures [40].

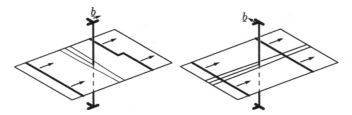

(a) Cutting of an edge dislocation. Depending on the orientation of the edge dislocation, a kink forms in the cutting dislocation

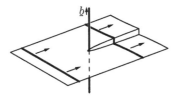

(b) Cutting of a screw dislocation. A jog forms in the cutting dislocation

Fig. 6.32. Cutting of dislocations of various types and orientations (after [40]). The type and orientation of the moving dislocation is not determined, here. Depending on its type and orientation, a kink or jog is created in the immobile dislocation

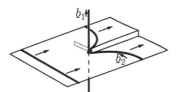

Fig. 6.33. Screw dislocation with a edge-type segment. The only way to move the edge dislocation segment in the original slip direction is by incorporating or emitting vacancies

6.4 Strengthening mechanisms

Plastic deformation of metals is mainly determined by the mobility of dislocations. To design engineering materials with high strength, dislocation movement has to be impeded. In this section, we want to discuss possible mechanisms to do this by different obstacles and to see what amount of strengthening (or hardening, as it is also called) can be achieved.

6.4.1 Work hardening

As explained above, dislocations are obstacles for other dislocations. The more dislocations there are in a metal, the higher is its yield strength. Dislocation sources, like the Frank-Read source or others described in section 6.2.8, create new dislocations during plastic deformation and serve to increase the dislocation density. This hardens the material, a process called *work hardening, strain*

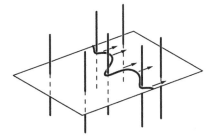

Fig. 6.34. A dislocation crosses a configuration of other dislocations acting as obstacles

hardening, or sometimes *strengthening by cold-working.* The dislocation density can increase to a value of $10^{16}\,\mathrm{m}^{-2}$.

Work hardening is the reason why the flow curve of metals increases in the plastic regime (see chapter 3). If the material is unloaded after plastic deformation, the stress-strain curve follows a line parallel to the elastic line. If the load is raised again, the yield strength has increased and the stress-strain curve follows the same line as on unloading. The strain until the material starts to neck or fracture is reduced; the material has lost ductility.

The influence of the dislocation density on the strength of a metal can be estimated: Consider a dislocation line moving through an array of dislocations perpendicular to it as sketched in figure 6.34. Let the distance between the dislocation obstacles be 2λ. If the dislocations were insurmountable, they would have to be by-passed with the Orowan mechanism. As they can be cut instead, the necessary stress is smaller than the Orowan stress. This results in $\tau_{\mathrm{cut}} = k_{\mathrm{d}}\,Gb/2\lambda$, with $k_{\mathrm{d}} \approx 0.1\ldots0.2$.

The spacing between the dislocation lines is determined by the dislocation density ϱ. If we simply assume all dislocations to be parallel and arrayed in a regular way, each penetration point in a plane perpendicular to the dislocation occupies an area of $2\lambda{\cdot}2\lambda$. The dislocation density is the number of penetration points per unit area i.e., $\sqrt{\varrho} = 1/2\lambda$. Inserting this into the equation given above, we get

$$\Delta\sigma_{\mathrm{d}} = k_{\mathrm{d}}MGb\sqrt{\varrho} \tag{6.20}$$

as the contribution of the dislocations to the strength of the material. Here we used the Taylor factor M introduced in section 6.2.6 to convert from shear to tensile stresses.

The contribution of work hardening can, according to equation (6.20), amount to several hundred megapascal. If we compare two materials (for instance, a low- and a high-strength steel) that differ strongly in their yield strength, the absolute contribution of work hardening is similar for both. Relative to the initial yield strength, the high-strength material thus has a smaller amount of hardening than the low-strength material. In section 3.2.3, it was explained that this causes a lower elongation without necking (ductility).

The strength of a material can thus be increased by simply deforming it plastically. This is used during rolling or wire drawing. Table 6.4 shows

Table 6.4. Effect of work hardening on the yield strength $R_{p0.2}$ and fracture strain $A_{11.3}$ [4] (true strain φ)

alloy	$\varphi/\%$	$R_{p0.2}/\text{MPa}$	$A_{11.3}/\%$
Al 99.5	0	20 . . . 55	35
	30	90	4
	50	130	3
AlMg 3	0	80	17
	20	190	4
	65	250	2

the increase in strength and the decrease in ductility for different states of deformation of pure aluminium and an aluminium alloy.

One advantage of work hardening is that it is simple to achieve and is often a by-product of the manufacturing process, for instance in deep drawing of steel sheets for car body parts. However, increasing the dislocation density also decreases the ductility, so work hardening is only suitable for materials with high ductility. Another disadvantage is that the strengthening is lost at high temperatures (for instance during welding) due to recovery (see section 6.2.8).

6.4.2 Grain boundary strengthening

Grain boundaries are barriers for the movement of dislocations. As the crystal orientation in the neighbouring grain is different, a dislocation cannot simply enter it. The stress field of the dislocation may initiate dislocation movement in the neighbouring grain, but if the slip systems are less favourably oriented there, a larger stress is needed to move dislocations than in the first grain.

If a slip system is activated in a crystal, several dislocations are moving on one slip plane in the same direction and can pile up at a grain boundary. Thus it is plausible, as will be explained below, that the strength of metals increases with decreasing grain size. This strengthening mechanism is called *grain boundary strengthening* or *strengthening by reduction of the grain size*.

The amount of grain boundary strengthening can be estimated using some simplifying assumptions. Consider a system of m dislocations, piled up at a grain boundary and being numbered starting at the grain boundary (see figure 6.35). This configuration may have been created by a dislocation source within the crystal that created several dislocations on the same slip plane. On each of these dislocations, the external stress τ acts to push it forwards, reduced by a frictional stress τ_i in the lattice (see section 6.2.9), resulting in an effective stress τ^*. In addition, there is a forward-pushing stress on each dislocation, caused by the interaction with the dislocations behind it. The forward acting stress τ_f on the jth dislocation is thus

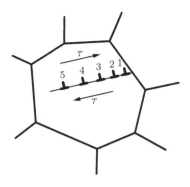

Fig. 6.35. Pile-up of dislocations at a grain boundary

$$\tau_{\mathrm{f}}^{(j)} = \tau^* + \sum_{k=j+1}^{m} \tau^{(jk)}, \tag{6.21}$$

where $\tau^{(jk)}$ is the stress that dislocation k exerts on dislocation j. For the backward-pushing stress τ_{b} we find analogously

$$\tau_{\mathrm{b}}^{(j)} = \sum_{k=1}^{j-1} \tau^{(jk)}. \tag{6.22}$$

Two dislocations exert the same, oppositely oriented stress on each other: $\tau^{(jk)} = -\tau^{(kj)}$. Another stress acts on the first dislocation at the grain boundary, namely the obstacle stress $-\tau'$ created by the grain boundary that causes the pile-up. In equilibrium, forward- and backward-pushing stresses must be the same on each dislocation: $\tau_{\mathrm{f}}^{(j)} + \tau_{\mathrm{b}}^{(j)} = 0$. Summing over all dislocations results in

$$\sum_{j=1}^{m} \tau_{\mathrm{f}}^{(j)} = -\sum_{j=1}^{m} \tau_{\mathrm{b}}^{(j)},$$

$$\sum_{j=1}^{m} \left(\tau^* + \sum_{k=j+1}^{m} \tau^{(jk)} \right) = \tau' - \sum_{j=1}^{m} \sum_{k=1}^{j-1} \tau^{(jk)}$$

$$= \tau' - \sum_{j=1}^{m} \sum_{k=j+1}^{m} \tau^{(kj)}.$$

Using the condition $\tau^{(jk)} = -\tau^{(kj)}$, we can eliminate both double sums to get

$$m\tau^* = \tau'. \tag{6.23}$$

The number m of dislocations piled up in a grain is on the one hand proportional to the diameter of the grain, on the other it is also proportional to the stress τ^*, for the larger the stress, the smaller is the equilibrium distance between the dislocations.[7] Introducing a proportionality constant k, we find from equation (6.23)

$$\tau' = k(\tau^*)^2 d.$$

The retarding stress τ' cannot exceed a critical value τ_c. If the applied shear stress is larger than this, a slip system in the neighbouring grain will be activated, which then starts to flow. If $\tau' = \tau_c$, the external stress takes the value

$$\tau = \tau_i + \frac{\sqrt{\tau_c}}{\sqrt{kd}}. \tag{6.24}$$

Here we again separated the contribution of the frictional stress. Grain boundary strengthening thus contributes to the material's strength with an amount that is proportional to the inverse of the square root of the grain size. If we convert from the shear stress on the slip plane to the tensile stress by using the Taylor factor (see section 6.2.6) and introduce a new proportionality constant k_{HP}, the amount of grain boundary strengthening is

$$\Delta\sigma_{gbs} = \frac{k_{HP}}{\sqrt{d}}. \tag{6.25}$$

This is the *Hall-Petch equation*, containing the *Hall-Petch constant* k_{HP}. Its value is $3.5\,\mathrm{N/mm^{3/2}}$ for copper, $12.6\,\mathrm{N/mm^{3/2}}$ for titanium, and $22\,\mathrm{N/mm^{3/2}}$ for a low-alloy steel [55]. Figure 6.36 shows the dependence of the yield strength of a low-alloy steel on the grain size.

Strengthening by grain boundaries has another cause, already discussed in section 6.2.6: During plastic deformation of a polycrystal, neighbouring grains have to deform so that neither material overlaps nor gaps are created. Therefore, more slip systems have to be activated near the grain boundary to enable compatible deformation of the grains. Generally, some of these are more difficult to activate and thus require a higher stress. This effect is already included in the measured values of the Hall-Petch constant.

Grain boundary strengthening has the advantage that the ductility of the material does not decrease with decreasing grain size and increasing strength. One disadvantage is that, at elevated temperatures, grain boundaries soften and constitute a weak point of the material. This will be discussed further in chapter 11. Fine-grained materials are thus advantageous only in the low-temperature regime.

In a material cooled from the melt, the grain size is determined mainly by the cooling rate. To produce a fine-grained material, the cooling rate must be large, but this is technically difficult to achieve. Fine-grained materials are therefore usually produced in another way, by *recrystallisation*.

---

[7] This argument shows that m increases with τ^*. It is not so easy to show that the dependence is actually a proportionality.

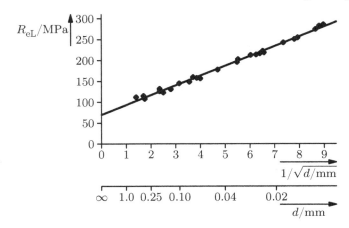

Fig. 6.36. Dependence of the lower yield strength on the grain size in a low-alloy steel at room temperature (after [68])

Here the material is heavily deformed at first, increasing the dislocation density to values of the magnitude $10^{15}\,\mathrm{m}^{-2}$ (see section 6.4.1). Due to the elastic distortion around the dislocations, the amount of stored elastic strain energy in the crystal is large. If the temperature is raised, the material recovers (see section 6.2.8) by re-ordering the dislocations and annihilating some of them, thus slightly reducing the dislocation density and the stored energy.

Because of the large amount of stored elastic energy, the deformed state is thermodynamically unstable. Favourably oriented regions, for instance near grain boundaries or inclusions, serve as starting points or *nuclei* for the formation of new, undeformed grains. During recrystallisation, these nuclei grow by moving their boundaries into the deformed material. The newly created grain now has a low dislocation density and thus a smaller amount of stored elastic energy. As the boundary between the nucleus and the already existing grains is, like every grain boundary, a region of high energy, growing can only occur if the increase in grain boundary energy is compensated by the decrease of the stored elastic strain energy from the decrease of the dislocation density. The higher the dislocation density, the easier the nuclei can grow. A large initial dislocation density finally produces a fine-grained structure, for the rate of activation of grains is large.

To produce fine-grained metals, they are at first heavily deformed (by rolling, for example) and are then heat treated in a way leading to recrystallisation. By controlling the amount of deformation and the heat treatment temperature, the resulting grain size can be adjusted rather precisely.

6.4.3 Solid solution hardening

Another important way of strengthening metals is to alloy them with elements that are dissolved in the crystal lattice and form a solid solution. Such atoms

(a) Substitutional solid solution   (b) Interstitial solid solution

Fig. 6.37. Different types of solid solutions

(a) Smaller atom   (b) Larger atom

Fig. 6.38. Different sizes of dissolved atoms in solid solutions

elastically distort the crystal and can thus interact with the stress field of a dislocation and impede its movement.

Atoms in solid solutions can be situated on two different kinds of lattice sites: They can either sit at the same position as the original atoms, thus substituting one atom by another (*substitutional solid solution*), or they can be placed in interlattice positions between the original atoms, forming an *interstitial solid solution*. Figure 6.37 sketches both cases. An interstitial solid solution can only form when the dissolved atoms are much smaller than those of the host atoms. Carbon in iron provides an example.

Substitutional atoms act as obstacles for dislocation movement by different mechanisms. Most important is the elastic distortion of the lattice (figure 6.38) that interacts with the distortion around the dislocation. If, for example, the dissolved atoms are larger than the host atoms, they produce compressive stresses in their vicinity. An edge dislocation trying to enter this region with its own compressive region will thus be repelled and needs additional energy to move on. If the dislocation approaches the dissolved atom with its tensile region, it will be attracted, and thus it becomes difficult to detach the dislocation from the solid solution atom, pinning the dislocation. Smaller substitutional atoms behave in the opposite way.

A further interaction between the dislocation and the solid solution atom is due to the different strength of the atomic bond between the dissolved atom and its neighbours, resulting in a locally changed elastic modulus in the vicinity of the solid solution atom. The line tension of the dislocation thus either increases or decreases when it approaches the atom, causing another obstacle effect known as *modulus interaction*.

Short-range order interaction [40] (sometimes called configurational interaction or Fisher effect) can also occur. If, for example, the binding energy

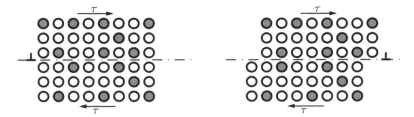

Fig. 6.39. Short-range order interaction during dislocation slip. If it is energetically unfavourable for the grey-coloured atoms to be on adjacent lattice sites, the slip shown is hampered

between host atom A and dissolved atom B is larger than that between B atoms, it is energetically favourable to surround a B atom with A atoms. If this short-range order is disturbed by slip, resulting in two neighbouring B atoms, additional energy is needed (see figure 6.39). Dislocation movement is thus impeded.

Experimentally, the following relation between the contribution to strengthening $\Delta\sigma_{\mathrm{sss}}$ and the concentration c of the impurity atoms is found:

$$\Delta\sigma_{\mathrm{sss}} \propto c^{n} . \tag{6.26}$$

The exponent n takes values of about 0.5. This is plausible because the spacing of the obstacles decreases approximately with $\sqrt{c}$ as we will see in section 6.4.4 (equation (6.28)).[8]

From what has been said so far, it might be presumed that it is best to choose atoms with strongly differing radii as substitutional atoms to achieve a large strengthening contribution. This, however, is only partly true because the solubility of atoms decreases with increasing difference in the radii. For instance, 100% nickel can be dissolved in copper because the radius difference is only 2.7%, but copper can only dissolve 10% aluminium with a radius difference of 12%. In general, a difference in the radii of less than about 15% is required for good solubility. The solubility is also larger if the elements are chemically similar and have the same crystal structure.

A different case is the *interstitial solid solution*, for instance of carbon or nitrogen in steel. Here a large difference in the radii is required because it enables the dissolved atoms to sit in an interstitial position.

Substitutional solid solution strengthening has the advantage of being rather temperature insensitive. With increasing temperate, for instance during welding, the solubility of the atoms does not decrease, but increases, so strengthening at room temperature is not impaired. As long as the dissolved atoms diffuse only slowly through the crystal and thus cannot move along

---

[8] There, the volume fraction of precipitates f_{V} is used, which is equivalent to the concentration c of the solid solution.

Table 6.5. Influence of solid solution strengthening on the yield strength $R_{p0.2}$ and the fracture strain $A_{11.3}$ in the annealed condition [18]. The numbers in the material names state the approximate content of the particular alloying element in percent

alloy	$R_{p0.2}$/MPa	$A_{11.3}$/%
Al 99.5	20…55	35
AlMg 1.5	45	20
AlMg 2.5	60	17
AlMg 3	80	17

with the dislocations, $\Delta\sigma_{sss}$ will be significant even at elevated temperatures. Nickel-base superalloys strengthened with tungsten, molybdenum, and rhenium are an example.

Another advantage of solid solution strengthened and thus single-phase alloys is their good corrosion resistance. This is due to the absence of *localised galvanic cells*. Localised galvanic cells are formed by two contacting phases of different chemical composition with a different position in the electrochemical series i.e., one of them being less noble than the other. If the material is attacked by a corrosive medium, the less noble metal can be dissolved.

> However, it has to be noted here that some solid solution strengthened alloys are supersaturated at room temperature. This is the case, for example, for the alloy AlMg 4.5 Mn. If the material is cooled too slowly from elevated temperatures, precipitation reactions can occur (see section 6.4.4). If the precipitated particles are semi-coherent or incoherent, as in the case above, and thus have a large nucleation barrier, heterogeneous nucleation at grain boundaries occurs preferredly. This frequently leads to embrittlement and larger sensitivity to intercrystalline corrosion.

Table 6.5 shows the effect of solid solution strengthening for aluminium. If we compare it to table 6.4, we see the favourable combination of strength and ductility.

One major disadvantage of solid solution strengthening is that those atoms that would yield a large effect due to their large radius difference have only a limited solubility (see above). Thus, this method can usually achieve only a moderate strengthening. The same is usually the case for interstitial solid solutions, for relatively large interstitial atoms possess a limited solubility, also. One famous exception is carbon in ferritic steels. Because the face-centred cubic γ phase can dissolve several percents of carbon at high temperature, it is possible to 'freeze in' these large contents when cooling to the body-centred cubic α phase, although the elastic lattice distortion is large (*hardening*, see also section 6.4.5).

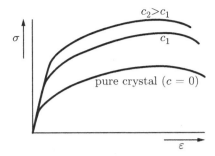

Fig. 6.40. Effect of solid solution atoms on stress-strain curves

Yield point phenomenon and strain ageing

As already discussed, the interaction between dissolved atoms and a dislocation can cause pinning of the dislocation. Because the dissolved atoms can move through the lattice by diffusion, they can pin dislocations even if these do not move. This is especially so for interstitial atoms, for they have a large diffusivity. This is the cause of the apparent yield point of some metals and for the so-called *Portevin-Le-Châtelier effect* (PLC) as we will see now.

If the diffusivity of the dissolved atoms is negligible, as assumed in the previous section, dislocations can be pinned only when they move due to external stresses. In this case, the presence of the dissolved atoms causes strengthening (figure 6.40).

If we increase the diffusivity (for example by raising the temperature), we encounter the *yield point phenomenon* i. e., we find an upper and a lower yield strength (see section 3.2). Solid solution atoms diffuse into the distorted regions near the dislocations line while the material is stress-free. If external stress is applied, the dislocation has to be 'teared off' its pinning points. The stress required for this defines the upper yield strength R_{eH} of the material. After the dislocation has left its pinning points, it is more mobile than before, and the yield strength reduces (*lower yield strength* R_{eL}). Deformation localises in this region, with only a few grains participating. Dislocations pile up at the grain boundaries, thus increasing the stress in the neighbouring grain, allowing dislocations there to become mobile as well. Narrow bands of localised deformation, so-called *Lüders bands,* form within the material. This alternation between local hardening by dislocation pile-up and removal of this deformation obstacle by tearing off dislocations in the neighbouring grain causes a strongly serrated flow curve. Apart from these fluctuations, there is no hardening. Only after the Lüders bands have spread throughout the material and all dislocations are removed from their pinning points does the yield strength increase beyond the lower yield strength by work hardening.

If the load is removed and the material stored for some time, the dissolved atoms diffuse again to the dislocations and thus re-anchor them. If the material is deformed again, an upper and lower yield strength are again encountered. This is called *strain ageing*. The temperature and time required for this ageing

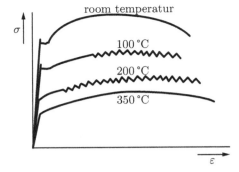

Fig. 6.41. Schematic illustration of the dependence of the stress-strain curve on the temperature for steel

process depend on the diffusivity of the dissolved atoms and thus on the alloy. In many steels, strain ageing can occur at room temperature.

The yield point phenomenon is often not desired because it causes an inhomogeneous plastic deformation. In deep drawing of metal sheets for car bodies, for instance, the surface of the sheets becomes rough (so-called 'orange peel'). Counter measures are therefore needed. Steel sheets, for instance, are usually temper rolled before the deep drawing process to tear off the dislocations by minor plastic deformation.

> For the case of steels, we want to estimate typical diffusion path lengths of carbon in the iron matrix. The *mean diffusion distance* is $x \approx \sqrt{Dt}$, where t is the time available and D is the diffusion coefficient or diffusivity. The diffusion coefficient can be calculated from the Arrhenius law:
>
> $$D = D_0 \exp\left(-\frac{Q}{RT}\right) = 0.2\,\mathrm{m}^2/\mathrm{s} \cdot \exp\left(-\frac{103\,\mathrm{kJ/mol}}{RT}\right) ,$$
>
> where Q is the *activation energy* of diffusion and D_0 is the *diffusion constant*. At room temperature, the diffusion coefficient is $D \approx 2 \times 10^{-19}\,\mathrm{m}^2/\mathrm{s}$. The diffusion distance covered in 24 hours is thus 130 nm. If the dislocation density has a typical value of $10^{12}\,\mathrm{m}^{-2}$ to $10^{14}\,\mathrm{m}^{-2}$, the mean distance between the dislocations is, according to section 6.4.1, 100 nm to 1000 nm. The atoms can thus easily travel to the dislocations within a day.

If the diffusivity is increased further (by further heating), the diffusion speed of the atoms becomes so large that they can move to a dislocation as soon as it stops.[9] Tearing-off and re-pinning of the dislocation alternate, causing a serrated flow curve (figure 6.41), the *Portevin-Le-Châtelier effect*. Another consequence of this effect is that when the strain rate is increased, the dissolved atoms may not be fast enough anymore to catch up with the dislocation and

---

[9] Dislocations usually travel quickly for a certain distance and then pause for some time.

pin it. In this case, the yield strength of the material becomes smaller when the strain rate increases, in contrast to the usual behaviour (see section 6.3.2), and the serration of the flow curve vanishes.

If the diffusivity (or the temperature) is raised even further, the speed of the dissolved atoms is so high that they simply accompany the dislocation during its movement. In this case, there is neither an apparent yield point nor serrated flow.

6.4.4 Particle strengthening

Many alloys comprise not only one phase, but several. Among these are precipitation hardened and dispersion strengthened materials. They contain fine particles of the second phase with sizes far below a micrometre. The particles are thus strong obstacles for dislocation movement (see section 6.3.1), resulting in high-strength materials. These materials have to be distinguished from metals with a coarse two-phase microstructure with large spacing between the particles and, consequently, a small Orowan stress. Nevertheless, significant strengthening can also occur in this case. Both classes of materials will now be explained. Metals can also be strengthened by adding fibres; this will be covered in chapter 9.

Coarse two-phase materials

Coarse particles of a second phase influence all bulk properties of a material. If, for example, Young's modulus of the second phase is larger than that of the matrix, load will be transferred from the matrix to the particle if the material is stressed elastically, and the stiffness increases. In the context of fibre- and plate-shaped particles this will be discussed further in chapter 9. The amount of load transfer depends strongly on the shape of the particles and their arrangement.

Other physical properties behave similarly to Young's modulus. If, for example, copper is used to dissipate heat from a ceramic structure, the coefficients of thermal expansion of the copper alloy and the ceramic should not differ too much to reduce thermal stresses at the interface. Adding tungsten particles to copper reduces its coefficient of thermal expansion so that it becomes closer to that of the ceramic.

The wear resistance of a material can also be improved by adding particles of a second phase. If the particles are hard, the softer matrix material will wear off first, resulting in the hard particles sticking out of the surface and then determining the wear properties. This method is used, for example, in aluminium cylinder liners in combustion engines whose wear resistance is improved by silicon precipitates, or in cast iron (figure 6.42).

Coarse particles also influence the plastic properties. If their Young's modulus is larger than that of the matrix, load transfer reduces the stress in the matrix, increasing its yield strength. For this to be possible, the strength of

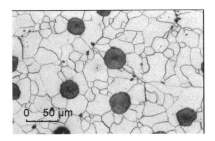

Fig. 6.42. Two-phase microstructure of grey cast iron. Graphite grains are embedded in the light ferritic phase

the particles must be larger than that of the matrix. In this case, there are further strengthening effects, discussed in the following.

To discuss these effects in detail, we consider the example of a tensile load on a matrix material that is strengthened by spherical particles arrayed regularly (for instance, in a cubic structure). If the volume fraction of the particles is not too large, it is possible to find a deformation path at 45° to the external load that does not cut through any of the particles. If the matrix were perfectly plastic, it could deform along this path, and the strength would not increase. If, however, the matrix hardens, hardening is increased compared to the unreinforced material because plastic deformation is restricted to parts of the matrix, resulting in larger strains there. This effect does therefore not increase the yield strength, but the amount of hardening.

If the particles are elongated, for instance in the shape of fibres or small plates, finding a deformation path at 45° to the external load may be impossible even at small volume fractions. In this case, a more intricate deformation pattern has to evolve. This obstruction of the deformation causes a marked strengthening effect with load transfer to the particles. Accordingly, elongated particles are especially efficient as long as their strength is high. This will be discussed in detail in chapter 9.

Coarse particles of a second phase embedded in a matrix of the first phase can also contribute to strengthening by their interaction with dislocations. Although they can be easily by-passed by the Orowan mechanism, dislocation loops around the particles are created, thus increasing the dislocation density and leading to more work hardening. This effect is used in manufacturing deep-drawn car body parts from dual-phase steels that contain martensitic islands in a matrix of ferrite (see also section 6.4.5). Due to their small initial yield strength but large amount of work hardening, these materials are easy to deform and nevertheless exhibit high strength after deformation has finished.

Although coarse second-phase particles do not strengthen a material as efficiently as fine-grained particles, they offer a multitude of ways to influence material properties.

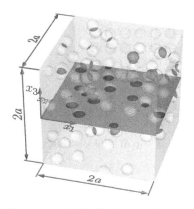
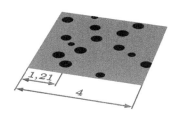

Fig. 6.43. Calculation of the average spacing of randomly distributed particles within one arbitrary plane. The presented example incorporates $N = 100$ particles with a radius of $r = 0.22\,\mu m$ in a cube with edge length $2a = 4\,\mu m$ ($f_V = 0.07$). Using equation (6.27) yields an average spacing of $2\lambda = 1.21\,\mu m$

Strengthening by fine particles

Small particles of a second phase, evenly distributed in the grains of the first phase, form a strong barrier to dislocation motion. This was previously discussed in section 6.3, and we saw there that there are two possible ways to overcome such obstacles, the Orowan mechanism and cutting of the particles. The mechanism actually occurring depends on the strength of the obstacles and on their distance. This strengthening mechanism is frequently called *precipitation hardening,* because the particles are usually created by a precipitation process, described below.

The effect of second phase particles not only depends on their radius r, but also on the number of the particles (measured as their volume fraction f_V) and their mean spacing 2λ on an arbitrarily chosen slip plane. These three quantities are not independent because, if the volume fraction is kept constant, the spacing of the particles increases when we increase their radius. To calculate the relation between them, we consider a cube with an edge length of $2a$ as shown in figure 6.43. Into this cube, we insert N particles with radius r. We want to calculate the mean distance 2λ between those particles intersecting the x_1-x_2-plane. Each sphere with a centre coordinate x_3 between $-r$ and r intersects the plane, so the probability that a randomly positioned sphere does so is $2r/2a$. On average, there thus are $N \cdot 2r/2a$ spheres in the plane. The number N of spheres can be calculated from their volume fraction

$$f_V = N\frac{4\pi r^3}{3(2a)^3}\,.$$

The mean area $(2\lambda)^2$ per sphere on the plane is given by the area $(2a)^2$ of the plane divided by the number of spheres intersecting it, $N \cdot 2r/2a$, yielding

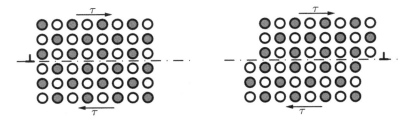

Fig. 6.44. Formation of an anti-phase boundary during slip of a dislocation through an ordered lattice structure containing two elements

$$(2\lambda)^2 = \frac{(2a)^2}{N \cdot 2r/2a} = \frac{1}{f_V}\frac{(2a)^3 4\pi r^3/3}{2r(2a)^3} = \frac{2\pi r^2/3}{f_V} \approx \frac{2r^2}{f_V}. \qquad (6.27)$$

Thus, the mean spacing 2λ of the particles is

$$2\lambda \approx \sqrt{\frac{2}{f_V}}\, r\,. \qquad (6.28)$$

The particles act as obstacles in a similar way to the solid solution atoms discussed in section 6.4.3. They elastically distort the crystal and thus interact with the dislocations, and they also change the line tension of a dislocation, for the elastic stiffness of the particle is usually different from that of the matrix. If the dislocation cuts through the particle, one part of the particle is sheared relative to the other. This increases the surface of the particle (cf. figure 6.25 on page 191). The additional surface energy has to be provided by the external work and thus raises the force needed to move the dislocation. If the particles have an ordered lattice structure containing more than one element (a superlattice), this order is disrupted by the passing dislocation (figure 6.44), forming a so-called anti-phase boundary (APB). The new configuration of the atoms has a higher energy than the old one.

It is difficult to quantitatively estimate these effects. Approximately, it can be assumed that the force F exerted by the obstacles increases with the particle radius as $F \propto r^{3/2}$.

Because the force on a dislocation is, according to equation (6.14), $F = 2\lambda b\tau$, the strengthening contribution of the particles is

$$\Delta\tau_{\mathrm{ps}} = \frac{F}{b}\frac{1}{2\lambda} \stackrel{.}{=} \frac{F}{b}\frac{1}{r}\sqrt{\frac{f_V}{2}} = k_{\mathrm{ps}}\sqrt{f_V r}\,, \qquad (6.29)$$

if we use the relation $F \propto r^{3/2}$ and introduce the constant k_{ps}. The effect of particle strengthening thus increases with the particle radius and the volume fraction.

This relation is valid only in the case that the particles are cut by the dislocations. With increasing particle radius, the force required for cutting the particles increases. On the other hand, if the volume fraction is kept fixed

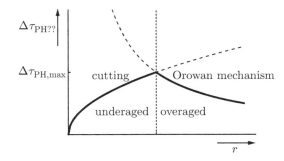

Fig. 6.45. Dependence of the strengthening effect on the radius of precipitates

at f_V, the spacing of the particles grows until the Orowan mechanism for overcoming the obstacles becomes easier because here the force decreases with increasing particle distance. According to equation (6.17), this contribution is

$$\Delta\tau_{\mathrm{ps}'} = \frac{Gb}{2\lambda} = k_{\mathrm{ps}'}\,\frac{\sqrt{f_V}}{r}\,. \tag{6.30}$$

Consequently, large radii are disadvantageous because the particles are overcome by the Orowan mechanism as the particle distance increases.

The total strengthening contribution of coherent or semi-coherent particles is the minimum of both contributions because the easier mechanism will be used by the dislocation. If we keep the volume fraction of the particles constant, the result is a curve that increases proportionally to the square root of the particle radius and reaches a maximum beyond which the curve drops again (figure 6.45). The particle radius is optimal at the intersection of the two curves i.e., the contribution to strengthening is maximised here. In general, this radius is between 10 nm and 100 nm; the particles have to be small to optimise strengthening. Since the particle size can be adjusted by the process of ageing as described below, the material state is often called *underaged* when particles are smaller than the optimum and *overaged* when they are larger.

In practice, it is usually better to choose particle radii slightly above the optimum value. If the radii are at or below the optimum, particles will be overcome by cutting. Continual cutting processes can destroy the particles, producing locally softened regions where deformation concentrates. This is especially problematic under cyclic loads and can cause a strong reduction of the life time of a component (see chapter 10).

Table 6.6 summarises the strength of some particle-strengthened aluminium alloys.

Precipitation hardening

Strengthening of materials by fine particles is frequently obtained by *precipitation hardening*. This process will now be explained, using the example of the alloy system aluminium-copper.

Table 6.6. Effect of particle strengthening on yield strength $R_{p0.2}$ and fracture strain A of aluminium alloys after EN 485. The specified values are minimum values for sheets with a thickness of more than 12.5 mm, according to the standard. The state remarks describe the heat treatment

alloy	state	$R_{p0.2}$/MPa	A/%
Al 99.5	O, annealed	20 . . . 55	35
AlSi 1 MgMn	O, annealed	≤ 85	16
	T4, naturally aged	110	13
	T61, artificially aged	200	12
	T6, artificially aged	240	8
AlCu 4 MgSi	O, annealed	≤ 145	12
	T4, naturally aged	250	12
AlCu 4 Mg 1	O, annealed	≤ 140	11
	T3, naturally aged	290	11
AlCu 4 SiMn	O, annealed	≤ 140	10
	T4, naturally aged	250	10
	T6, naturally aged	400	6

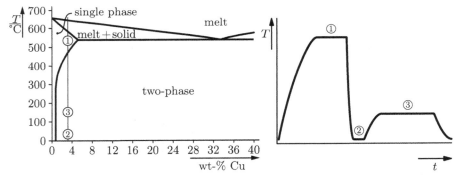

(a) Phase diagram of the alloy system Al-Cu   (b) Heat treatment for precipitation hardening

Fig. 6.46. Precipitation hardening of an aluminium-copper alloy

Figure 6.46(a) shows part of the phase diagram of the Al-Cu system. One prerequisite for precipitation hardening is the existence of a two-phase region where the matrix phase (in the example, aluminium with copper in solid solution) is in equilibrium with the precipitation phase (a copper-rich phase in the example), a so-called *miscibility gap* (see section C.3).

To create second-phase particles that strengthen the material significantly, their radius must be near the optimum as discussed above, so the particles must be sufficiently small and also be evenly distributed. This requires a precipitation reaction in the solid state since particle coarsening rates are too high in liquids. Such a reaction is possible if the solubility of the second element (copper in the example) decreases with decreasing temperature. If, for example, we heat an alloy with a copper content of 3.5 wt-% to a temperature of about 550°C (state ①) and keep it for a sufficiently long time at this temperature, all copper will be dissolved because a single-phase state is in equilibrium at this temperature. This step is therefore called *solution heat treatment* or *solutionising*. If we quench the system sufficiently fast (state ②), time is not sufficient to precipitate the second phase. The system is still single-phase, but oversaturated with copper. A slight warming of the system (to temperatures of about 150°C, state ③), often referred to as *artificial ageing*, raises the diffusivity of the copper atoms sufficiently so that they can form a second phase in relatively short time (usually a few hours). Figure 6.46(b) sketches a possible heat treatment. Because of the ageing step required to produce precipitation hardened alloys, they are frequently also called *age hardened.*

In some alloys, for example Al-Cu-Mg alloys, this final step is not necessary because the diffusivity of the atoms is sufficient already at room temperature. The particle size is in this case usually below the optimum and cannot be adjusted precisely. The advantage of this so-called natural ageing is that one step in the heat treatment can be dispensed with.

Because a fine distribution of the particles is required for particle strengthening, it is best if a large number of precipitates can form. Therefore, a large number of initial nuclei of the second phase are required. An initial nucleus, formed by random fluctuations, will be stable if its further growth will decrease its energy[10]. Two effects are crucial for this: On the one hand, precipitating the particle decreases the energy because during quenching the system has moved into the miscibility gap (see section C.3) and is oversaturated. This gain in energy is proportional to the volume of the particle. On the other hand, energy has to be expended to create the energetically costly interface between particle and surrounding matrix. The total change ΔQ in energy is thus

$$\Delta Q = \frac{4}{3}\pi r^3 Q_V + 4\pi r^2 \gamma, \tag{6.31}$$

where Q_V is the specific energy gain due to decomposition of the oversaturated phase in two phases, and γ is the specific interface energy of the boundary. For an oversaturated phase, $Q_V < 0$. Figure 6.47 shows how ΔQ depends on the radius r of the nucleus. ΔQ is maximal at a critical radius r^*. If a nucleus with a radius of r' forms by random fluctuations, it will grow if further growth reduces the energy i.e., if $r' > r^*$. If we differentiate equation (6.31) and set the derivative equal to zero, the critical radius can be calculated as $r^* = -2\gamma/Q_V$. Therefore, the larger r^*, the less probable is the formation of a growing nucleus and the fewer nuclei will form in the material.

---

[10] To be more precise, its free enthalpy, see appendix C.2.

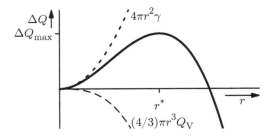

Fig. 6.47. Dependence of the energy gain ΔQ on the particle radius r

Precipitation hardening is therefore most efficient if the *nucleation barrier* is small. As we saw, the energy of the interface between particle and matrix is crucial for this. If it, and thus the nucleation barrier, is large, the rate of nucleation is small, and few particles are created in any time interval. Thus, there is ample time for the particles to grow before the oversaturation is relieved. Accordingly, the precipitates are rather large. Therefore, alloys forming incoherent particles with a large value of the interface energy are not suitable for precipitation hardening. This is the case in Al-Mg alloys. Other aluminium alloys, like Al-Cu, Al-Mg-Si, and Al-Zn-Mg form coherent particles with low interface energy and can be precipitation hardened.

The energy difference Q_V also plays an important role in determining the precipitation process. Its absolute value becomes larger with increasing oversaturation of the solid solution. If it is small, for example when the temperature is only slightly below the limiting temperature between the two-phase and the single-phase region, the nucleation barrier also becomes large and, as before, large precipitates form. Therefore, it might seem that ageing should be performed at low temperatures. However, in this case diffusion rates are low, and complete precipitation of the second phase and particle growth to the optimum size cannot be achieved within a realistic time-scale. This explains why natural ageing of aluminium alloys does not lead to the same strength levels as artificial ageing (see table 6.6).

Frequently, the solubility of the second element in the matrix phase cannot be neglected so that some of these atoms are dissolved in the matrix and increase the strength additionally by solid solution hardening.

The main advantage of precipitation hardening is that a large strengthening effect can be achieved. One disadvantage is that the heat treatment is rather complex and has to be fine-tuned and controlled precisely to adjust the desired particle radius. Furthermore, precipitation-hardened alloys are sensitive to high temperatures. If they are locally overheated, for instance by welding, the precipitates may coarsen or even dissolve. After re-precipitation, the strength near the weld may be large. Residual stresses produced on cooling may then be sufficient to cause cracks in the material. Precipitation-hardened alloys are thus less weldable than solid solution strengthened alloys.

Fig. 6.48. Fracture surface of an austenitic chrome-nickel steel X 5 CrNi 18-10 (EN number 1.4301) caused by intercrystalline corrosion

A further problem is connected to corrosion. If a large number of particles have precipitated on the grain boundaries, localised galvanic elements can form (see page 206) and cause intercrystalline corrosion. If the precipitates are less noble than the matrix, they are attacked. This is especially dangerous if they completely cover the grain boundaries. If the precipitates are more noble than the matrix, oxidation or corrosion of the matrix material is accelerated in their vicinity and the material also dissolves near the grain boundaries.

A similar effect is observed in stainless steels. At chromium contents of more than about 13%, steel becomes corrosion-resistant in weakly corrosive media because a passivating chromium oxide (Cr_2O_3) layer separates the reaction partners. If a steel is alloyed with a chromium content slightly above this value, precipitation of chromium carbides can reduce the chromium content in the vicinity of the precipitates below the required value, and the matrix becomes sensitive to corrosion. As long as these chromium-depleted regions do not overlap, this is uncritical, for corrosion cannot destroy the whole material. At grain boundaries, however, it is more probable that these regions overlap, for two reasons: On the one hand, chromium carbide forms preferentially at the grain boundaries due to a reduced nucleation barrier; on the other hand diffusion is faster there, increasing the size of the depleted region compared to the inside of the grain. Both effects increase the corrosion-sensitivity of the areas adjacent to the grain boundaries and can lead to intercrystalline corrosion. A steel with such corrosion-sensitive grain boundary regions is called *sensitised*. Figure 6.48 shows a fracture surface resulting from such a corrosion process.

dispersion strengthening

Precipitation-hardened alloys can be very strong, but their service temperature is limited. For example, long-time application of precipitation-hardened aluminium alloys at temperatures above 200°C is impossible due to excessive coarsening of the precipitates. To use high-strength materials at high temperatures, another method of particle strengthening can be achieved by

Fig. 6.49. Micrograph of an aluminium reinforced with aluminium-oxide particles [124]

introducing incoherent particles into the matrix. This method is called *dispersion hardening* because particles are distributed in the matrix similar to a dispersion. Suitable dispersoids should be thermodynamically stable even at high temperatures and contain at least one element that is not soluble in the matrix, thus avoiding coarsening of the particles. One example is aluminium containing aluminium oxide particles (Al_2O_3), inserted by powder metallurgical means (figure 6.49).

One advantage of dispersion strengthened alloys is their high temperature resistance. Furthermore, particles cannot be overcome by cutting because cutting is only possible in coherent or semi-coherent particles. Theoretically, very high strength levels can be achieved at very small particle radii. In practice, however, it is very difficult to distribute such fine particles as evenly as nature does when precipitates form. Another disadvantage of dispersion strengthening is that distributing the particles in the matrix is a complex and expensive process.

6.4.5 Hardening of steels

Steels are strengthened by a particular mechanism that is discussed in this section. This mechanism frequently is called *hardening* without further specification, in contrast to e. g., work hardening.

Iron can exist in three different phases: The body-centred *ferritic* α phase, the face-centred cubic austenitic γ phase and also the body-centred cubic δ phase that is completely irrelevant for our discussion.

Pure iron has a phase transformation between the α and the γ phase at a temperature of 912°C. If iron is cooled sufficiently slowly from above this temperature, the face-centred cubic phase transforms to the body-centred one. The phase transformation is diffusive, for it proceeds by atoms exchanging their places and forming nuclei of the α phase which then grow by incorporating more atoms from the γ phase.

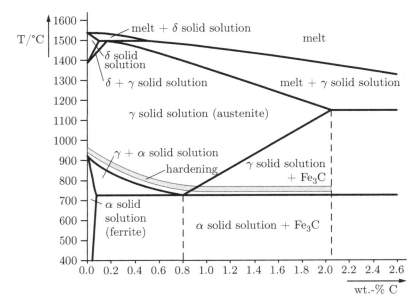

Fig. 6.50. Section of the phase diagram iron-carbon

Two opposing effects are involved in this process: The driving force for the transformation becomes larger when the temperature decreases. This driving force is the energy difference between the two phases[11] because the α phase has a lower energy than the γ phase below the transformation temperature (see also page 215). The more the γ phase is undercooled, the greater the driving force for the transformation.

On the other hand, the diffusivity of the atoms becomes exponentially smaller when the temperature is reduced. If the material is undercooled strongly, it becomes more and more difficult for the atoms to move to their new positions. The transformation speed is determined by these two opposing effects. It has a maximum at a temperature of about 700°C (figure 6.51) and strongly decreases with decreasing temperature. If we were to undercool an iron crystal abruptly (with a cooling rate of about 10^5 K/s) from the γ phase to room temperature, the γ phase would be frozen in a *metastable state* because no diffusion is possible. The energy difference to the α phase, which is thermodynamically stable, would, however, be huge.

In practice, it is not possible to undercool an iron crystal in such a way. Instead of the diffusive phase transformation, a diffusion-less transformation, also known as *martensitic transformation,* takes place, without exchange of atomic positions. Instead, the atoms only slightly rearrange themselves to

---

[11] More precisely, the difference in free enthalpy, see appendix C.2.

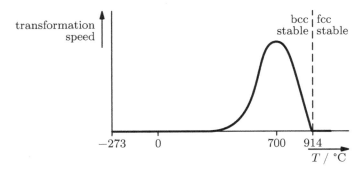

Fig. 6.51. Dependence of the transformation speed from face-centred to body-centred cubic structure of iron on the temperature (after [9])

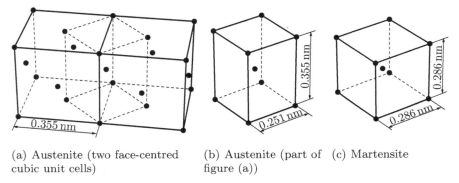

(a) Austenite (two face-centred cubic unit cells)

(b) Austenite (part of figure (a))

(c) Martensite

Fig. 6.52. During the martensitic transformation, the face-centred cubic crystal is compressed in one direction and elongated in the others

produce the favourable body-centred cubic structure from the face-centred crystal.

This process can most easily be visualised by comparing the unit cells of the face-centred and the body-centred phases as it is shown in figure 6.52. If we look at two neighbouring unit cells of the face-centred cubic structure, we see a distorted body-centred unit cell between them that is compressed along two of its axes and dilated in the third direction. A slight shift of the atoms in such a distorted cell can thus transform the lattice to a body-centred cubic lattice without any atoms having to exchange their places. The transformation is thus diffusion-less. The energy for this process is provided by the lower energy of the thermodynamically stable α phase. Because differently oriented grains deform unequally in different spatial directions, considerable residual stresses can be produced and distort the component.

> This visualisation of the diffusion-less transformation according to Bain is not totally correct. X-ray diffraction experiments have shown that, in reality, the (111) planes of the face-centred cubic lattice become the

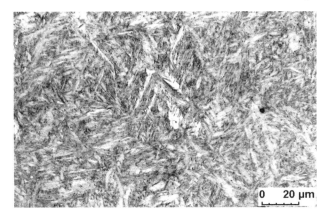

Fig. 6.53. Micrograph of a hardened C 60 steel. The martensite needles can be observed

(011) planes of the body-centred lattice. A more detailed account of the process is given by *Fujita* [52], but to understand the principle of the process, these fine details are irrelevant.

The transformation begins at numerous sites within the crystal and proceeds almost with the sonic speed. The movement of the atoms is coordinated, with atoms adjacent to already transformed regions shifting to the new structure. The orientation of the new lattice is not random, but depends on that of the γ lattice. The α phase usually grows mainly in two dimensions, forming lens-shaped structures. If a metallographic image of such a structure is taken, the so-called martensite needles can be seen (figure 6.53). Because several transformations occur simultaneously within each grain, the resulting microstructure is very fine. The fine microstructure produced by the martensitic transformation strengthens the material, for the large number of interfaces causes strengthening similar to grain boundaries (see section 6.4.2). Furthermore, the dislocation density increases markedly due to the mechanical distortion.

Technically, it is almost impossible to harden pure iron in this way because the extreme cooling rates needed are very difficult to achieve. However, if carbon is added, it significantly reduces the required cooling rates, making the production of a martensitic microstructure feasible.

In addition, the carbon further strengthens the alloy considerably. The solubility of carbon is much smaller in the body-centred than in the face-centred crystal structure because the interstitial spaces are smaller. If γ iron with a sufficient amount of dissolved carbon ($> 0.008\,\text{wt-}\%$) is quenched, the carbon remains dissolved in the body-centred lattice. The carbon atoms strongly distort the body-centred cubic cell to a tetragonal one (figure 6.54). This distorted lattice structure has an extreme strength when highly oversaturated because the stress field cannot be passed by dislocations. As a rule of thumb,

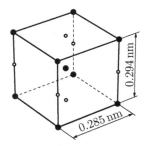

Fig. 6.54. Distorted martensitic unit cell. The dissolved carbon can reside at the marked positions

it can be stated that at least 0.2 % carbon are needed to achieve significant strengthening by martensite transformation. The martensitic structure, however, is then extremely brittle.

To reduce the brittleness of the martensite, a subsequent heat treatment, called *tempering,* can be used. Tempering reduces part of the lattice distortion by precipitating finely distributed carbide particles (Fe_3C), similar to precipitation hardening. As a result, the strength of the steel decreases, but its ductility increases substantially. The heat treatment cycle consisting of quenching and subsequent tempering is often also referred to as 'quench and draw', 'draw the temper', or 'harden and temper'.

This possibility of creating an extremely oversaturated solid solution by martensitic transformation is the reason for the immense technical importance of carbon steels. In addition, the austenitic phase can be stabilised at room temperature by the addition of other alloying elements, allowing the design of a vast number of different alloys with varying properties (see, for example, the book by *Honeycombe* [68]).

Martensitic transformations can also occur in other alloys. Of special importance are *shape memory alloys.* The most commonly used are based on nickel and titanium. In these alloys, a reversible martensitic phase transformation can occur that will be briefly described here.

A typical shape memory alloy starts in its austenitic phase. When the material is strained, it transforms to the martensitic phase, oriented in a way to produce an elongation in the direction of the load. When the load is removed, the austenitic phase forms again because it the thermodynamically stable one. Thus it is possible to produce reversible deformations with *pseudo-elastic* strains of several percent, much more than usually possible for elastic strains in metals. For this reason, the phenomenon is also called *superelasticity.* One advantage of pseudo-elasticity is that the stress in the pseudo-elastic regime is almost independent of the strain.

One application are tooth braces, containing a wire-shaped spring made of a shape memory alloy that exerts a force on the teeth to be repositioned. Due to the strain-independence of the stress, the wire does not need to be readjusted as soon as a tooth has moved. Tubes to stabilise blood vessels, so-called *stents* can also be constructed in this way. Here, a superelastic tube is compressed and placed inside a catheter. In the compressed form, the tube can easily be moved through the vessels to the designated position. The catheter is then removed and the stent unfolds, preventing collapse of the vessel.

Because the stability of the phases also depends on temperature, a shape memory alloy can also deform when the temperature is changed. To achieve this, the alloy is at first heated to a temperature well within the austenitic regime (usually several hundred degree centigrade). At this temperature, it is formed to the desired shape and then quenched to room temperature where martensite is stable. If the material is deformed now, deformation in the martensitic phase occurs not by dislocation movement, but by reorientation of lattice planes, a process called twinning that is described in the next section. After re-heating to the austenitic phase, the original crystal lattice forms again, and the material takes its old shape it has 'remembered', thus explaining the name of these alloys. This shape memory effect can also be used in medicine, for instance to precisely move catheters or endoscopes within the human body. Stents can also be constructed in this way by compressing them plastically at low temperatures. The body heat lets them take the desired shape after they have been positioned [22].

* 6.5 Mechanical twinning

Besides dislocation movement, there are other mechanisms of plastic deformation. These are the martensitic transformation we already discussed, diffusion creep at high temperatures (to be covered in chapter 11), and finally the so-called *twinning*. *Mechanical twinning* usually contributes only slightly to plastic deformation and is in general more difficult to activate than dislocation movement. Therefore, it will be discussed only briefly.

Mechanical twinning mainly occurs at low temperatures and in metals with a small number of slip systems i. e., when slip of dislocation is difficult. The hexagonal metals show a greater tendency to form twins. Mechanical twinning is a shear deformation of the lattice in which atoms are shifted parallel to the so-called *twinning plane,* as shown in figure 6.55. The orientation of the twinning plane is determined by the crystal lattice. In contrast to dislocation movement, single atoms can travel over large distances. The further away an atom is from the twinning plane, the further it moves. In a polycrystalline metal, large displacements of single atoms are not possible so that usually two adjacent twinning planes form, producing a small twin band. The atoms usually shear by a small angle only, whereas the orientation of the unit cell can change by much larger values. For instance, in hexagonal crystals the orientation can change by about 87° at a shear of about 7°. The shear angle

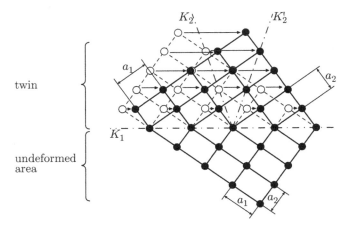

(a) Formation of a twin. The atoms above the twin plane K_1 shear by the angle $\angle(K_2, K_2') = 37°$ when using the given ratio $a_1 : a_2$. Thereby, the lattice rotates by a noticeable larger value of $\alpha = 71°$. Shear angles in true crystal structures normally are significantly smaller

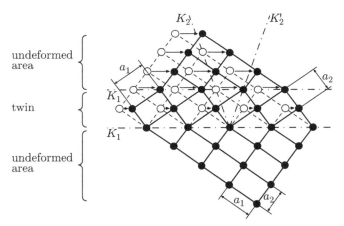

(b) Formation of a twin band. Two parallel, opposite twin planes form simultaneously, resulting in the upper part of the crystal moving along the twin plane in parallel without rotation. The distance is proportional to the width of the twin band

Fig. 6.55. Illustration of mechanical twinning

Fig. 6.56. Deformation twins in ferritic steel. The twinning planes can be observed as lines within the grains, some marked by arrows

and the rotation of the unit cell are also determined by the crystal lattice. Figure 6.5 shows mechanical twins in a ferritic steel (S 235 JR).

In hexagonal metals, which glide preferentially in the basal plane (cf. section 6.2.4), the rotation of the lattice by twinning can lead to a more favourable orientation of the crystal and thus increase the deformability by subsequent dislocation movement.

7

Mechanical behaviour of ceramics

Ceramics feature a high elastic stiffness, high strength – especially under compressive loads –, good resistance against many chemicals, and a high temperature stability. This latter point is only valid for crystalline ceramics, whose high-temperature behaviour will be investigated in chapter 11. Amorphous ceramics (glasses) do not have a melting point, but soften when the temperature is increased and behave then like a viscous fluid, with decreasing viscosity at increasing temperature. The softening temperature is considerably below the melting temperature of typical crystalline ceramics. Window glass, for example, can be deformed at temperatures of several hundred degree centigrade. Because this behaviour of ceramics is similar to that of amorphous thermoplastics, subject of chapter 8, it will not be covered further in this chapter.

Besides the positive properties enumerated above, ceramics also have a distinct disadvantage: They are brittle, which poses problems not only during service, but also in manufacturing ceramic components. This brittle behaviour of ceramics is – as already explained in section 1.3 – due to the nature of their chemical bonds. Ceramics usually fail by brittle fracture, so their strength is determined by initial cracks already present in the material. As these are usually produced during manufacturing, the chapter starts with a brief survey of these processes. Next, we will discuss the mechanisms that influence crack propagation in ceramics and thus determine their strength. Because the size of the initial cracks is stochastically distributed, statistic methods are required to analyse the strength of ceramics. They are the topic of section 7.3. To safely design components, the size of the largest crack can be restricted by applying a load once that is larger than the expected service loads. This so-called proof test will be covered in section 7.4. Finally, we will discuss methods for strengthening ceramic materials.

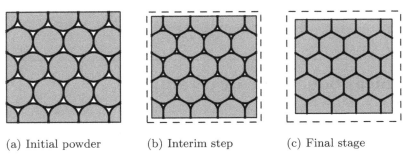

(a) Initial powder   (b) Interim step   (c) Final stage

Fig. 7.1. Sintering (after [9])

7.1 Manufacturing ceramics

Engineering ceramics are almost exclusively made from powders because they cannot be cast in the liquid state due to their high melting temperature.[1] The powder is formed to a near-net shape *green body,* for instance by *cold compaction* in the dry state. Alternatively, ceramics can be shaped by mixing them with a plasticiser or a liquid solvent. One example for this type of process is *slip casting* where ceramic powder is mixed with a fluid (usually water or alcohol) and poured into a porous ceramic mould. The capillarity of the mould dehumidifies the ceramic powder. After drying, the ceramic is sufficiently rigid to be taken out of the mould. In *injection moulding,* the ceramic powder is mixed with a polymer binder and processed similarly to an injection moulded polymer. However, the larger abrasion of the ceramic powder has to be taken into account.

Next, the green body is compacted at high temperature, whereupon the ceramic powder particles, previously only loosely bound mechanically or by binders, bind chemically. Standard processes for this are *sintering, hot pressing,* and *hot isostatic pressing* (HIP) [126]. While sintering is done without external load, the other processes superimpose uniaxial or hydrostatic stresses.

Figure 7.1 sketches a sintering process. During compaction, material diffuses to the contact area between the particles, driven by the energy gained in reducing the surface. The contact areas round off and become larger. One consequence of this process is that the component shrinks by as much as 30% to 40%.[2]

Because ceramics sinter only slowly due to their high melting temperature, a high sintering temperature alone is frequently not sufficient to produce dense

---

[1] Ceramic glasses are an exception because they can be processed in the viscous state when softened at elevated temperature.

[2] This means a reduction of the length of the component in each direction of about 10% to 15%.

ceramics without significant porosity between the former powder particles. Sintering aids, like magnesium oxide (MgO) for silicon nitride (Si_3N_4), can be added that produce a liquid phase at the sintering temperature and thus facilitate the sintering process. One disadvantage is that this phase, the so-called *glassy phase,* is amorphous and thus reduces the strength at elevated service temperatures (creep strength, see chapter 11).

7.2 Mechanisms of crack propagation

Because dislocations are completely immobile in ceramics at room temperature due to the directed atomic bonds and the complex crystal structures (see section 1.3), ceramics can in general not deform plastically. Failure can occur only by cleavage fracture, usually with initial cracks growing and propagating. The pores remaining after compaction are defects acting as initial cracks and thus cause failure by crack propagation. As there is no plastic deformability, it cannot unload these initial cracks by evening out stress concentrations or dissipate energy during crack propagation. Therefore, the fracture toughness of ceramics is comparably small. This is also reflected in the toughness-strength diagram 5.11 on page 146. Because of the crack sensitivity of ceramics, even small defects can determine the strength – the pre-cracks formed during manufacturing are thus crucial for the mechanical behaviour. The theoretical strength of a perfect ceramic without any defects is technically irrelevant.

Usually, ceramics always contain cracks of different sizes with different orientations. The strength of the ceramic is determined by the cracks with the lowest failure strength. Under tensile loads, cracks can, depending on their orientation, be loaded in all modes, I, II, or III (cf. section 5.1.1), under compressive loads only in mode II or III, for the stress component perpendicular to the crack surface closes the crack. Because the fracture toughness is much smaller for mode I than for modes II or III, ceramics under tensile loads usually fail in this mode and are thus more sensitive to tensile than to compressive stresses. The compressive strength of ceramics is usually 10 to 15 times larger than its tensile strength.[3]

The fracture toughness is primarily determined by the strength of the chemical bonds within the ceramic because this determines the energy needed to create fresh surface. Beyond that, other effects within ceramics can occur that impede crack propagation because they require additional energy and thus increase the fracture toughness. The basic mechanisms are discussed in this section; in section 7.5, we will see how they can be utilised to strengthen ceramics.

---

[3] This property of ceramics is exploited in the design of ferroconcrete (steel-reinforced concrete), for example, see section 9.1.1.

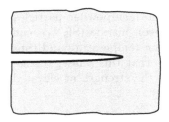   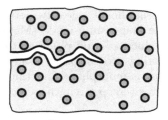

(a) Without particles   (b) With particles

Fig. 7.2. Deflection of a crack by appropriate particles in a ceramic

7.2.1 Crack deflection

When the crack can be deflected from its straight path, the surface of the crack per advanced distance becomes larger, thus requiring additional energy for crack propagation and increasing the fracture toughness. This can be achieved in several ways, often by adding particles.

One possible mechanism is the *modulus interaction,* already discussed in another context in the chapter on metals (section 6.4.3). If the particles have a larger Young's modulus than the matrix, the matrix is partly unloaded in the vicinity of the particles, and the stress available to propagate the crack is reduced. The crack is deflected away from the particle (see figure 7.2). If Young's modulus of the particles is smaller than that of the matrix, the stress is raised in the vicinity of the particles, and the crack is attracted by the particle. If the crack cannot penetrate the particle, the crack must proceed along its boundary. In all these cases, the crack path becomes longer.

Another way to deflect cracks are *residual stresses* caused by the particles. Compressive stresses reduce the force opening the crack tip and thus repel cracks. Such residual stresses can stem from differences in the coefficient of thermal expansion of the particles or from phase transformations on cooling from the sintering temperature.

7.2.2 Crack bridging

When the two opposed crack surfaces interact during crack propagation, the energy dissipation during crack propagation can be increased or the crack tip can be partially relieved. This kind of interaction can occur in coarse-grained microstructures with intercrystalline crack propagation. In this case, the crack surfaces can contact and rub against each other when the crack is opened (see figure 7.3) or can even be geometrically clamped, so that the crack cannot open at all. Fibres or particles are another crack bridging mechanism. Fibres will be discussed in chapter 9.

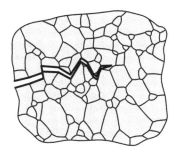

Fig. 7.3. Coarse-grained microstructure in which the crack surfaces are in contact and dissipate energy by sliding on each other

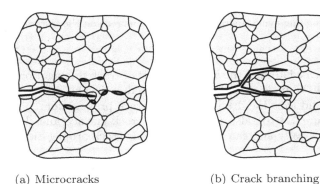

(a) Microcracks   (b) Crack branching

Fig. 7.4. Examples of microcracks and crack branching

7.2.3 Microcrack formation and crack branching

The stress concentration near the crack tip can create microcracks at weak points in the ceramic. Examples are unfavourably oriented grain boundaries (perpendicular to the largest principal stress as drawn in figure 7.4(a)), grains with a cleavage plane perpendicular to the largest principal stress, or regions containing residual stresses.

Microcrack formation raises the fracture toughness because it increases the energy dissipation. This can be understood by looking at the stress-strain diagram of a volume element that is passed by the crack tip (see figure 7.5): When the volume element approaches the crack tip, its load increases, and microcracks form. These reduce the stiffness of the volume element, causing a reduction in the slope of the stress-strain curve. On unloading (when the volume element moves away from the crack tip), the unloading curve is not the same as the loading curve. The shaded area in the diagram is the energy dissipated in this process; this additional energy has to be provided during crack propagation.

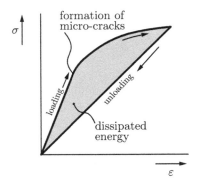

Fig. 7.5. Stress-strain diagram of a volume element during microcrack formation. During the loading and unloading cycle, energy is dissipated, increasing the fracture toughness

If microcracks have formed around the crack tip, further crack propagation is hampered also because Young's modulus is locally reduced. This reduces the stress in this region and thus the driving force for crack propagation.[4]

Crack branching (figure 7.4(b)) also increases the crack surface and decreases Young's modulus locally and can thus also impede crack propagation.

7.2.4 Stress-induced phase transformations

So-called *stress-induced phase transformations* can produce additional compressive residual stresses during crack propagation and thus increase the crack-growth resistance K_{IR}. This is caused by particles in the matrix that can increase their volume by a phase transformation. Initially, the particles have to be in a metastable state which is thermodynamically unfavourable, but cannot transform to the thermodynamically stable phase because a nucleation barrier has to be overcome for this, similar to the process in precipitation hardening (see section 6.4.4).

If a sufficiently large tensile stress is applied, for instance, at the crack tip, it may need less energy to transform the particles to the phase with the greater volume than to deform them elastically (figure 7.6). This case can also be understood by considering the stress-strain diagram of a volume element passed by the crack tip (figure 7.7). The phase transformation starts when the elastic energy is sufficiently large. Because the particle was metastable prior to the transformation, the transformation proceeds even when the stress becomes smaller due to the volume increase. During the transformation, tensile

---

[4] This argument is only valid when the microcracks are restricted to the region around the crack tip. If the whole material is cracked, the global stiffness is reduced and fracture toughness decreases, see also section 7.5.3.

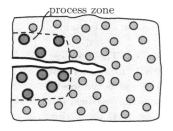 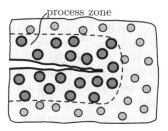

(a) Before transformation   (b) After transformation

Fig. 7.6. Unloading of a crack by a phase transformation of particles. In the process zone, residual compressive stresses are superimposed to the external tensile stress field

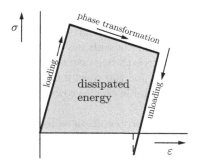

Fig. 7.7. Stress-strain diagram of a volume element, showing stress-induced phase transformations (after [46]). Similar to the formation of microcracks (figure 7.5), energy is dissipated during the loading and unloading cycle and thus increases the fracture toughness

stresses in circumferential and compressive stresses in radial direction around the particles are superimposed to the external load. After unloading, part of the volume increase remains and compressive residual stresses are generated that reduce the stress on the crack and thus may partly or totally close it.

Because of the tensile stresses in circumferential direction around the particles, microcracks can form and – as described in the previous section – cause further dissipation of energy (see also section 7.5.3).

> To be more precise, the stress-induced transformation is based on a reduction of the free enthalpy, defined in equation (C.4). The phase with the larger volume is thermodynamically stable, but, as already stated, a nucleation barrier has to be overcome to transform the particle. If hydrostatic tensile stress is added, the free enthalpy of the phase with

the larger volume decreases more strongly, according to equation (C.4), thus increasing the driving force for the transformation that enables the particle to overcome the nucleation barrier.

In some metals, an analogous behaviour is observed: The stainless steel X 5 CrNi 18-10, which is austenitic (face-centred cubic) at room temperature, is only metastable. Thus, the ferritic phase is thermodynamically stable, but the transformation does not occur because the driving force is too small. Under mechanical load, for instance during forming, a martensitic transformation can take place in parts of the component, easily detectable by the component becoming ferromagnetic locally.

7.2.5 Stable crack growth

In the previous sections (7.2.2 to 7.2.4), we encountered several mechanisms (crack bridging, microcracking, stress-induced phase transformations) that may increase the crack-growth resistance of a material. They all have in common that the resistance initially grows during crack growth because a process zone forms near the crack tip and are thus examples for the mechanism discussed in section 5.2.5. If conditions are appropriate, stable crack growth may thus occur in a certain stress range.

The crack-growth resistance increases during crack propagation as long as the process zone grows. For example, if friction of the crack surfaces (figure 7.3) occurs, the crack-growth resistance initially increases because the contact area of the surfaces grows. If the crack propagates further, a stationary state is reached because parts of the surface far away from the crack tip will not touch anymore when the crack has opened too much. Then, the crack-growth resistance stays constant because for every newly formed region of fracture an equally large region is lost further away from the crack tip. When energy is dissipated within the material, as it happens in microcracking or stress-induced phase transformations, the process zone initially grows, for initially there are no microcracks or transformed particles near the crack tip. Only after a stable equilibrium is reached does the crack-growth resistance remain constant.

* 7.2.6 Subcritical crack growth in ceramics

In section 5.2.6, it was explained that ceramics may, under certain conditions, exhibit subcritical crack propagation which can be quantified using the *crack-growth rate* da/dt. Figure 7.8 shows crack growth curves for a glass in different environments. Frequently, the crack growth curve is a line when a log-log scale is used (region ①). In some cases, a plateau follows (region ②), and, finally, the crack-growth rate rapidly increases shortly before reaching the fracture toughness K_{Ic} (region ③).

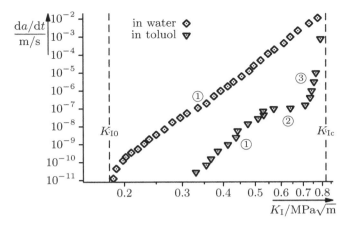

Fig. 7.8. Dependence of the crack-growth rate on the stress intensity factor of soda-lime glass in water and toluene (phenylmethane) at 25°C [104]

Often, a power law is used to describe the crack-growth rate in region ①:

$$\frac{da}{dt} = AK_{\mathrm{I}}^{n} = A^{*}\left(\frac{K_{\mathrm{I}}}{K_{\mathrm{Ic}}}\right)^{n} \tag{7.1}$$

where n, A, and A^{*} are temperature-dependent material parameters [103].[5] The unit of A depends on the exponent n and can thus contain fractional exponents. To avoid this, A^{*} can be used which has always the unit of a velocity because $K_{\mathrm{I}}/K_{\mathrm{Ic}}$ is dimensionless. As already mentioned in section 5.2.6, the dependence of the crack-growth rate on the stress intensity factor is sometimes very strong, resulting in a large value of the exponent n in equation (7.1).

If a time-dependent stress $\sigma(t)$ is applied, the life time of the component can be calculated by integration of equation (7.1). If the load is static, it is

$$t_{\mathrm{f}} = B^{*}\sigma^{-n} = B\sigma_{\mathrm{c}}^{n-2}\sigma^{-n}, \tag{7.2}$$

where B^{*} and B depend on the material and, via the geometry factor Y, also on the geometry. σ_{c} is the so-called *inert strength*, the load required to break the specimen in a chemically inert environment where it would fail not by subcritical crack growth, but by fracture at K_{Ic}. Because the stress exponent n is large, there is a strong dependence of the time to failure on the applied stress. In table 7.1, some examples for the stress exponent n and the prefactor are summarised.

The effect of time-dependent loads on subcritical crack growth will be discussed in section 10.3. In exercise 24, an example for designing ceramic components against subcritical crack growth is given.

---

[5] This shape of the crack growth curve da/dt is very similar to the crack-growth rate da/dN in cyclic loading of metals, to be discussed in section 10.6.1.

Table 7.1. Exemplary parameters of subcritical crack propagation [104]. For the specimen geometry investigated, $Y = 1$ holds true; the parameter B^* thus contains only material parameters

ceramic	medium	$T/°C$	n	$\lg(B^*/(MPa^n h))$
Al_2O_3	water	20	52.2...67.6	121.1...162.7
Al_2O_3	conc. saline solution	70	20	45.5
$Si_3N_4 + 5.5\% Y_2O_3$		1 100	37	106.5
		1 200	30	84.2
$Si_3N_4 + 2.5\% MgO$		1 000	26	69.8
		1 100	22.6	61.8

7.3 Statistical fracture mechanics

Because ceramics cannot compensate for inner defects by plastic deformation, the statistical scatter of defect sizes causes a large scatter in the mechanical properties, different from metals and polymers. Therefore, it is usually not sufficient to simply state a failure load. Because it is not feasible to measure the size and position of every single defect within a component and thus to predict its strength exactly (deterministically), the statistics of the defect distribution is considered, and, using the methods of *statistical fracture mechanics,* a failure or survival probability is calculated.

The objective of this section is to describe the probability of failure of a ceramic component analytically, using statistical fracture mechanics. Simplifyingly, we assume that defects with a certain defect size are distributed homogeneously in the material and that crack propagation at only one of them will cause complete failure. Initially, we will also assume a constant stress σ within the component.

The *probability of failure* $P_f(\sigma)$ states the probability of the component failing when the stress σ is applied. If, for instance, in a batch of (macroscopically) identical specimens, the probability of failure is $P_f(200\,MPa) = 0.3$, 30% of the specimens will fracture when we try to apply a load of $200\,MPa$. The value is not be understood in such a way that 30% of the specimens will fail exactly at this stress value, but at stresses lower or equal to it.

If defects were not statistically distributed, the behaviour of the material would be deterministic: It would fail at a critical stress σ_0 and the probability of failure would discontinuously change from 0 to 1. In reality, there is always a probability that the material will bear larger loads or will fail at smaller ones, and the 'edge' at σ_0 is rounded off.

7.3.1 Weibull statistics

Usually, ceramics fail as soon as a crack starts to propagate. Therefore, their strength is determined by the stress value at which the first, and thus critical, crack starts to grow. The probability of failure is thus given by the probability that the critical crack has a certain failure stress. If loads are tensile, one of the cracks that are at least partially loaded in mode I will govern the strength.

> To describe the probability of failure, a statistical approach is needed that takes into account the statistical distribution of the density and size of the cracks [104].
>
> If we consider a component with homogeneous stress distribution and known number of defects, the size distribution of the defects can be used to determine the failure probability. This is equal to the probability that at least one crack has the critical crack length. As the critical crack determines failure, only the largest defects are relevant. The probability of finding a large defect eventually becomes smaller with increasing defect size; therefore, different defect size distributions will look similar in the relevant region. The details of the defect size distribution are thus not important.
>
> Because the number of defects differs between different components, the defect density distribution must also be taken into account by using it to accumulate the probability of failure for all possible numbers of defects.

From these statistical considerations, it can be shown that the failure stress σ of the critical defect is distributed according to the so-called *Weibull distribution*. From this, the probability of failure can be calculated as[6]

$$P_{\mathrm{f}}(\sigma) = 1 - \exp\left[-\left(\frac{\sigma}{\sigma_0}\right)^m\right] . \tag{7.3}$$

The parameter m, called the *Weibull modulus,* quantifies the scatter of the strength values and is thus a measure of how strongly the edge in a plot of the probability of failure is rounded off as shown in figure 7.9 for some examples. As can be seen, a larger Weibull modulus reduces the scatter of the failure stress. For $m \to \infty$, there is no scatter anymore, and σ_0 is equal to the fracture stress. The Weibull modulus is a material parameter; the reference stress σ_0 depends on the material and the specimen volume. Equation (7.3) is only valid when a constant, homogeneously loaded specimen volume is considered. The influence of the volume will be discussed from the following page onwards. Table 7.2 gives a synopsis of some values for the Weibull modulus in different materials.

Frequently, a linearised representation of the probability of failure is used. For this purpose, equation (7.3) is re-written as follows:

$$1 - P_{\mathrm{f}} = \exp\left[-\left(\frac{\sigma}{\sigma_0}\right)^m\right] ,$$

---

[6] To correctly describe the probability of failure, the volume of the component must also be taken into account, see equation (7.6) below.

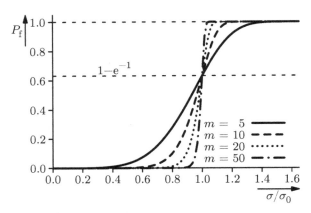

Fig. 7.9. Dependence of the failure probability on the stress for several Weibull moduli

Table 7.2. Weibull modulus m and reference stress σ_0 of some ceramics (after [58, 64, 104]). V_0 is the reference volume in equation (7.6). For comparison, the Weibull modulus of cast iron and steel is specified

material	m	σ_0/MPa	V_0/mm^3
SiC	$8\ldots 27$	$250\ldots 600$	1
Al_2O_3	$8\ldots 20$	$100\ldots 600$	1
Si_3N_4	$8\ldots 9$	$750\ldots 1\,350$	1
ZrO_2	$10\ldots 15$	$200\ldots 500$	1
cast iron	≈ 40		
steel	≈ 100		

$$\frac{1}{1 - P_\mathrm{f}} = \exp\left[\left(\frac{\sigma}{\sigma_0}\right)^m\right],$$

$$\ln\frac{1}{1 - P_\mathrm{f}} = \left(\frac{\sigma}{\sigma_0}\right)^m,$$

$$\ln\left(\ln\frac{1}{1 - P_\mathrm{f}}\right) = m\ln\frac{\sigma}{\sigma_0}. \tag{7.4}$$

If we now plot $\ln\big(\ln[1/(1 - P_\mathrm{f})]\big)$ as a measure of the probability of failure versus $\ln(\sigma/\sigma_0)$ as a measure for the applied stress, a linear equation with slope m through the origin results as sketched in figure 7.10.

The probability of failure in equation (7.3) depends on the material volume. This is plausible if we assume that a single defect of critical size will cause the component to fail, for, if the volume and thus the number of defects is increased, the probability of a critical defect being present increases as well. This will now be shown using an example. It is useful to consider the *probability of survival* P_s instead of the probability of failure, defined as

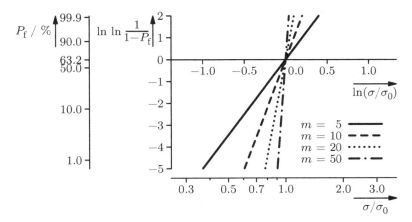

Fig. 7.10. Plot of the Weibull equation in linearised representation

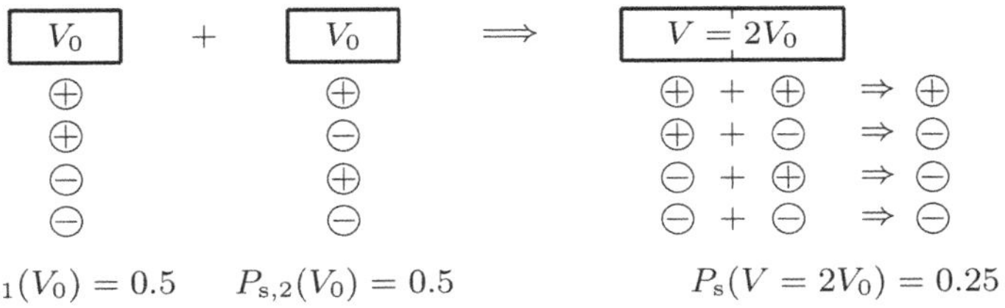

$P_{s,1}(V_0) = 0.5 \qquad P_{s,2}(V_0) = 0.5 \qquad\qquad P_s(V = 2V_0) = 0.25$

Fig. 7.11. Demonstration of the volume dependence of the survival probability. The circled symbols denote the chances for survival and failure of the component, respectively ($\oplus$: survival, $\ominus$: failure)

$$P_s = 1 - P_f \,.$$

Consider two specimens with the probabilities of survival $P_{s,1}(V_0) = P_{s,2}(V_0) = 0.5$ and volumes V_0. If we join them to a single specimen of volume $V = 2V_0$ and enumerate all possibilities of survival or failure (see figure 7.11), the probability of survival is $P_s(V = 2V_0) = 0.25$.

To analyse the volume effect in general, we consider a specimen of volume V, loaded homogeneously with a load σ. If we imagine it to be composed of n partial volumes V_0 with a probability of survival $P_{s,i}(V_0)$, the specimen survives if all partial volumes do. According to the laws of probability, the probability of survival of the total volume is

$$P_s(V) = \prod_{i=1}^{n} P_{s,i}(V_0) = \left[P_{s,i}(V_0)\right]^n \,.$$

Using $n = V/V_0$ yields

$$P_s(V) = \left[P_{s,i}(V_0)\right]^{V/V_0} \,. \tag{7.5}$$

For the example from figure 7.11, we get the same result from this equation as calculated above: $P_{\mathrm{s}}(V = 2V_0) = 0.5^2 = 0.25$.
With

$$P_{\mathrm{s},i}(V_0) = \exp\left[-\left(\frac{\sigma}{\sigma_0}\right)^m\right], \quad i = 1\ldots n,$$

and $(\mathrm{e}^x)^y = \mathrm{e}^{xy}$, we get from equation (7.5)

$$P_{\mathrm{s}}(V) = \exp\left[-\frac{V}{V_0}\left(\frac{\sigma}{\sigma_0}\right)^m\right],$$

$$P_{\mathrm{f}}(V) = 1 - \exp\left[-\frac{V}{V_0}\left(\frac{\sigma}{\sigma_0}\right)^m\right]. \tag{7.6}$$

V_0 is called the *reference volume*. This general equation must replace equation (7.3) when the volume dependence is to be taken into account correctly. It is called the *Weibull equation*.

So far, we always assumed that the whole component was homogeneously loaded with a constant stress σ. In practice, this is seldom the case, for instance in bending problems or at stress concentrations in notches. If the component is loaded with different stresses σ_i in its partial volumes V_i, the probability of survival is

$$P_{\mathrm{s}}(V) = \prod_{i=1}^n P_{\mathrm{s},i}(V_i, \sigma_i) = \prod_{i=1}^n \exp\left[-\frac{V_i}{V_0}\left(\frac{\sigma_i}{\sigma_0}\right)^m\right]$$

$$= \exp\left[-\sum_{i=1}^n \frac{V_i}{V_0}\left(\frac{\sigma_i}{\sigma_0}\right)^m\right].$$

When taking the limit of infinitely small volumes, we have to replace the sum by an integral

$$P_{\mathrm{s}}(V) = \exp\left[-\frac{1}{V_0}\int\left(\frac{\sigma(x)}{\sigma_0}\right)^m \mathrm{d}V\right]. \tag{7.7}$$

This is the general form of the Weibull equation for arbitrarily loaded components.

When failure of a ceramic is not caused by volume defects, but by surface defects, the probability of failure depends on the surface. Instead of summing or integrating over partial volumes, partial surface elements have to be used in this case [103].

Sometimes, another material parameter σ_1 is introduced in equation (7.6)

$$P_{\mathrm{f}}(\sigma, V) = \begin{cases} 1 - \exp\left[-\dfrac{V}{V_0}\left(\dfrac{\sigma - \sigma_1}{\sigma_0}\right)^m\right] & \text{for } \sigma \geq \sigma_1, \\ 0 & \text{for } \sigma < \sigma_1. \end{cases} \tag{7.8}$$

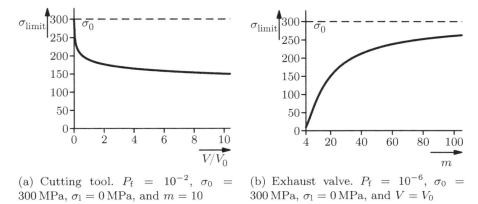

(a) Cutting tool. $P_{\mathrm{f}} = 10^{-2}$, $\sigma_0 = 300\,\mathrm{MPa}$, $\sigma_1 = 0\,\mathrm{MPa}$, and $m = 10$ (b) Exhaust valve. $P_{\mathrm{f}} = 10^{-6}$, $\sigma_0 = 300\,\mathrm{MPa}$, $\sigma_1 = 0\,\mathrm{MPa}$, and $V = V_0$

Fig. 7.12. Permissible stress for the examples

This parameter allows for a lower limit stress σ_1 below which the probability of failure is zero. The underlying distribution is called *three-parameter Weibull distribution*. In the following, we will usually assume $\sigma_1 = 0$, however.

If the Weibull modulus m and the reference stress σ_0 are known, components can be designed using the Weibull equation (7.6) or (7.8). The engineer chooses a probability of survival (or failure) to be met by the component (e. g., $P_{\mathrm{f}} \le 10^{-5}$: one in $100\,000$ components may fail). Putting this into equation (7.8) and solving for σ yields the maximum allowed stress

$$\sigma_{\mathrm{limit}} = \sigma_1 + \sigma_0 \sqrt[m]{-\frac{V_0}{V}\ln(1 - P_{\mathrm{f}})}\,. \tag{7.9}$$

Examples

A ceramic cutting tool is to be manufactured from a material with parameters $\sigma_0 = 300\,\mathrm{MPa}$, $\sigma_1 = 0\,\mathrm{MPa}$ and $m = 10$. The probability of failure was chosen to be $P_{\mathrm{f}} = 10^{-2}$. To calculate the maximum allowed stress for different tool volumes, we can use equation (7.9):

$$\sigma_{\mathrm{limit}} = 300\,\mathrm{MPa} \times \sqrt[10]{-\frac{V_0}{V} \times \ln 0.99} = 189.4\,\mathrm{MPa} \times \sqrt[10]{\frac{V_0}{V}}\,.$$

Figure 7.12(a) plots the dependence of the allowed stress on the tool volume.

As a second example, the dependence of the allowed stress on the Weibull modulus is calculated for a ceramic exhaust valve. The design parameters are $P_{\mathrm{f}} = 10^{-6}$, $\sigma_0 = 300\,\mathrm{MPa}$, $\sigma_1 = 0\,\mathrm{MPa}$, and $V = V_0$. From equation (7.9), we find

$$\sigma_{\mathrm{limit}} = 300\,\mathrm{MPa} \times \sqrt[m]{-\ln(1 - 10^{-6})} = 300\,\mathrm{MPa} \times \sqrt[m]{10^{-6}}\,.$$

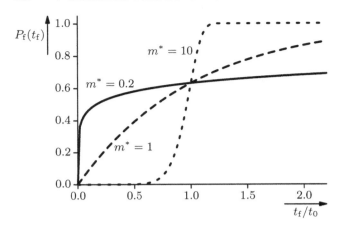

Fig. 7.13. Failure probability depending on the loading duration. Depending on the Weibull modulus m^*, characteristic curves arise

Figure 7.12(b) shows the dependence of the stress on the Weibull modulus. As the figure illustrates, the allowed maximum stress σ_{limit} is much smaller than σ_0 due to the high component reliability required[7]. Decreasing the scatter in the failure stress (increasing m) significantly increases the allowed stress.

* 7.3.2 Weibull statistics for subcritical crack growth

If a ceramic can fail by subcritical crack growth, its life time when loaded with a certain stress can be calculated from equation (7.2). In section 7.2.6, we used a deterministic approach to do so, but by now we have learned that the failure stress values scatter and the probability of failure follows a Weibull distribution $P_f(\sigma)$. The time to failure, depending on the failure stress, must therefore also be distributed stochastically.

The failure probability after a certain time t_f can be calculated when equation (7.2) is solved for σ, putting the result into equation (7.6):

$$P_f(t_f) = 1 - \exp\left[-\frac{V}{V_0}\left(\frac{t_f}{t_0(\sigma)}\right)^{m^*}\right]. \tag{7.10}$$

Here, $m^* = m/(n-2)$ is the Weibull modulus for the life time and $t_0(\sigma) = B^* \sigma^{-n}$ the reference time. This equation is completely analogous to equation (7.6), but another Weibull modulus and a reference time instead of a reference stress have to be used.

The Weibull modulus m^* characterises the failure type of the ceramic [2]. If $m^* < 1$, failure usually occurs shortly after applying the load (so-called *infant failure*, see figure 7.13). If $m^* = 1$, the probability of failure is the

---

[7] For example, if $m = 10$, the allowed stress is only 75 MPa.

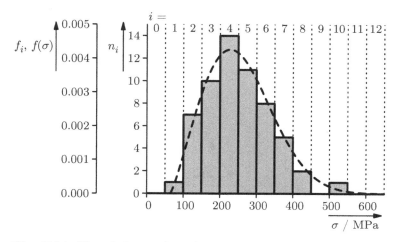

Fig. 7.14. Plot of the number of specimens n_i that have failed within the stress interval i of width $\Delta\sigma$. The total number of specimens is $N = 60$. The 'discrete probability density' f_i results from equation (7.11). As approximation, a Weibull distribution according to equation (7.13) with the parameters $V/V_0 = 1$, $m = 2.2$, $\sigma_1 = 65\,\text{MPa}$, and $\sigma_0 = 215\,\text{MPa}$ has been used

same at each time interval in each specimen. If $m > 1$, there is a certain most probable time of failure t_0.

$*$ 7.3.3 Measuring the parameters σ_0 and m

To measure the failure probability, a large number of experiments are performed on identical specimens, measuring the failure stress σ_i or failure time t_i of each. In both cases, the determination of the parameters (σ_0 and m or t_0 and m^*, respectively) is done in the same way. In the following, we use the example of the failure stress.

One method to determine the distribution of failure stresses is to divide the stress region containing the failure stress values into intervals of width $\Delta\sigma$ as shown in figure 7.14. For each stress interval i, we count the number of specimens that failed at stress values within it. The probability that another specimen will also fail in this stress interval is given by n_i/N. Normalising this by the interval width $\Delta\sigma$ yields the 'discrete probability density' f_i:

$$f_i = \frac{n_i}{N\Delta\sigma} \, . \tag{7.11}$$

The values of f_i are shown as columns in figure 7.14.

The 'discrete probability density' f_i can be approximated by a function $f(\sigma)$ which can be determined from the probability of failure P_f. If we increase the stress σ by $\Delta\sigma$, a fraction of $P_f(\sigma + \Delta\sigma) - P_f(\sigma)$ of all specimens

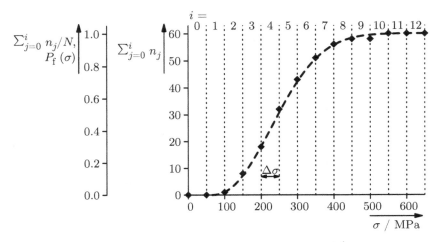

Fig. 7.15. Plot of the accumulated number of specimens $\sum_{j=0}^{i} n_j$ failed until the end of the current stress interval. The normalised number results from dividing by the total specimen number N; the normalised number approaches one for a large stress. The resulting failure probability P_f is plotted for the parameters $V/V_0 = 1$, $m = 2.2$, $\sigma_1 = 65\,\text{MPa}$, and $\sigma_0 = 215\,\text{MPa}$, according to figure 7.14

will fail in this stress interval. Normalising this number by the stress interval $\Delta\sigma$ as we did in equation (7.11) and taking the limit $\Delta\sigma \to 0$ yields

$$f(\sigma) = \lim_{\Delta\sigma \to 0} \frac{P_f(\sigma + \Delta\sigma) - P_f(\sigma)}{\Delta\sigma} = \frac{dP_f}{d\sigma}. \tag{7.12}$$

Thus, the probability density $f(\sigma)$ is the derivative of the probability of failure with respect to the stress. Using equation (7.8) for the probability of failure results in the *Weibull distribution* [58, 146]

$$f(\sigma) = \begin{cases} \dfrac{V}{V_0} \dfrac{m}{\sigma_0} \left(\dfrac{\sigma - \sigma_1}{\sigma_0}\right)^{m-1} \exp\left[-\dfrac{V}{V_0}\left(\dfrac{\sigma - \sigma_1}{\sigma_0}\right)^{m}\right] & \text{for } \sigma \geq \sigma_1, \\ 0 & \text{for } \sigma < \sigma_1, \end{cases} \tag{7.13}$$

shown as dashed line in figure 7.14. As before, V is the volume of the components for which the probability of failure is to be determined, and V_0 is the volume of the test specimen used to determine the parameters σ_0, σ_1, and m.

The measured values from figure 7.14 can also be used to determine the probability of failure when all specimens are counted that have fractured at stresses below those at the end of each interval. This is shown in figure 7.15.

Another way to determine the Weibull modulus m and the reference stress σ_0 considers each measured value directly, without grouping it in intervals in a histogram. We assign an estimate of the failure probability to each measured value of the fracture stress, starting with a very small value for the smallest

Table 7.3. Estimated failure probabilities $\tilde{P}_{f,i}$ for 12 measurements on Al_2O_3

i	σ_i/MPa	$\tilde{P}_{f,i}$
1	234.0	0.042
2	257.4	0.125
3	273.0	0.208
4	273.8	0.292
5	275.3	0.375
6	276.9	0.458
7	280.8	0.542
8	288.6	0.625
9	290.2	0.708
10	296.4	0.792
11	312.0	0.875
12	335.4	0.958

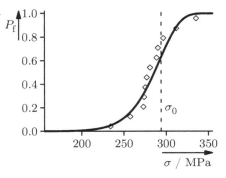

Fig. 7.16. Plot of the failure probabilities for the experimental data in table 7.3. Additionally, the curve calibrated to the experimental data using equation (7.3) is included

measured fracture stress σ_1 because only a small fraction of all specimens had failed at this stress level. Similarly, the largest stress σ_N is assigned a value close to one, for no specimen remains intact. To be more precise, the following estimate is used at each fracture stress value:

$$\tilde{P}_{f,i} = \frac{i - 0.5}{N} \, . \tag{7.14}$$

Here, N is the number of specimens and i its index, sorted from the smallest to the largest value of the fracture stress. Table 7.3 shows an example with values for Al_2O_3. This corresponds to plotting the normalised number of specimens broken until the current stress value has been reached, but corrected in such a way that the value for the first specimen is as far from zero as that of the final specimen is from one. Figure 7.16 shows a plot of this for the values from table 7.3.

The objective is now to use equation (7.3) to approximate the measured values by adapting the parameters m and σ_0.[8] Using the linearised form (7.4), approximations for m and σ_0 can be calculated analytically. For this, we rewrite the equation as follows:

$$\ln\left(\ln\frac{1}{1 - P_f}\right) = m \ln\frac{\sigma}{MPa} - m \ln\frac{\sigma_0}{MPa} \, . \tag{7.15}$$

If we now plot $\ln\big(\ln[1/(1 - \tilde{P}_{f,i})]\big)$ versus $\ln(\sigma_i/MPa)$ for all measured values, a simple linear regression or a graphical method can be used to fit equation (7.15) to the values. This is shown in figure 7.17. In figure 7.16, the transformation back to the σ-P_f coordinate system has been performed. The method is also used in exercise 22.

---

[8] It is not necessary to use the volume-dependent Weibull equation (7.6) because we determine the parameters for the specimen volume V_0.

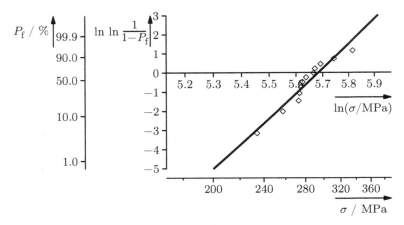

Fig. 7.17. Determination of the parameters $m = 13.1$ and $\sigma_0 = 294\,\text{MPa}$ using experimental data

∗ 7.4 Proof test

The large scatter of the mechanical properties of ceramics is problematic for the engineer because failure may occur even at small loads. If such a failure cannot be avoided, the design has to be extremely conservative and it is thus almost impossible to exploit the advantages of ceramics.

One way to overcome this problem is the *proof test*. Here, the component is momentarily loaded with a proof stress σ_p that is larger than the largest stress expected in service. All components with a strength σ smaller than σ_p will fail the test.

As long as no subcritical crack propagation (section 5.2.6) or fatigue (section 10.3) occurs, the tested component is now fail-safe in service if σ_p is never exceeded. But even if it is, the proof test reduces the probability of failure at a certain stress level. This change in the probability of failure is now calculated.

Before the proof test, the probability of failure P_f is given by equation (7.3).[9] After the proof test, only components with a failure stress $\sigma > \sigma_\text{p}$ remain. For them, the probability density $g(\sigma)$ is calculated by cutting off the original probability density $f(\sigma)$ at σ_p (see figure 7.18). However, $g(\sigma)$ has to be normalised again to ensure that the failure probability

$$G_\text{f}(\sigma) = \int_{\sigma_\text{p}}^{\sigma} g(\sigma)\mathrm{d}\sigma$$

becomes 1 for $\sigma \to \infty$. To do so, G_f is calculated from the quotient of the two areas shown in figure 7.18. A_1 is the area below $f(\sigma)$ between the proof

---

[9] It is sufficient to use the equation without volume dependence because the tested component is identical to the one in service.

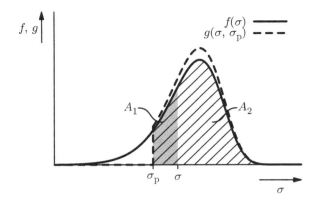

Fig. 7.18. Probability density before (f) and after (g) a proof test at σ_p [104]. The marked areas A_1 and A_2 are used to calculate the distribution function G_f after the proof test

stress σ_p and the current stress σ, whereas A_2 is the total area below the curve $f(\sigma)$ for values larger than σ_p. Thus, G_f is

$$G_\mathrm{f}(\sigma) = \frac{A_1(\sigma)}{A_2} = \frac{\int_{\sigma_\mathrm{p}}^{\sigma} f(\sigma)\mathrm{d}\sigma}{\int_{\sigma_\mathrm{p}}^{\infty} f(\sigma)\mathrm{d}\sigma}\,.$$

Because $f(\sigma)$ is the derivative of $P_\mathrm{f}(\sigma)$, the integrals can be calculated as follows:

$$G_\mathrm{f}(\sigma) = \frac{\left\{1 - \exp\left[-\left(\frac{\sigma}{\sigma_0}\right)^m\right]\right\} - \left\{1 - \exp\left[-\left(\frac{\sigma_\mathrm{p}}{\sigma_0}\right)^m\right]\right\}}{1 - \left\{1 - \exp\left[-\left(\frac{\sigma_\mathrm{p}}{\sigma_0}\right)^m\right]\right\}}$$

$$= 1 - \exp\left[-\left(\frac{\sigma}{\sigma_0}\right)^m + \left(\frac{\sigma_\mathrm{p}}{\sigma_0}\right)^m\right]\,. \tag{7.16}$$

This equation states the probability of failure for components tested with a proof stress σ_p. It is only valid for $\sigma > \sigma_\mathrm{p}$; otherwise, the probability of failure is zero. Figure 7.19 shows the probability of failure before ($P_\mathrm{f}(\sigma)$) and after ($G_\mathrm{f}(\sigma)$) the proof test with a linear scale and in the linearised form. Below σ_p, the probability of failure is zero. But even above the proof stress, it is smaller than before the proof test because removing the failed specimens has changed the normalisation.

If we perform a proof test for the example from section 7.3.3 ($\sigma_0 = 294\,\mathrm{MPa}$, $m = 13.1$), using a proof stress of $\sigma_\mathrm{p} = 220\,\mathrm{MPa}$, 2.2% of the components will fail the test. For the remainder, the new probability of failure is

$$G_\mathrm{f}(\sigma) = 1 - \exp\left[-\left(\frac{\sigma}{294\,\mathrm{MPa}}\right)^{13.1} + 0.022\right]\,.$$

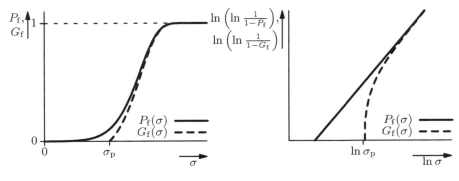

(a) Linear scale

(b) Linearised representation according to figure 7.10 [104]

Fig. 7.19. Failure probability before (P_f) and after (G_f) the proof test with σ_p

Below 220 MPa, the probability of failure is zero. If the proof stress is exceeded, the probability of failure is reduced, for example from 3.9 % to 1.8 % at a stress of $\sigma = 230$ MPa.

It is crucial to simulate the in-service stress distribution of the component as closely as possible during the proof test. This is not always feasible, for example in the case of thermal stresses simulated by mechanical ones. To gain sufficient safety, a large value of the proof stress can be chosen, although this has the disadvantage of producing unnecessary scrap parts.

7.5 Strengthening ceramics

Similar to metals, strengthening ceramics is of great technical importance. The methods used for metals (section 6.4) cannot be used for ceramics because they are all based on impeding dislocation movement, a phenomenon that is completely irrelevant for the failure of ceramics. In this section, we will discuss some methods appropriate for strengthening ceramics.

There are two approaches: In the first, we try to reduce the size of defects or initial cracks in the material, thus increasing the critical stress according to equation (5.4). The fracture toughness remains unchanged. In the other approach, the energy dissipation during crack propagation is increased, also increasing the fracture toughness. The basic mechanisms have already been discussed in section 7.2. Frequently, particles or fibres are added to the ceramic. Ceramics strengthened by particles are called *dispersion-strengthened ceramics* and will be discussed here; fibre-reinforced ceramics are one topic of chapter 9. Table 7.4 summarises the mechanical properties of some technically important ceramics.

Table 7.4. Mechanical properties of some ceramics [142]. σ_{B} is the failure stress in a four-point-bending experiment

ceramic	E GPa	K_{Ic} MPa$\sqrt{\mathrm{m}}$	σ_{B} MPa	m
SSN (Si_3N_4, sintered)	290...330	5...8.5	700...1 000	10...15
RBSN (Si_3N_4, reaction bonded)	80...180	1.8...4	200...330	14...16
ATI ($Al_2O_3 \cdot TiO_2$)	10...50	3...5	15...100	10...12
PSZ (ZrO_2)	200...210	5.8...10.5	500...1 000	20...25
aluminium oxide (Al_2O_3)	220...380	4...5.5	230...580	10...15
ZTA (Al_2O_3 + ZrO_2 particles)	380	4.4...5	400...480	10...15

7.5.1 Reducing defect size

We already saw in section 7.1 that the defect size distribution of compacted ceramics crucially depends on the manufacturing process. Methods aiding the compaction process, for instance by adding a compressive stress or sintering aids, should therefore increase the strength. However, all these methods have their disadvantages as well: hot isostatic pressing, for example, is much more expensive than sintering, whereas sintering aids cause the formation of an amorphous glassy phase on the grain boundaries and thus may impair the mechanical high-temperature behaviour.

Another important parameter is the size of the ceramic powder used. It should be chosen as fine as possible for two reasons: Fine powder enlargens the specific surface of the green body and thus improves sintering. In addition, the size of the cavities between the powder particles scales with their size. Both effects reduce the pore size after compaction. Table 7.5 shows this effect for Al_2O_3. It can be clearly seen that the physical properties are not changed significantly by decreasing the powder size, but the strength improves considerably.

7.5.2 Crack deflection

We saw in section 7.2.1 that a crack can be deflected by adding particles that the crack cannot penetrate. One example for a ceramic strengthened this way is sintered silicon nitride (SSN, Si_3N_4) which contains elongated, rod-shaped crystallites with a large aspect ratio. Due to the sintering process, they are

Table 7.5. Comparison of the properties of Al_2O_3 with the purity 99.9% produced by two different powder sizes [101]

		fine	coarse
powder size	µm	1...6	15...45
density	g/cm^3	3.96	3.99
Young's modulus	MPa	366 000	393 000
tensile strength	MPa	310	206
bending strength	MPa	551	282
compressive strength	MPa	3 790	2 549

surrounded by a very thin fringe which may be either in a glassy or crystalline state. During crack propagation, the crack cannot penetrate the Si_3N_4 crystallites. Instead, it has to grow along the thin fringe and to wind around the rod-shaped grains [60]. Due to this mechanism, the fracture toughness of SSN is surprisingly large for an unreinforced ceramic (see table 7.4). Because of their high wear and corrosion resistance and their strength, silicon nitride ceramics are used in bearings, valves, cutting tools, and in apparatus engineering [103].

> Silicon nitride can crystallise in two different structures, the α and β phase. Both are hexagonal, but the unit cell is about twice as large in c direction in the β as in the α phase [25]. Manufacturing starts with powders made of α-Si_3N_4 grains. During sintering, they transform to particles of the thermodynamically more stable β phase. The particles are elongated because they grow preferentially in one direction. To achieve complete compaction, sintering aids are added that are responsible for the formation of the glassy layer along the grain boundaries.

> Alternatively, Si_3N_4 can be manufactured as reaction bonded silicon nitride (RBSN). This process starts with silicon powder that reacts with nitrogen in a nitrogen atmosphere. This produces a somewhat porous ceramic, containing both α and β phase. The strength of this material is much smaller than that of sintered silicon nitride due to the different grains and the larger defect size caused by the increased porosity (see table 7.4).

In fibre-reinforced ceramics, deflection of the crack is an important means of increasing fracture toughness as we will see in chapter 9. One example for a biological material with a fracture toughness that is increased by a similar mechanism as sintered silicon nitride is mother-of-pearl (see section 9.4.4).

Residual stresses near particles may, as explained in section 7.2.1, also deflect a crack. One example is silicon nitride reinforced with titanium nitride [20]. Due to the difference in thermal expansion of the two materials, the matrix is stressed compressively near the particles, impeding crack propa-

Fig. 7.20. Microcracks around a particle (after [137])

gation. The fracture toughness K_{Ic} of silicon nitride can be increased by 11 % when 30 wt-% of titanium nitride are added. The titanium nitride particles also improve the wear resistance.

7.5.3 Microcracks

The formation of microcracks during crack propagation can increase the fracture toughness due to additionally dissipated energy (see section 7.2.3). Frequently, microcracks are created in the manufacturing process. Depending on their distribution, they may either increase or decrease fracture toughness. Both can be advantageous, depending on the requirements of the application.

One case where the fracture toughness increases is sketched in figure 7.20. Here, microcracks have formed near a particle during manufacturing, either due to differences in the coefficient of thermal expansion of matrix and particle or to a phase transformation of the particle. It is crucial that the microcracks reduce the elastic modulus *locally* [19, 137], resulting in an attraction of the crack by the particle. When the crack approaches the particle, the crack tip is partially unloaded because microcracks in its vicinity open as well. Locally concentrated microcracks thus cause crack deflection with an increase of the crack surface and also crack branching at the particles. One example is aluminium oxide reinforced with zirconium oxide (ZrO_2) particles that can undergo a phase transformation during manufacturing. As we will see below, the particles have to be sufficiently large for this to be possible. On the other hand, they must not be too large because otherwise the microcracks themselves would reduce the strength. This is one principal problem of strengthening by microcracks – frequently, the fracture toughness increases, but the strength does not.

If microcracks are not restricted to the vicinity of the crack tip or the particles, but are homogeneously distributed throughout the volume, they globally reduce Young's modulus. In a stress-controlled situation, the stored elastic strain energy increases according to section 2.4.1, thus increasing the energy release rate $\mathcal{G}_I$ during crack propagation (see equation (5.10)). Crack propagation can occur at smaller stresses, and the fracture toughness decreases with decreasing Young's modulus according to equation (5.16). For strain-controlled applications, however, a smaller Young's modulus – caused, for example, by microcracks – is advantageous because the reduction of the stress due to the smaller Young's modulus is larger than the decrease in fracture toughness K_{Ic}. From equations (5.15) and (5.16), using also $\sigma_c = E\varepsilon_c$, we find

$$\varepsilon_c = \sqrt{\frac{\mathcal{G}_{Ic}}{E\pi a}} . \tag{7.17}$$

The critical strain ε_c has increased because of the reduced stiffness.

One example for a ceramic with homogeneously distributed microcracks is aluminium titanate (ATI, $Al_2O_3 \cdot TiO_2$). Due to its rhombohedral crystal structure, its thermal expansion is strongly anisotropic. In two directions, we find positive coefficients of thermal expansion, but in the third direction the coefficient is negative; there is 'thermal contraction'. During cooling from the sintering temperature, this causes large residual stresses which lead to microcracking. This is the reason for the small Young's modulus and the small fracture toughness and strength shown in table 7.4. Macroscopically, the coefficient of thermal expansion of ATI is very small and, together with the small Young's modulus, makes the material very resistant against thermal shock [142].

7.5.4 Transformation toughening

Transformation toughening increases the crack-growth resistance by producing compressive residual stresses in the material during crack propagation. These are caused by stress-induced phase transformations, described in section 7.2.4. To achieve this, particles are added to the matrix that perform a phase transformation that results in a larger volume of the particles when a sufficient tensile stress is applied.

Zirconium oxide (ZrO_2) is especially suited for transformation toughening. Pure zirconium oxide solidifies at a temperature of 2680°C in a cubic lattice that is stable down to a temperature of 2370°C (cf. figure 7.21). Here, ZrO_2 transforms to a tetragonal phase. On further cooling, the monoclinic phase becomes stable at 1170°C. This usually occurs by a martensitic, diffusionless phase transformation (see section 6.4.5) in which the volume grows by 3% to 5% and the material is sheared by 1% to 7%. This process is almost impossible to suppress when manufacturing pure zirconium oxide and causes residual stresses that are so large as to damage the material excessively by crack formation. Accordingly, pure zirconium oxide cannot be used for load-bearing applications.

Transformation toughening occurs when the zirconium oxide is in the metastable, tetragonal phase. This can be achieved by stabilising the tetragonal phase with another oxide. For example, by adding yttrium oxide (yttria, Y_2O_3) to zirconium oxide, the transformation temperature can be reduced to

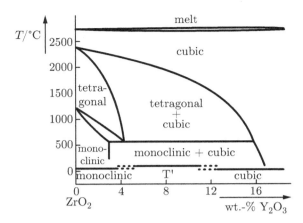

Fig. 7.21. Phase diagram of the system ZrO_2-Y_2O_3 (after [26])

550°C (see figure 7.21).[10] To transform from the tetragonal to the monoclinic phase, an energy barrier must be overcome because sufficiently large nuclei of the monoclinic phase have to be formed, in analogy to the nucleation barrier in precipitation hardening (see section 6.4.4). This nucleation barrier is so large in stabilised zirconium oxide that it cannot be surmounted without an additional driving force.

The necessary driving force can be provided by external stresses. If a hydrostatic tensile stress is superimposed, energy would be required to deform the tetragonal phase elastically. This increases the energy of the tetragonal phase compared to the less dense monoclinic phase, thus reducing the nucleation barrier. External stresses, for instance near a crack tip, can thus induce the phase transformation. Residual stresses developing during cooling can also reduce the nucleation barrier by this mechanism. These residual stresses increase with increasing grain size so that the grain size must be sufficiently small to avoid the phase transformation during cooling.

> The influence of the grain size on the transformation process can be explained with a simple model [46]: Consider a corner inside a grain of tetragonal zirconium oxide, embedded in a matrix with a different coefficient of thermal expansion. Due to thermal stresses, a stress singularity will form in this edge, and the stresses near the edge will thus be very large. In a certain region near the corner, the stress is sufficiently large to initiate the phase transformation. To grow, this transformed region must be sufficiently large. The size of the region in which the necessary stress is reached is proportional to the grain size. Therefore, the transformation can occur during cooling in large grains, but not in small ones.

---

[10] Additionally, compressive stresses can be imposed during cooling to favour the tetragonal phase energetically.

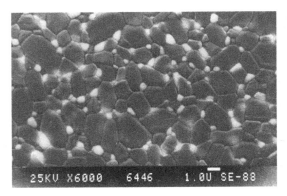

Fig. 7.22. Grain structure of zirconium oxide (light-grey) in aluminium oxide (dark-grey). The horizontal bar has a length of 1 μm. Courtesy of CeramTec AG, Plochingen, Germany

In the case of partially stabilised zirconium oxide (partially stabilised zirconia, PSZ), the amount of the added oxide is so large that the material is still in the single-phase cubic region at the sintering temperature of about 1800 °C. By rapid cooling, this state is 'frozen in'. A subsequent heat treatment at about 1400 °C causes the formation of tetragonal precipitates in the cubic matrix. This method, similar to precipitation hardening of metals (see section 6.4.4), has the advantage that the size of the tetragonal particles can be controlled precisely. Maximum strength is achieved if the particle size is slightly below the critical value described above.

Because strengthening increases linearly with the amount of tetragonal phase, a 'fraction' of 100 % of tetragonal phase gives the highest strength. Zirconium oxide with almost 100 % tetragonal phase is called TZP, *tetragonal zirconia polycrystals*. In this case, the amount of added oxide must be limited so that the phase is still tetragonal during sintering.[11] The amount must, on the other hand, not be too small because then the transformation of the tetragonal grains to the monoclinic state could not be suppressed. Such a complete stabilisation of zirconium oxide can be accomplished, for example, by adding 4 wt-% yttrium oxide. By using extremely fine powders and low sintering temperatures (typically about 1400 °C), a small grain size of about 1 μm is ensured to avoid transformation to the monoclinic phase.

TZP is among the ceramics with the highest strength and fracture toughness. It is used, for example, in ceramic hammers or as ceramic cutting edges in knifes. TZP is not only used as a bulk material, but also as reinforcement in other ceramics. As an example, figure 7.22 shows aluminium oxide (alumina) reinforced with zirconium oxide (zirconia) particles, called zirconia-toughened alumina (ZTA) [104, 137]. If the size of the zirconium oxide particles is larger

---

[11] Usually, this can not be ensured completely and a small fraction of cubic phase remains.

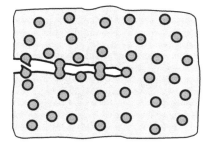

Fig. 7.23. Hampering crack propagation by ductile particles (after [149])

than the critical value or if the compressive stresses during cooling are not sufficient, the phase transformation can occur during cooling. In this case, there is no transformation toughening, but the other mechanisms to increase the crack-growth resistance discussed above (crack deflection, microcracking) take effect. ZTA is used, for example, as cutting tool for machining metals.

7.5.5 Adding ductile particles

Ductile particles can cause crack bridging because a crack penetrating them has to do additional work to deform the particles plastically before they finally break. This is depicted in figure 7.23. It is reasonable to choose particles that attract the crack, for example by using particles with a smaller Young's modulus. This strengthening mechanism has the main disadvantage that the ceramic cannot be used at high temperatures because the metallic phase then looses its strength and will be oxidised when it is reached by the crack tip. So far, this mechanism is therefore of no technical importance. One possible application could be in medical engineering where hydroxyapatite (see section 9.4.4) is frequently used as implant material. The fracture toughness of hydroxyapatite can be significantly increased by adding ductile platinum particles [30].

Mechanical behaviour of polymers

There is a multitude of polymers with widely differing properties, making them suitable for different applications like rubber tyres, crash helmets, food packagings, or plastic bags. This multitude is due to the fact that polymers consist of organic chain molecules whose structure can be controlled within wide margins (see also chapter 1).

As already discussed in section 1.4, we distinguish amorphous and semi-crystalline thermoplastics, elastomers and duromers. All explanations given in the following relate to amorphous thermoplastics, unless noted otherwise. The peculiarities of the other groups are discussed separately.

The mechanical properties of amorphous thermoplastics are mainly determined by the intermolecular bonds between the chains, not by the covalent bonds within them. These intermolecular bonds are, depending on the chemical composition, van der Waals, dipole, or hydrogen bonds. The different strengths of these bonds cause the wide spectrum of mechanical properties. In addition, the geometry of the molecules is important because they have to move against each other, especially so in plastic deformation.

The bonds between the chains are weaker than covalent or metallic bonds and may be overcome by thermal activation even at room temperature. Thus, as we will see in detail in section 8.1, polymers are in their high-temperature regime even at room temperature. Their deformation is therefore time-dependent, and it is not always easy to distinguish elastic and plastic deformations. The mechanical properties of polymers are the subject of sections 8.2 to 8.4. Methods to improve the mechanical properties of polymers are discussed subsequently. The chapter closes with a brief discussion of the sensibility of polymers against environmental influences.

8.1 Physical properties of polymers

8.1.1 Relaxation processes

Amorphous thermoplastics consist of covalently bonded chain molecules, bonded to each other by intermolecular interactions. At low temperatures (of a few

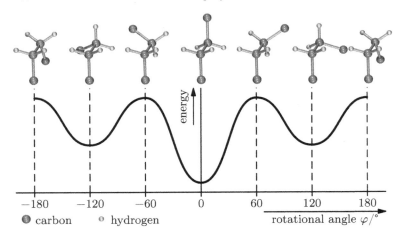

Fig. 8.1. Sketch of the potential energy of the bond between two carbon atoms in the backbone (after [79]). Only the hydrogen atoms of the two middle carbon atoms are shown

kelvin), the chain molecules are fixed in their positions, and the intermolecular bonds are strained under mechanical loads.

At elevated temperatures, the behaviour of polymers is much more complex because thermally activated rearrangements and movements within and between the chains can occur, which are frequently reversible. These processes are mainly responsible for the physical and mechanical properties of polymers. They are called *relaxation processes* and are the topic of this section. Their name is due to the fact that they may cause a relaxation i. e., a reduction of applied stresses, as we will see later.

In an amorphous polymer, there are a large number of such relaxation mechanisms which are determined by the chemical structure of the polymer and are active at different temperatures. In the following, we will discuss some examples.

A simple example of a relaxation process is a rotation along the axis of a molecular chain, for example, in polyethylene. Although a single C-C bond can rotate freely, this is not true for the bond along the molecular chain because the attached hydrogen atoms and the chain itself impede the rotation. Figure 8.1 schematically shows the energy of the bond as a function of the rotational angle in a molecular chain. As can be seen, the energy is smallest when the hydrogen atoms on neighbouring carbon atoms are in opposite positions. In addition, there are two local minima at rotational angles of $\pm 120°$ with slightly larger energies. A rotation from one of the local minima to another thus needs to overcome an energy barrier and may be thermally activated.

A single rotation along the chain of a polymer as shown here, cannot occur in a polymer because a large chain segment would have to be moved as a whole. This is not possible because neighbouring molecules would get in the

Fig. 8.2. Rotation of carbon atoms in a chain molecule (crankshaft relaxation, after [79])

way. However, it is possible to simultaneously move several atoms along the chain as shown in figure 8.2 for the example of a polyethylene molecule. Due to its shape, this process with four moving atoms is called *crankshaft relaxation*. The energy barrier of this rotation is about 60 kJ/mol [79]. If the thermal energy is sufficiently large, rotations like this can occur. In polyethylene, this is the case at temperatures above −100°C.

If a component made of polyethylene is loaded mechanically, the molecules can rearrange with this mechanism and thus enable an additional deformation. Due to the thermal activation necessary for this, the process requires some time. Therefore, the deformation is time-dependent. This will be described further in section 8.2.

Side groups along the chain can also cause relaxation processes. Polymethylmethacrylate (PMMA, see figure 1.23, page 27) provides an example. The monomer of PMMA contains one methyl group CH_3 directly bonded to the chain (often called the 'backbone') and a $COOCH_3$ group, comprising a carboxyl group (COO) and another methyl group. The methyl group bonded to the carboxyl group is rather mobile, and its rotation can be thermally activated at temperatures of only 6 K. The other methyl group directly attached to the main chain is impeded in its movement by the other neighbouring groups and starts to become mobile only above approximately −170°C. The carboxyl group is even larger and is also constrained in its movement by its methyl group. Furthermore, it is polar so that electrostatic interactions help to fix it in its position. A relaxation of a side group additionally bonded by dipole or hydrogen bonds can only occur if the temperature is sufficiently large to overcome this binding energy and to move the side group to another position. The carboxyl group of PMMA thus can only relax at temperatures above 20°C.

> The single relaxation processes of a polymer are frequently denoted with Greek letters. The α relaxation is the relaxation process with the highest temperature, the one with the next-lower temperature is called β relaxation and so on. This is a bit confusing because physically identical relaxation processes in different polymers may bear different names.

Larger chain segments of about 50 units may also become mobile at sufficiently large temperatures. Thermally activated collision processes between

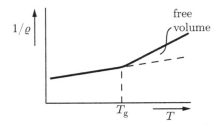

Fig. 8.3. Specific volume $1/\varrho$ of an amorphous polymer as function of the temperature [9]

the atoms increase the distance between the chains and may enable these larger segments to slide past each other. This larger mobility causes the so-called *glass transition* described in the next section.

In crystalline regions, relaxation processes may also occur. Many of the mechanisms described above, for example the crankshaft relaxation, are not possible in crystalline regions due to the larger packing density. Instead, the mobility of free chain ends or rearrangements of faults in the lattice may provide relaxation mechanisms.

8.1.2 Glass transition temperature

The physical properties of a polymer are different in the amorphous and crystalline regions and are therefore covered separately. The *glass transition temperature* discussed in this section is only relevant for the amorphous regions.

To understand the glass transition temperature, we need to look at the so-called *specific volume,* the reciprocal of the density. Figure 8.3 shows the dependence of the specific volume on the temperature for an amorphous polymer. At low temperatures, the specific volume grows with the temperature because of thermal expansion as explained in section 2.6 for the case of metals and ceramics. A temperature raise is accompanied by an increase in the energy of the molecules, widening the bonds between the chains. At a certain temperature, the *glass transition temperature* (or *glass temperature* for short) T_{g}, the curve of the specific volume has a kink and the specific volume grows more strongly with temperature than before. This additional volume is called *free volume.*

Microscopically, this larger increase implies that the distance between the molecular chains grows more strongly than at small temperatures. This is again due to thermal activation which is now large enough to overcome the intermolecular bonds and increase the mobility of the molecules. In this context, it is frequently said that the bonds 'melt' when the glass temperature is reached. This does not mean that the bonds between the chains are broken; but the thermal energy is sufficient that rearrangements of the molecules are possible without an external stress, momentarily breaking and re-forming the bonds between the chains. The activation energy for this process is completely provided by thermal activation. The situation is analogous to the melting of

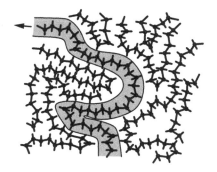

Fig. 8.4. Movement of a chain molecule through a 'tunnel' formed by the surrounding molecules (after [9])

a solid. At temperatures beyond the melting temperature, strong interatomic bonds still exist, otherwise the material would be a gas, not a liquid. But thermal activation allows to break and re-form these bonds frequently enough for the atoms to be able to move freely past each other. Similarly, the thermal movement of the chain molecules is less restricted by intermolecular bonds above the glass temperature. Increasing the temperature, and thus the kinetic energy, also increases the distance between the molecules strongly.

The mobility of the chain molecules is much larger than below the glass temperature, enabling the molecules to relax along large chain segments as described in the previous section. The viscosity at a temperature above the glass temperature is nevertheless much larger than in a molten metal, for the molecules are entangled, and sliding them past each other is geometrically constrained. For the polymer to behave like a liquid, molecules that were initially close have to be able to separate by a large distance i. e., a molecule must be drawn out of the assembly of the other molecules. This can be visualised by imagining the molecule to move (or 'reptate' like a snake) through a 'tunnel' formed by the surrounding molecules (see figure 8.4). Because of the complex shape of the tunnel, the molecule must be able to rotate along the covalent bonds on the backbone, for otherwise the movement is geometrically constrained. Due to the increase in the free volume with temperature, the mobility of the molecules increases with temperature and the viscosity thus decreases.

8.1.3 Melting temperature

The properties of the crystalline regions are clearly distinct from those of the amorphous parts. In crystalline regions, the chain molecules are more strongly bound. On the one hand, this is due to the higher density of intermolecular bonds, on the other hand, the bond lengths are smaller because the chain molecules are more regularly arranged. This denser packing hinders most of the relaxation processes described in section 8.1.1. Therefore, the simple spring model of the atomic bond that was explained in section 2.3 is valid to higher temperatures than in the amorphous regions.

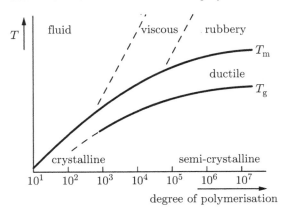

Fig. 8.5. Dependence of the glass and melting temperature as well as the mechanical behaviour of a crystalline or semi-crystalline polymer on the degree of polymerisation (after [27]). At very low degrees of polymerisation, complete crystallinity is possible

If the thermal energy is of the order of the binding energy, the intermolecular bonds melt when the temperature is raised, as described above, and the crystalline regions become liquid. The melting temperature and the glass transition temperature are both characterised by the thermal energy being sufficiently large to enable rearrangements of the chain molecules which are held together by their intermolecular bonds. If, upon cooling, the temperature falls below the melting temperature, the molecules rearrange from a irregular to a regular, crystalline structure. There is no such rearrangement in the amorphous regions at the glass temperature. This rearrangement in the crystalline regions releases energy as heat, resulting in a latent heat at the melting temperature. It is mainly the chemical composition of a polymer that determines whether it becomes crystalline or amorphous on cooling as we will discuss further in section 8.5.2.

Because long-chained polymers can never be fully crystalline (see section 1.4.2), the properties of amorphous and crystalline regions are superimposed in semi-crystalline polymers. On heating, the glass temperature is reached first and the melting temperature second. Frequently, the glass temperature is about 60 % of the melting temperature (measured in kelvin) because the bonds in the crystalline regions are stronger due to their more favourable geometry.

The values of the melting and the glass temperature also depend on the length of the chain molecules (the degree of polymerisation). This can be understood because the free ends of a chain molecule are more weakly bound and thus more mobile. Short-chained polymers with a large number of free ends thus increase the free volume more strongly than large-chained ones, easing the rearrangement of the molecules. Figure 8.5 depicts the dependence of the melting and the glass temperature on the degree of polymerisation. In a technical polymer, the molecules are never all of the same length. Therefore, its melting temperature is not as distinctly defined as it is in a pure metal.

The degree of polymerisation also has an influence on the mechanical properties – with increasing chain length, amorphous thermoplastics become more viscous even above the glass temperature. This is due to the folding and entanglement of the single molecules (see section 1.4.2), which makes sliding the molecules past each other the more difficult, the longer the molecules are. In section 8.3.1, we will see that this entanglement can even cause rubbery behaviour. Above their glass, but below their melting temperature, semi-crystalline polymers are ductile; a fact discussed in more detail in section 8.4.2.

Duromers and elastomers also exhibit a glass transition temperature with melting of the non-covalent intermolecular bonds, with localised sliding of the molecules being consequently easier. However, due to the covalent bonds between the molecules, it is not possible to pull single molecules out of the molecular network. Thus, these materials never become viscous liquids, but always remain solid. It is not possible to heat them to temperatures where the covalent intermolecular bonds are molten because they will decompose before.

8.2 Time-dependent deformation of polymers

Young's modulus of polymers is about two orders of magnitude smaller than that of metals and ceramics (see table 2.1), whereas the yield strength is smaller by only about one order of magnitude. Therefore, polymers can exhibit much larger elastic strains without deforming plastically. When components made of polymers are designed, this large elastic deformation has to be taken into account.

Both the elastic and the plastic behaviour of polymers are time-dependent even at room temperature; polymers are thus viscoelastic and viscoplastic. In this section, we discuss the time-dependent deformation behaviour phenomenologically and explain how thermal activation of relaxation processes causes the time-dependence of deformation.

8.2.1 Phenomenological description of time-dependence

If the stress in a polymer is raised abruptly from zero to a value σ clearly below the yield strength and then kept constant, the polymer answers with a time-dependent strain $\varepsilon(t)$. The strain increases instantaneously to a value ε_0 (see figure 8.6), as in the case without time-dependent elastic behaviour, but then it further increases with time. We define the *time-dependent Young's modulus at constant stress* $E_c(t)$ as

$$E_c(t) = \frac{\sigma}{\varepsilon(t)} \, . \tag{8.1}$$

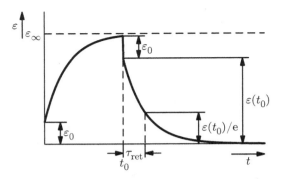

Fig. 8.6. Time response of the strain in a retardation experiment at constant stress for a viscoelastic material

Often, the time-dependent Young's modulus is denoted as *creep modulus*[1]. The deformation, and thus $E_c(t)$, approaches a constant value if the loading time becomes large. If the load is removed after a time t_0, the strain decreases instantaneously by the time-independent strain ε_0 and then reduces slowly to zero. The *retardation time* τ_{ret} is defined as the time needed to reduce the time-dependent part of the strain by a factor $1/e$ (figure 8.6).

At small strains, polymers are *linear viscoelastic:* An increase in stress causes a proportional increase in strain. At larger strains, this is not true anymore.

If we prescribe the strain instead of the stress of a component, the stress in it decreases with time (*stress relaxation*). Similar to the retardation process, the stress increases instantaneously, but then it decreases with time and approaches a constant value. Similar to the creep modulus, we can define the *relaxation modulus*

$$E_r(t) = \frac{\sigma(t)}{\varepsilon} \tag{8.2}$$

and the *relaxation time* τ_{rel}. The relaxation modulus and the relaxation time can never be larger than the creep modulus and retardation time, respectively. This is discussed further in exercise 25.

Phenomenologically, linear viscoelastic behaviour can be described using a simple mechanical model, called the *Kelvin model* or, occasionally, *Voigt model*. In this model, the behaviour of the material is described by a parallel connection of a spring element and a dashpot (or damping) element (see figure 8.7(a)). A constant load strains the spring, but the friction within the dashpot element provides a large initial resistance to the strain, causing the strain to increase with time. This model describes the behaviour of a purely viscoelastic material.

---

[1] The name 'creep modulus' is somewhat misleading because it denotes viscoelastic behaviour, although the term 'creep' is usually used for viscoplastic behaviour only.

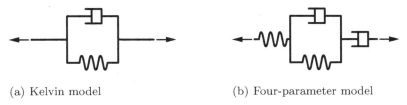

(a) Kelvin model   (b) Four-parameter model

Fig. 8.7. Mechanical models of viscoelastic and viscoplastic materials, built as systems containing spring and dashpot elements

As we already saw, in reality, the behaviour of a polymer is never purely viscoelastic. There is always an instantaneous elastic contribution to the deformation (without any time-dependence) and, at elevated temperatures, a plastic deformation which is irreversible. Similar to the elastic properties, the plastic properties of a polymer also strongly depend on time. Thus, polymers are viscoplastic i.e., they creep.[2]

This behaviour can be described using the so-called *four-parameter model*, containing a spring and a dashpot element in series with a Kelvin model (see figure 8.7(b)). The stiffness and the damping parameters of the elements are temperature-dependent. At small temperatures (well below the glass temperature), the elastic behaviour dominates, rendering the material essentially linear-elastic and in most cases brittle. If the temperature is raised, the behaviour is viscoelastic; if it is raised further (clearly beyond the glass temperature), the material becomes a viscous liquid as discussed in section 8.1.2, with the dashpot element determining its behaviour.

This simple model is only qualitatively, but not quantitatively, correct. It also ignores the stress-dependence of the deformation: If the stress is sufficiently large, a polymer can deform plastically even below the glass temperature. In other words, as in the case of metals, the material has a yield strength.

The time-dependence is important for technical applications. Figure 8.8 shows *isochronous stress-strain curves* of polymethylmethacrylate (PMMA) with a glass temperature of about 100°C. These curves are obtained in *retardation experiments,* where the strain in a specimen kept at constant stress after a fixed loading time is measured. Different from 'ordinary' stress-strain curves, each value of the stress requires its own experiment. The deformation becomes larger the longer the loading time is.

In the diagram, we can see the linear viscoelastic region at small strains, with stress and strain being proportional, but with the slope of the curve being time-dependent. This slope is the creep modulus defined above, which

---

[2] In this book, we use the term 'creep' only for the time-dependent plastic deformation. In the context of polymer science, the time-dependent *elastic* deformation is also frequently called 'creep', but this is not done here, except for using the term 'creep modulus'. Creep (i.e., viscoplasticity) can also occur in metals and ceramics at high temperatures and will be discussed in chapter 11.

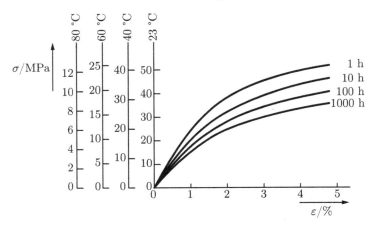

Fig. 8.8. Isochronous stress-strain curves of amorphous polymethylmethacrylate with a glass temperature of approximately 100°C (after [41])

decreases with increasing loading time. At strains larger than approximately 0.5 %, we observe a deviation from linear behaviour. The material becomes non-linear viscoelastic and, at even larger strains, flows viscoplastically. If the strain is increased further, plastic behaviour dominates, and the slope of the curves decreases. If the loading time is increased in this region, the strain increases proportionally i. e., the strain rate is almost constant as a function of stress. Viscoelastic effects can occur at temperatures well below the glass temperature (in this case, 80 °C below T_g) and thus have to be considered when designing with polymers.

Curves similar to those in figure 8.8 are obtained when the strain is kept fixed instead of the stress. In this case, the stresses decrease with time (relaxation) as explained above. For technical applications, retardation curves are usually more important because in most cases the load on the component is known and it has to be checked whether the permissible maximum strain is exceeded.

The time-dependence of deformation renders the parameters measured in simple tensile tests (see section 3.2) much less important than they are in metals, for instance. Although they do describe the behaviour at short-termed loads, time-dependent parameters – obtained, for example, from isochrones – have to be used to design polymer components. Viscoelastic and viscoplastic effects can be neglected only if strains and loading times both are small.

8.2.2 Time-dependence and thermal activation

Relaxation processes, explained in section 8.1.1 above, play an important role in the deformation of polymers. As they are thermally activated at sufficiently high temperatures, the probability of a relaxation process increases exponentially with the temperature and also with the time available for the process.

To discuss this in detail, we consider a segment of the polymer chain that has to overcome an energy barrier Q of a relaxation process (see section 8.1.1) to slide past a neighbouring segment and enable the deformation. The probability P to overcome this barrier by thermal activation is, according to appendix C.1, $P \propto \exp(-Q/kT)$, with the temperature T and Boltzmann's constant k.

Analogous to section 6.3.2, an external stress σ eases overcoming the barrier. If this stress acts on a molecular segment of cross section A and if the width of the barrier is d^*, the external stress does the work $W = \sigma A d^*$ when the segment overcomes the barrier. The probability to overcome the barrier is thus

$$P^+ \propto \exp\left(-\frac{Q - \sigma A d^*}{kT}\right).$$

If the barrier is overcome, the back-reaction can also occur. However, it is less probable than the forward-reaction because work against the external stress has to be done. Its probability is

$$P^- \propto \exp\left(-\frac{Q + \sigma A d^*}{kT}\right).$$

The total probability to overcome the barrier is the difference of both contributions:

$$P = P^+ - P^- \propto \exp\left(-\frac{Q}{kT}\right)\left[\exp\left(\frac{\sigma A d^*}{kT}\right) - \exp\left(-\frac{\sigma A d^*}{kT}\right)\right].$$

From this, we obtain the strain rate (using $\sinh x = (\mathrm{e}^x - \mathrm{e}^{-x})/2$)

$$\dot{\varepsilon} = \dot{\varepsilon}_0 \exp\left(-\frac{Q}{kT}\right) 2 \sinh\left(\frac{\sigma A d^*}{kT}\right). \tag{8.3}$$

Here, $\dot{\varepsilon}_0$ is a constant parameter.

In the following, we will use this equation to study the time-dependence of the elastic and plastic behaviour.

Time-dependence of elastic deformation

For small stresses, we can use the approximation $\sinh x \approx x$ in equation (8.3) so that the strain rate is proportional to the applied stress. In this case, the behaviour is linear and viscous. As stresses are small, the deformation is not plastic, but elastic, for there is a restoring force corresponding to the spring element in figure 8.7(a), whereas equation (8.3) describes the dashpot element of the Kelvin model. The behaviour is thus linear viscoelastic. At larger stresses, deviations from linearity occur, although the behaviour is still viscoelastic.

From what we discussed so far, it seems plausible to assume that the behaviour of a polymer at small temperatures and large times is similar to that at larger temperatures and smaller times. In fact, it is possible to find conversion factors, allowing to extrapolate the creep or relaxation modulus from one temperature to another. This will be discussed in a similar way in section 11.1 for the creep of metals.

To discuss this conversion in detail, we consider the creep modulus i.e., the time-dependent Young's modulus at constant stress. We assume that it was measured at a temperature T_1 with a loading time t_1. According to the ideas presented above, the creep modulus at temperature T_2 should have the same value if we change the loading time to t_2.

If the viscoelastic behaviour is dominated by a relaxation process with activation energy Q, the value of t_2 can be calculated easily: At constant stress, the strain rate is given by equation (8.3):

$$\dot{\varepsilon} = A \exp\left(-\frac{Q}{kT}\right) . \tag{8.4}$$

For the strain to be identical at a different temperature, the product of loading time and strain rate must be constant:

$$t_1 \dot{\varepsilon}_1 = t_2 \dot{\varepsilon}_2 \quad \Rightarrow \quad \frac{t_1}{t_2} = \frac{\exp\left(-\frac{Q}{kT_2}\right)}{\exp\left(-\frac{Q}{kT_1}\right)} = \exp\left[-\frac{Q}{k}\left(\frac{1}{T_2} - \frac{1}{T_1}\right)\right] . \tag{8.5}$$

If the activation energy Q is known, this equation can be used to calculate t_2 directly. Otherwise, one additional experiment at temperature T_2 (or another temperature) is required to determine the activation energy.

In deriving this result, we assumed that the viscoelastic properties are determined by a single activation energy. In real-world polymers, this is usually not the case. Nevertheless, the equation can be used approximately as long as the same relaxation processes are involved. It is not possible, however, to extrapolate to temperatures so low that some relaxation processes do not occur, or to temperatures so high that additional mechanisms are activated.

Near the glass temperature and at larger temperatures, this conversion between temperature and time is not valid anymore because not the relaxation processes, but the sliding of chain molecules past each other determines the elastic behaviour. Although this process is thermally activated as well, the dependences are more complicated because raising the temperature also increases the free volume and thus has an additional effect on the movement of the molecules. Therefore, the conversion factor changes to

$$\frac{t_1}{t_2} = \exp\frac{\ln 10 \cdot C_1(T_2 - T_1)}{C_2 + (T_2 - T_1)} , \tag{8.6}$$

with C_1 and C_2 being constants that are approximately the same for all amorphous polymers. If we choose $T_1 = T_g$, their values are $C_1 = 17.5$ und $C_2 = 52\,\text{K}$. This equation is called *Williams-Landel-Ferry equation*, or WLF-*equation* for short. If we plot the creep modulus at temperature T_1 versus time in a double-logarithmic scale, a change to temperature T_2 shifts the whole curve horizontally by $a_T = \log(t_1/t_2) = \log(t_1/\text{h}) - \log(t_2/\text{h})$. a_T is frequently called the *shift factor*.

Time-dependence of plastic deformation

At all technically relevant temperatures, polymers deform by creep. To describe the time-dependence of plastic deformation, we again exploit equation (8.3). In contrast to the viscoelastic deformation, there is no restoring force in viscoplasticity. Equation (8.3) is thus used to describe the dashpot element connected in series in the four-parameter model from figure 8.7(b).

For large stresses, we can use the approximation $2 \sinh x \approx \exp x$ in equation (8.3). This yields

$$\dot{\varepsilon} = \dot{\varepsilon}_0 \exp\left(-\frac{Q}{kT}\right) \exp\left(\frac{\sigma A d^*}{kT}\right). \tag{8.7}$$

A simple form of this equation is obtained by solving for σ/T:

$$\frac{\sigma}{T} = \frac{k}{Ad^*}\left(\frac{Q}{kT} + \ln\frac{\dot{\varepsilon}}{\dot{\varepsilon}_0}\right). \tag{8.8}$$

If we plot the quotient of the stress σ in the polymer and the temperature versus the logarithm of the strain rate, points at constant temperature should fall onto a common line. Figure 8.9 plots this for polycarbonate, thus confirming our considerations. This kind of plot is called *Eyring plot*. If there is more than one type of obstacle, the slope of the curve is not constant anymore. If, for instance, one type of obstacle is dominant up to one temperature, but another is more important at higher temperatures, the Eyring plot contains a kink because both obstacles differ in the activation volume Ad^*.

8.3 Elastic properties of polymers

8.3.1 Elastic properties of thermoplastics

The elastic behaviour of polymers is mainly determined by the intermolecular bonds between the chain molecules, not by the covalent bonds within. For elastomers and duromers, the covalent bonds linking the chains are also relevant. In the following, we will start by discussing the elastic properties of thermoplastics and afterwards study the influence of cross-linking.

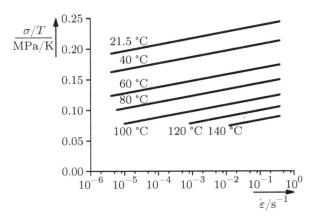

Fig. 8.9. Eyring plot of polycarbonate (after [97])

Figure 8.10(a) schematically shows the temperature dependence of Young's modulus in an amorphous thermoplastic, measured at a typical, constant loading time (for example, one second). Increasing the loading time would cause a reduction of Young's modulus. As can be seen from the figure, the stiffness strongly decreases at temperatures close to the glass temperature. The elastic behaviour will therefore be discussed separately in the temperature regimes below and above the glass temperature.

Energy elasticity

The elasticity of thermoplastics below their glass temperature is mainly due to the energy needed to displace atoms from their equilibrium position. On unloading, the atoms return to their original position which has the lowest energy. For this reason, this behaviour is called *energy elasticity*. It is mostly the weak, intermolecular van der Waals, dipole, or hydrogen bonds that are strained. The covalent bonds do not contribute significantly to the elastic properties. Their stiffness is so large that they nearly cannot be strained elastically as long as the other bonds can deform. Only if the chain molecules are aligned in parallel, as in polymer fibres like aramid (kevlar), the covalent bonds determine Young's modulus which can then take very large values of up to 440 GPa.

We already saw in section 2.6 that Young's modulus is approximately proportional to the melting temperature and thus to the binding energy. For amorphous polymers, the relevant temperature is the glass transition temperature because this is the temperature where the bonds melt. The rather low values of the glass temperature (listed in table 1.3) thus also explain why Young's modulus of polymers is smaller than for the other material classes.

The strong decrease of Young's modulus at the glass temperature (see figure 8.10(a)) will be discussed in the next section. More interesting in the

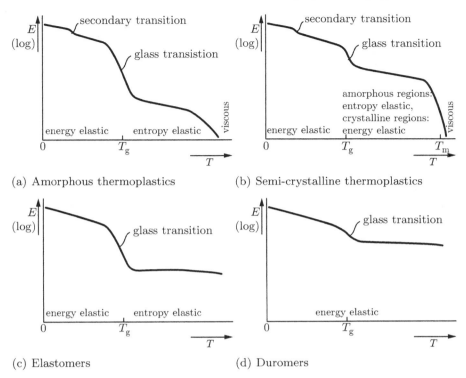

Fig. 8.10. Temperature dependence of Young's modulus in different types of polymers (after [19]). Because a logarithmic scale is used, the reduction of Young's modulus appears to be smaller than it is in reality. More explanations in the text

present context is the fact that even below the glass temperature, there may be temperature values at which Young's modulus decreases markedly by about a factor of approximately 2. These so-called *secondary transitions* are caused by relaxation processes which enable a limited mobility of the chain molecules and thus cause a stress relaxation by movement of molecule segments. Because such rearrangements always require overcoming some activation energy, they become more probable if the loading time increases. They are responsible for the viscoelastic behaviour of polymers.

As the activation energies of different relaxation processes differ, their relaxation time also differs. This is the reason why the simple spring-and-dashpot model from section 8.2.1 cannot be used to make quantitative predictions. This would require coupling several such elements [97] with relaxation times chosen to fit their respective processes.

We already saw in section 8.1.1 that the activation energy of some relaxation processes is so low that it can be overcome by thermal activation already at temperatures as low as a few kelvin. At room temperature, their relaxation

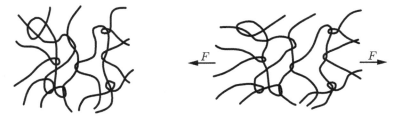

Fig. 8.11. Elastic deformation of a polymer above the glass temperature. The molecules are straightened between the entanglement points

times are thus very short (of the order of 10^{-8} s). Relaxation is almost instantaneous so that these processes contribute to the initial deformation of the polymer when the load is applied.

If the stressed polymer has deformed viscoelastically by relaxation, the deformed configuration has a higher energy than the initial one. Upon unloading, the molecules return to their initial positions. This process again requires thermal activation and is therefore time-dependent as well.

Entropy elasticity

If the temperature exceeds the glass temperature, Young's modulus strongly decreases. From what has been said so far, it could be surmised that the polymer should deform like a viscous liquid if heated beyond the glass temperature, exhibiting viscosity, but no elasticity. This, however, is not the case.

The reason for this is the strong entanglement of the chain molecules. As discussed in section 1.4.2, the chain molecules are strongly folded. Different chain molecules are thus 'tied together' like a knot in many places. On loading, the molecules are straightened. Directly above the glass temperature, they cannot slide past each other because this movement is hampered by the surrounding molecules (see figure 8.11). The molecules thus straighten between their entanglement points. During sliding, energy barriers have to be overcome because the chain molecules are straightened, rotate, and because side groups have to move. Due to the higher temperature and the larger distance between the molecules, this process is much easier at temperatures above the glass temperature. The deformation of the material is still time-dependent due to the required thermal activation.

If the load is removed, there is no force on the straightened molecules, so there seems to be no reason why they should return to their initial position. Because of the stochastic thermal movements of the molecules in the polymer, the molecule will probably return from the straightened to a folded geometry because there are a lot more possibilities for a folded molecule than for a straightened one. The arbitrary thermal collisions with the surrounding molecules thus fold up the molecule again. Thus, there is a thermodynamic driving force because the entropy of the molecule is larger in the folded than in the straightened state. This behaviour is therefore called *entropy elasticity*. The entanglement points between the chain molecules remain in a fixed position on the molecules during elastic deformation so that the molecule returns to its initial shape. In contrast to the deformation below the glass temperature, it is not the smaller energy of the initial configuration that drives the return to this form, but its larger entropy. As before, the movement of the molecules is time-dependent.[3]

The viscoelasticity of amorphous polymers is most pronounced near the glass temperature in the transition regime between energy-elastic and entropy-elastic behaviour. At lower temperatures, only smaller parts of the molecules can slide past each other as explained above. As we approach the glass temperature, more and more sliding processes become possible. As the sliding processes can be more easily thermally activated the higher the temperature becomes, the relaxation time decreases. At temperatures well above the glass temperature, relaxation times are small, and the system returns quickly to its initial state.

So far, we only considered amorphous thermoplastics. Semi-crystalline thermoplastics show a different behaviour as shown in figure 8.10(b). Due to the stronger intermolecular bonds in the crystalline regions, their elastic stiffness is usually larger than that of amorphous polymers. The decrease in Young's modulus on reaching the glass temperature is smaller because only the amorphous regions become entropy-elastic, whereas the crystalline regions remain in the energy-elastic state. The cohesion between the crystalline and the amorphous regions is ensured because most chain molecules extend over several crystalline and amorphous regions.

8.3.2 Elastic properties of elastomers and duromers

Elastomers and duromers are characterised by additional covalent cross-links between the chain molecules. In the energy-elastic regime, these additional bonds do not influence the elastic properties significantly; Young's modulus only increases slightly.

At temperatures above the glass temperature, the additional bonds become important. Elastomers are entropy-elastic at these temperatures. The covalent bonds between the molecules increase the linking between them compared to thermoplastics where molecules are linked by geometric entanglement only. These additional links cannot be broken during sliding of the molecules and thus increase the effect of entropy elasticity. With increasing number of cross-links, the covalent bonds are loaded more heavily during elastic deformation so that Young's modulus increases with the cross-linking density as can be seen

---

[3] Above the glass temperature, there is always plastic deformation as well. If the loading time is sufficiently short, the plastic strain rate is small enough to be neglected; at larger loading times, it has to be taken into account (see also section 8.4.1).

from figure 8.10(c). Because the restoring force in entropy-elastic deformation is the entropy, which becomes more important the larger the temperature is (see equation (C.3)), Young's modulus of elastomers often increases with increasing temperature.

Contrary to metals and ceramics, the elastic strains in elastomers can become very large and attain values of several hundred percent. The reason is that the molecules are straightened during deformation, but the cross-links prevent the molecules from sliding past each other and thus inhibit plastic deformation. Upon unloading, entropy-elasticity completely restores the initial arrangement of the molecules. This behaviour is called *hyperelasticity*.

> During deformation of hyperelastic materials, large strains of 100% or more can occur. The material behaviour is strongly non-linear. Therefore, the theory of large deformations has to be used to describe the material behaviour (see section 3.1).
>
> The basis of the description is the energy of the deformation: Because it is elastic (i. e., reversible), energy is stored in the material and can be regained on unloading. Hyperelastic materials can therefore be described by specifying the energy density as a function of strain. The stress in the material can be calculated as the derivative of the energy density with respect to the strain. This description is useful for two reasons: On the one hand, the energy density in the material can be calculated using methods of thermodynamics, on the other hand, it ensures that the stored energy does not depend on the material history, but only on the current state of deformation. This is necessary because hyperelastic processes do not dissipate energy; it would be difficult to accomplish by defining a stress-dependent Young's modulus.

If the cross-linking density of a polymer is increased further, the entropy-elastic behaviour vanishes nearly completely because the large number of cross-links prevent the straightening of the molecules. For this reason, duromers show only a small decrease of Young's modulus with temperature (see figure 8.10(d)) caused by relaxation processes. They are energy elastic even above the glass temperature.

Table 8.1 lists the magnitude of Young's modulus for the different polymer groups as a function of their cross-linking density. This quantity is normalised by assigning a value of 1 to diamond in which all atoms contribute to the cross-linking.

8.4 Plastic behaviour

Polymer elasticity is determined by the reversible deformation of the chain molecules as we saw in the previous section. Polymers can also deform plastically, with chain molecules sliding past each other over large distances as

Table 8.1. Cross-linking density and Young's modulus of different types of polymers (cf. section 1.4.2)

type of material	cross-linking density	E/GPa	
thermoplastics	0	$0.1 \ldots 5$	(for $T < T_{\mathrm{g}}$)
elastomers	$10^{-4} \ldots 10^{-3}$	$0.001 \ldots 0.1$	(for $T > T_{\mathrm{g}}$)
duromers	$10^{-2} \ldots 10^{-1}$	$1 \ldots 10$	
diamond	1	$1\,000$	

sketched in figure 8.4 on page 261. The plastic behaviour of polymers strongly depends on the temperature because obstacles have to be overcome by thermal activation and because the size of the 'tunnels' in which the molecules move is determined by the specific volume (see section 8.1.2).

As in the previous section, we start by discussing amorphous thermoplastics and afterwards discuss how things change in semi-crystalline thermoplastics. Elastomers and duromers only allow for a small amount of plastic deformation because the cross-links prevent molecule sliding as explained above. Elastomers used above their glass temperature can be deformed with large elastic strains instead; duromers are brittle, with the covalent bonds between the chain molecules breaking in brittle failure.

8.4.1 Amorphous thermoplastics

We start this section by discussing the plastic behaviour of amorphous thermoplastics. The stated temperature regions are, due to the time-dependence of plastic deformation, valid for rather large strain rates (with testing times of a few seconds). Increasing the testing time i. e., decreasing the strain rate, is equivalent to increasing the temperature (see section 8.2).

Far below the glass temperature

At temperatures lower than about 80 % of the glass temperature T_{g}, the bonds between the molecules are so strong and the specific volume is so small that chain molecules cannot move by sliding. On loading, the molecules are straightened viscoelastically. If the load is raised further, as sketched in figure 8.12(a), brittle failure ensues, mainly breaking the intermolecular bonds.

Slightly below the glass temperature

At temperatures of about 80 % of the glass temperature T_{g}, amorphous thermoplastics have a limited ductility (see figure 8.12(b)). At these higher temperatures, the mean distance between the chain molecules is larger and enables

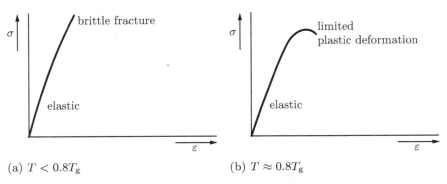

Fig. 8.12. Stress-strain diagrams of an amorphous thermoplastic at different temperatures [9]

them to partially overcome the binding forces, giving the molecules a limited mobility.

In contrast to metals, polymers do not work-harden because no new obstacles are created when the molecules slide past each other. Heat generated during deformation causes a local increase in temperature, further easing plastic deformation. This results in a local softening of the material, similar to a metal with an apparent yield point (see section 6.4.3). Only if the plastic strain becomes larger does some hardening occur because the molecules become aligned in the direction of the applied stress.

A typical microstructure of an amorphous thermoplastic loaded in tension slightly below the glass temperature is shown in figure 8.13(a). There are microscopically small, lens-shaped cavities, called *crazes*. They have a thickness of about 1 µm to 10 µm and a diameter of about 10 µm to 1000 µm and are bridged by fibrils. The fibrils comprise several chain molecules and have a diameter of approximately 10 nm to 100 nm. Their volume fraction within the craze is between 10 % and 50 %. Although the crazes do look crack-like, the strength of the material is only slightly reduced in this region compared to the strength of the undeformed material since the chain molecules within the fibrils are straightened and thus can bear a higher load. The thickness of a craze is almost independent of the applied stress, but it increases with increasing temperature. If the applied stress is large, a large number of small crazes form, if it is small, their number is smaller.

Usually, crazes are initiated at surface defects, for example scratches or impurities. Plastic deformation starts in these regions due to the slight stress concentration caused by these defects. Because the material softens as explained above, plastic deformation concentrates in this region, resulting in a slight local necking. This, in turn, causes the stress state to become triaxial and increases the hydrostatic tension. Small cavities with a diameter of a few nanometres form (figure 8.14). Because of the stress concentration, the material between the cavities is heavily loaded and deforms plastically,

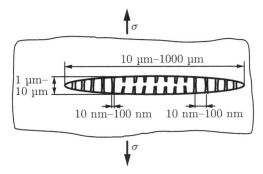

(a) Cross sectional view of a craze. The lens-shaped cavity and the fibrils can be observed

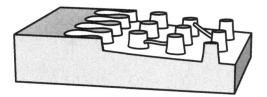

(b) Scaled partial view of the edge of a craze

Fig. 8.13. Microstructure of a craze (after [82, 128])

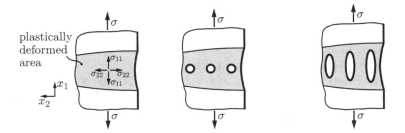

(a) Local necking   (b) Formation of cavities (c) Extension of cavities

Fig. 8.14. Development of a craze by formation of cavities (after [82])

straightening the molecules in this region. Fibrils between the cavities emerge and a craze is formed.

Despite the load-bearing capacity of the fibrils, there is a stress concentration near the edges of a craze, easing its further growth. The growth mechanism is a so-called meniscus instability: Near the edge of the craze, finger-shaped extensions evolve and contract, forming new fibrils (figure 8.15). Fibrils within the crazes initially elongate further by drawing other chain molecules from the bulk material. Cross-links between the fibrils may form if opposite

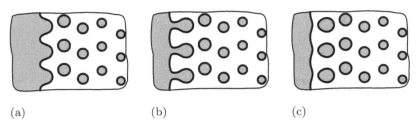

(a)   (b)   (c)

Fig. 8.15. Growth of a craze by a meniscus instability (after [82])

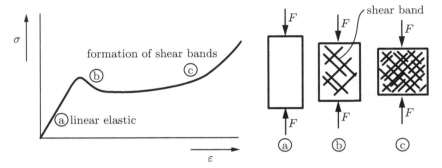

Fig. 8.16. Deformation of an amorphous thermoplastic under compressive load by formation of shear bands (after [9])

ends of a chain molecule are drawn into neighbouring fibrils (see figure 8.13(b)). Finally, fibrils in the centre of the craze break. The craze then grows continuously at constant load, rendering the plastic deformation time-dependent. This growth can eventually cause fracture of the polymer.

A polymer can deform not only by crazing, but also by forming *shear bands,* created at an angle between 45° and 60° [44, 82, 132] to the loading direction (figure 8.16). Formation of shear bands is especially important under compressive loads. Within the shear bands, large localised plastic deformations of 100 % or more can occur, whereas the deformation is very small outside of them. Shear band formation has not been studied as closely as crazing. A simple mechanical model is based on the shearing of chain molecules (figure 8.17). The shear stress component causes the chain molecules to either straighten or to form two kinks, resulting in a region with aligned chain molecules. If several shear bands converge, a crack can be initiated if one shear band reaches the already straightened molecules. Because these cracks are now loaded under shear where, according to section 5.1.1, the fracture toughness is larger ($K_{IIc} \gg K_{Ic}$), the fracture strain is significantly larger than under tensile loading.

Several factors determine whether a polymer deforms by shear bands or crazing. The crucial factor is that crazes, which are initiated by cavitation,

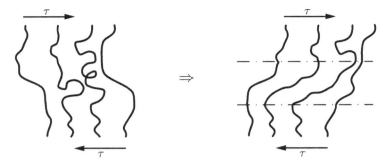

Fig. 8.17. Formation of a shear band by local stretching and contracting of the molecule chains (after [35, 132])

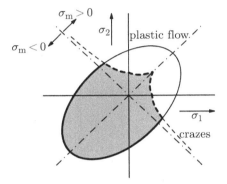

Fig. 8.18. Yield surface of an amorphous thermoplastic that can fail by crazing or formation of shear bands. If the hydrostatic tensile stress is sufficiently large, crazing occurs before shear bands (after [92, 132])

can only form under hydrostatic tensile stress.[4] The larger the hydrostatic tensile stress is, the stronger is the tendency for crazing. Figure 8.18 shows the yield surface of a polymer in plane stress, illustrating this.

The yield strength of polymers generally depends on hydrostatic stress because hydrostatic compression decreases the specific volume and thus hampers sliding of the molecules. This was already discussed phenomenologically in section 3.3.3. However, the criteria discussed there did not take crazing into account.

Apart from the multiaxiality of the stress state, the temperature and the loading time also play a role in determining the deformation mechanism. Large strain rates (and small temperatures) make shear band formation more difficult, thus favouring crazing.

---

[4] Even in a uniaxial stress state, there is a hydrostatic stress according to equation (3.25): $\sigma_{\text{hyd}} = \sigma/3$.

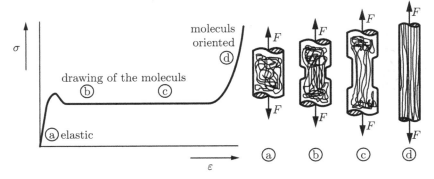

Fig. 8.19. Stress-strain curve of an amorphous thermoplastic closely below the glass temperature (after [9])

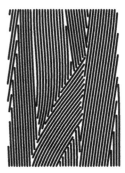

Fig. 8.20. Configuration of a drawn thermoplastic made of fibre bundles (after [35])

Close to the glass temperature

If the temperature approaches the glass temperature, the chain molecules become more and more mobile and may rearrange on loading. Figure 8.19 shows the stress-strain curve for this case. After the yield strength has been reached, the specimen starts to neck in some region because of local softening as explained in the previous section. If deformation continues, more and more chain molecules are drawn and straightened in parallel. The more pronounced the drawing of the chain molecules is, the more are the covalent bonds loaded, causing a local hardening. This eventually overcompensates for the reduction in cross section and forecloses further necking in this region. Instead, the necking region grows until the whole specimen comprises drawn chain molecules. In this process, strains can be as high as 300 %.

By drawing a thermoplastic, it is thus possible to manufacture a material with chain molecules arranged mainly in parallel. Figure 8.20 schematically shows the structure of such a polymer: The chain molecules are arranged in bundles, being parallel within them. The material deforms by sliding of these fibre bundles. As the fibre bundles are very long, even a small interfacial strength between them is sufficient to exploit the strength of the covalent bonds.[5] The force on a fibre bundle can become so large that the covalent bonds break. The fracture surface splices, exposing the fibre bundles. Fibres manufactured this way have a very large stiffness and strength compared to amorphous thermoplastics.

High-strength polymer fibres with drawn chain molecules can be produced by spinning. Aramid fibres with drawn molecules, for example, can have a Young's modulus in fibre direction of up to 450 GPa and an axial tensile strength of 4700 MPa. These fibres are frequently used in composites (see chapter 9).

Above the glass temperature

If the temperature significantly exceeds the glass temperature, the chain molecules can easily slide past each other because the strong increase in the specific volume (see section 8.1.2) and the melting of the intermolecular bonds strongly increases the mobility of the molecules. During plastic deformation, thermoplastics behave similar to highly viscous liquids. Their strength is therefore very low.

8.4.2 Semi-crystalline thermoplastics

The bond strength between the chain molecules is higher in the crystalline regions of a semi-crystalline thermoplastic than in the amorphous regions because of the smaller bond length. This increases Young's modulus and also the strength, even at temperatures above the glass temperature.

Plastic deformation starts by lengthening the amorphous regions (see figure 8.21(b)). At larger strains, the crystalline regions rotate the chain molecules into the loading direction (figure 8.21(c)). On further deformation, the crystalline regions separate into different blocks (figure 8.21(d) and (e)). In those crystalline regions where the molecules are directed transversely to the loading direction, the molecules may also rearrange to a vertical orientation, not by rotating block-wise, but by forming new layers in the vertical direction.

One problem of semi-crystalline thermoplastics is that impurities and short-chained molecules are concentrated in the amorphous regions because they are pushed from the crystalline regions on crystallisation. The interface between amorphous and crystalline regions is therefore weak and cracks may initiate there.

> Nature frequently uses polymers for load-bearing applications as well (see also section 9.4.4). One particularly interesting example for a biological polymer is the silk of spiders or some insects, for example the

---

[5] This will be explained in detail in section 9.3.2 for the case of fibre composites.

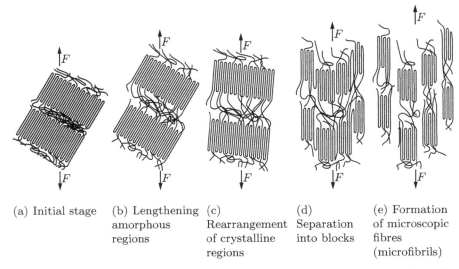

(a) Initial stage   (b) Lengthening   (c)   (d)   (e) Formation
                     amorphous   Rearrangement   Separation   of microscopic
                     regions   of crystalline into blocks   fibres
                                        regions   (microfibrils)

Fig. 8.21. Stages of plastic deformation of a semi-crystalline thermoplastic (after [44, 82])

larvae of the silk moth *Bombyx mori* [144]. Silks are made of protein fibres, spun to strings with a diameter between about $2\,\mu m$ and $10\,\mu m$ in spider silk and $10\,\mu m$ and $50\,\mu m$ in silk of the silk moth.

Proteins are polymers comprising amino acids as monomers (see also section 9.4.4). Their structure is similar to polyamide (see figure 1.23), with the chain 'R' consisting of a carbon atom with a side group. 20 different amino acids commonly exist in nature, resulting in a huge number of possible protein structures. In contrast to technical polymers, the structure of a protein i.e., the sequence of its constituting amino acids, is defined exactly. This sequence determines how the protein molecule folds up to form a three-dimensional structure.

Most silks are semi-crystalline polymers. Due to the exactly defined three-dimensional structure, proteins can be aligned exactly in the crystalline regions, with different side chains precisely interlocking. The silk of the silk moth, for example, contains large regions of two alternating amino acids, one of them (glycin) with a very small, the other (alanine or serine) with a slightly larger side group. These side groups interlock as shown in figure 8.22 and thus create crystalline regions with very high strength.

Silk properties vary strongly, depending on their structure. A single spider can possess up to seven different types of silks which may be used, for example, as dragline, for orb-spinning, or to encase prey or the eggs in an egg cocoon. Each type of silk is produced by its own silk gland.

The mechanical properties of spider dragline silk are especially well-studied, mostly for the common garden spider *Araneus diadematus* and

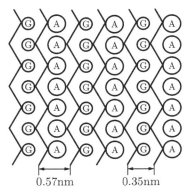

0.57nm   0.35nm

Fig. 8.22. Crystalline structure of the proteins in the silk of the silk moth. The side groups of neighbouring molecule chains interlock and cause a high strength. In each protein, the amino acids glycin and alanine with different side groups alternate. In the amorphous regions (not shown here), other amino acids are used, inhibiting this crystalline structure

the golden silk spider *Nephila clavipes*. Draglines are almost always produced by a spider during moving, serving as a safety rope for the case that it falls during climbing. As the dragline must not break, its fracture toughness has to be large. The tensile strength of a dragline can take values of up to 1.1 GPa, approximately one third of that of aramid fibres. Their fracture strain is about 30%, much larger than in aramid (with a fracture strain of about 2.7%). Thus, they can absorb large amounts of energy without breaking. Because it is impossible to perform crack propagation experiments with these microscopic specimens, the fracture toughness is characterised by measuring the energy absorption to fracture. This can take values of 1.5×10^8 J/m^3, more than four times higher than in aramid fibres.

Similar silks are also used during orb spinning in those lines that run radially outwards from the centre. The circumferential lines, forming the viscid net for capturing prey, are made from a completely different type. These have to absorb large amounts of energy to prevent prey hitting the orb from bouncing back and they have to cling to the prey to retain it. To achieve this, their fracture strain is especially large, with values of up to 800% and a tensile strength of about 500 MPa. The energy to fracture can be as high as 10^9 J/m^3, larger than in any other known material.

Spider silks are produced in silk glands in which the constituting proteins are dissolved in water [85]. During drawing of the thread, the tensile stress straightens and aligns the molecules although the material is still dissolved in a liquid. Shortly before it leaves the gland, the crystalline regions are arranged as discussed above and water is removed. Although no chemical reaction occurs at this stage, the silk is not soluble in water after it has left the gland. How the spider accomplishes this is not known. To achieve the correct microstructure, it is crucial that the silk is drawn from the gland because this is required to align the molecules. Silks cannot be pressed or squirted out of the gland.

The silk of the silk moth has been technically used for thousands of years. Because of their excellent mechanical properties and also because of their biocompatibility, spider silks are especially attractive for many technical applications, for example for wound dressing, sutures, or in microtechnics. Contrary to the larvae of the silk moth, spiders are highly territorial and have a strong tendency for cannibalism so that it is impossible to keep many of them in a confined space. Nowadays, it is tried to manufacture spider silk biotechnologically, using bacteria to produce the silk proteins.

8.5 Increasing the thermal stability

The thermal stability of polymers is inferior to that of metals and ceramics. Near the glass temperature, the stiffness and strength of amorphous thermoplastics strongly decrease. Above the glass temperature, viscous flow is the dominant deformation mechanism in amorphous thermoplastics. Therefore, amorphous thermoplastics can only be used at service temperatures markedly below their glass temperature in load-bearing applications because Young's modulus strongly decreases before the glass temperature is reached (see figure 8.10(a)). Semi-crystalline polymers can also be used above their glass temperature. Their strength is smaller here than below T_g, but their ductility increases. Elastomers are always used above the glass temperature, because they are rubbery only in this temperature regime. Duromers can be used below or above the glass temperature, depending on the application. Although the stiffness is smaller above the glass temperature, they do not flow viscously above T_g and are thus still serviceable.

Due to this temperature dependence, any means of increasing the thermal stability are of extreme importance, especially as they usually also increase the strength and stiffness. One can either increase the glass or, in a semi-crystalline polymer, the melting temperature, or the volume fraction of the crystalline regions. This will be discussed in the following.

8.5.1 Increasing the glass and the melting temperature

On reaching the glass temperature, the mobility of the chain molecules becomes large enough to allow them to slide past each other, as we saw in section 8.1.2. Figure 8.4 on page 261 visualises the movement of a chain molecule during sliding. It shows that, in the process, the molecule has to move through an intricately shaped tunnel formed by the surrounding molecules. If this movement can be impeded, the glass temperature will increase.

Because of the intricate shape of the tunnel, sliding through it is only possible if the chain molecule can rotate. One way to reduce the mobility of the molecules is thus to impede these rotations. In principle, the carbon-carbon bond can rotate freely (see also section 8.1.1), enabling the molecule to twist slide through the tunnel. If the rotation is impeded, sliding is impeded as well.[6] There are several ways to achieve this. If, for example, the simple carbon-carbon bond is replaced by a more complicated structure with less rotatable bonds, the glass temperature increases markedly. Among the polymers with the highest glass temperature are the polyimides (see table 1.3), with a molecular backbone formed not from a single carbon chain, but from a link between a benzene ring, two amide groups, and a carbon chain. As the ring-shaped part of the backbone is rigid, a rotation is not possible here, and the mobility is strongly reduced. The more mobile carbon chains between the rings serve to reduce the brittleness of the material and improve its processibility.

If large side groups are added to the molecule, they can also impede rotation. On the one hand, these side groups cannot penetrate each other and thus make sliding the molecule through the tunnel formed by the surrounding molecules more difficult. On the other hand, the energy required to rotate the side groups increases with their size (see figure 8.1) because the side groups on a single molecule interfere with each other due to their spatial extension and their electrostatic repulsion.[7] The glass temperature thus increases with the size of the side groups. If we compare the glass temperature of polyethylene ($-110°C \ldots -20°C$) without side groups, polypropylene ($-20°C \ldots 0°C$) with a simple methyl group, and polystyrene ($100°C$), we immediately see how the size of the side group influences the glass temperature.

> If the side chains are long and flexible, they may also decrease the glass temperature. On the one hand, the number of freely movable chain ends increases and thus causes an effect similar to a reduction in the chain length (see section 8.1.3), on the other hand, the side groups tend to increase the distance between the chain molecules and thus reduce the bond strength. If the chains become very long, the glass temperature rises again because the side chains can be arranged regularly, similar to a semi-crystalline polymer [13].

Impeding the rotation is also the mechanisms responsible for the rather large glass temperature of polytetrafluor ethylene (PTFE, Teflon) with a value of

---

[6] The tendency to form crazes is also reduced in this way since crazing requires to draw some molecules from the bulk material into the fibrils [83].

[7] This was already discussed in section 8.1.1 for the side groups in PMMA.

Fig. 8.23. Spatial structure of a PTFE chain molecule. The strong repulsion between the fluorine atoms results in a twisted and rigid molecule

126°C. The large electron affinity of fluorine causes the fluorine atoms to be partially negatively charged. They thus repel each other and cause a twist in the molecule to maximise the distance between them (figure 8.23). In a rotation of the chain, the charged atoms approach each other and thus need additional energy, resulting in an increase in the glass temperature.

Furthermore, the glass temperature is affected by the bond strength between the chain molecules. To increase the glass temperature, the bonds can be made stronger. This can be achieved, for example, by adding polar side groups which can form stronger dipole bonds between the molecules. This is the reason why the glass temperature of polyvinyl chloride is larger than that of polyethylene. Replacing a single hydrogen atom by chlorine increases the glass temperature from between −110°C and −20°C to approximately +80°C. This is due to the dipole bond being much stronger than the van der Waals bond as explained in chapter 1.

Because the electron affinity of fluorine is even larger than that of chlorine and because each monomer of PTFE contains four fluorine atoms, it might be surmised that the glass temperature of PTFE is much larger than that of PVC. This, however, is not the case because the dipole bonds in PVC are in fact stronger than those in PTFE. One reason for this is that the dipole moments of neighbouring regions of PTFE cancel each other due to the spatial structure of the molecule (see figure 8.23) which possesses only negative charges on the outside. Furthermore, fluorine is a much smaller atom than chlorine, resulting in a shorter bond length between carbon and fluorine. The strength of a dipole is directly proportional to its length, giving the carbon bond with the chlorine atom a larger dipole moment. Altogether, PTFE behaves like a nonpolar molecule, making it suitable for low-adhesion coatings.

Hydrogen bonds can also strongly bond the molecules in a polymer. One example is polyamide (see table 1.3) which contains hydrogen bonds formed by the hydrogen atoms of the amino groups of neighbouring molecules.

All methods discusses so far can also serve to increase the melting temperature in a semi-crystalline polymer. However, to achieve a high crystallinity, the chain molecules must be sufficiently mobile to allow them to arrange in an ordered alignment. Stiffening the molecules therefore may decrease the crystallinity. A further problem is that a polymer made of stiff molecules has a large viscosity at high temperatures, making manufacturing processes like injection moulding more difficult.

8.5.2 Increasing the crystallinity

The strength of semi-crystalline polymers is larger than that of amorphous polymers. If the crystallinity of a polymer can be increased, the mechanical properties are improved accordingly.

The crystallinity of a polymer can be changed by the manufacturing process and by the structure of the chain molecules. Upon cooling from the melt, crystalline regions can only form if there is sufficient time to arrange the molecules in the energetically favourable more densely packed crystal structure. Crystallinity is thus a function of the cooling speed. If this speed is too high, the polymer is purely amorphous. This is analogous to the production of glasses or to precipitation and transformation processes (see section 6.4.4).

The crystallinity can also be increased by orienting the chain molecules under mechanical loads. This was already discussed in section 8.4.1 for the plastic deformation of an amorphous thermoplastic close to the glass temperature. By applying a tensile load, the fibres are drawn and straightened, forming crystalline fibres bundles.

The size of the side chains and thus the mobility of the chain molecules also influences the crystallinity. The more immobile the molecules are, the more difficult it is to arrange them in a closely packed and regular manner.

For this reason, polyethylene is well-suited to form high-strength fibres because the polymer chain is very mobile due to its simple structure and can be easily drawn in fibre direction. Depending on the straightening of the molecules, Young's modulus can reach values of up to 200 GPa [107]. In technically used fibres, values between 62 GPa and 175 GPa are characteristic (see also table 9.1). These are rather high values, especially so if the low density of slightly less than 1 g/cm^3 is taken into account. They are due to the covalent carbon bonds along the backbone.

Polymers with stiff chain molecules usually have a lower crystallinity than those made of mobile chains. Exceptions from this rule do occur, however: Due to its very stiff and straight chain molecules, PTFE can reach crystallinity values of up to 90 %. This is only possible if the cooling speed is very low; in technical applications, the crystallinity is therefore usually less.

The molecules can also be oriented during cooling from the melt by shearing the melt with high speeds because this will also align the molecules. This has to be kept in mind when designing polymer components manufactured by injection moulding to avoid a strong fluctuation of the crystallinity in the final component.

Aramid is one example for a polymer that can be manufactured with high crystallinity in this way. This is mainly used to produce aramid fibres. Due to the aromatic rings on the backbone (see figure 1.23), the molecule is extremely stiff. It thus does not fold up, but usually exists in rod-like form (similar to PTFE). Well above the melting temperature, these rods are disordered in the

(a) Isotactic; all side groups are positioned on the same side

(b) Syndiotactic; the side group positions alternate

(c) Atactic; random positions of the side groups

● carbon   ● hydrogen   ● chlorine

Fig. 8.24. Configuration of side groups in polyvinyl chloride (PVC)

melt. If the temperature falls below a certain ordering temperature (which is larger than T_m), the molecules start to align themselves in parallel because this increases their binding energy. They now possess a preferential orientation, although their positions are still unordered.[8] If fibres are drawn from the melt, the molecules in the fibres orient themselves in the drawing direction.[9]

Especially important in determining the crystallinity of a polymer is the *tacticity*, the way the side groups of the monomers are arranged within the chain. We distinguish the *isotactic* arrangement (figure 8.24(a)), in which all side groups are regularly arranged on one side of the chain molecule, the *syndiotactic* arrangement with alternating side groups (figure 8.24(b)), and the irregular *atactic* arrangement with random orientation of the side groups (figure 8.24(c)). The more regular the arrangement of the side groups is, the easier it is to arrange the chain molecules in a crystalline structure. If the structure is the same otherwise, isotactic polymers thus have the largest, atac-

---

[8] In this state, the molecules thus form a liquid crystal. Liquid crystals are characterised by molecules that have a preferential orientation but unordered positions like the molecules of a liquid.

[9] This is similar to the drawing of silk, see page 281.

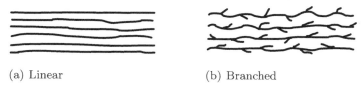

(a) Linear   (b) Branched

Fig. 8.25. Linear and branched chain molecules. Side chains impede the formation of crystalline structures

tic polymers the lowest crystallinity. The isotactic structure is also superior to the syndiotactic because the latter can only be crystalline if the side groups match and interlock exactly.

Although thermoplastics comprise un-linked chain molecules, they can nevertheless have a branched structure. In contrast to elastomers and duromers, these branches do not cause cross-linking between the chains, but they can inhibit a geometrically dense packing necessary to form crystalline regions. This is sketched in figure 8.25.

8.6 Increasing strength and stiffness

The most important methods to increase the strength or stiffness of a polymer are a direct consequence from what we saw in the previous section. They are:

- Increasing the bond strength, for example by adding polar side groups (section 8.5.1),
- impeding the sliding of the chains, for example by adding large side groups or by stiffening the chain molecule (section 8.5.1),
- increasing the crystallinity, for example during manufacturing or by using isotactic structures (section 8.5.2),
- orienting the chain molecules in load direction (section 8.4.1).

Table 8.2 contains a survey of the mechanical properties of different polymers. The effect of the different mechanisms to increase strength or stiffness can be clearly seen from the table. Low-density polyethylene (LDPE), containing branched polymer chains and thus possessing a low crystallinity of about 45%, has the lowest Young's modulus and the lowest tensile strength because it consists only of simple and mobile molecules with weak intermolecular bonds. Increasing the crystallinity to 75% and thus the density (creating high-density polyethylene, HDPE), markedly improves the properties because the larger crystallinity strongly increases the number and strength of the intermolecular bonds. If side groups are added, as it is done in polypropylene and polyvinyl chloride, the mechanical properties improve accordingly as already discussed.

Table 8.2. Typical properties of several polymers (after [44, 98]). Since the properties also depend on the degree of polymerisation and additives, the values specified only serve as guidelines

material	$R_\mathrm{m}/\mathrm{MPa}$	E/GPa	$\varrho/(\mathrm{g/cm^3})$
low-density polyethylene	15	0.3	0.92
high-density polyethylene	35	1.0	0.96
polypropylene	35	1.5	0.91
polyvinyl chloride	55	3.0	1.4
polyethyleneterephtalate	65	3.0	1.3
polymethylmethacrylate	70	3.3	1.2
polycarbonate	75	2.3	1.2
polyamide	80	3.5	1.2

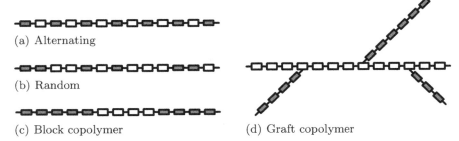

(a) Alternating

(b) Random

(c) Block copolymer   (d) Graft copolymer

Fig. 8.26. Different types of copolymers (after [19, 31])

One method of improving the strength or stiffness is particularly important in polymers: Combining them with other materials to form composites. This is the subject of chapter 9.

8.7 Increasing the ductility

As already mentioned in section 8.4.2, the ductility of semi-crystalline thermoplastics is increased if they are used above their glass temperature. Here, the amorphous regions are easily deformable, whereas the crystalline regions increase the strength. Semi-crystalline thermoplastics are thus well-suited for applications requiring an increased ductility.

Copolymerisation is another approach to improve the mechanical behaviour of polymers, especially their ductility. In copolymerisation, different monomers are used to form the chain molecules. There are several possibilities to arrange the monomers, shown in figure 8.26. The monomers can alternate (*alternating copolymerisation*) or can be arranged irregularly (*random copolymerisation*). In *block copolymers*, there are longer chain segments of one type

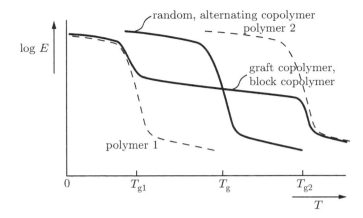

Fig. 8.27. Temperature dependence of Young's modulus of a copolymer. Alternating and random copolymers have a glass temperature between both basic polymers; graft and block copolymers behave like semi-crystalline thermoplastics (possessing two transition temperatures) [97]

alternating with those of the other, in *graft copolymers,* the main chain is formed by one monomer to which side chains made of the other are attached.

Copolymerisation is frequently used to decrease the glass temperature, with the intent of increasing the ductility. This is called *internal plasticisation.*[10]

Figure 8.27 shows the temperature dependence of Young's modulus of two polymers and their copolymers. One example of a copolymer is polybutadiene styrene, made of the monomers of polybutadiene and polystyrene. If the copolymerisation is alternating or random, the glass temperature is between that of the constituting polymers. This is easily understood because the larger side groups in polystyrene increase the glass temperature compared to that of polybutadiene, the more so, the higher the content of polystyrene.

In a graft or block copolymer, another effect occurs: The polymer contains regions rich in polybutadiene and rich in polystyrene. When the glass temperature of polybutadiene is reached, these regions start to melt, but the polystyrene-rich regions are still below their glass temperature. The overall behaviour is thus similar to that of semi-crystalline polymers due to the different melting temperature of the intermolecular bonds in the different regions. Therefore, copolymerisation causes a decrease in strength, but a significant increase in ductility.

Block copolymers can also increase the ductility by affecting crazing. As explained in section 8.4.1, crazes in amorphous polymers are initiated at surface defects because cavities can form there more easily. These crazes

---

[10] *Plasticisers* are molecules added to a polymer to reduce its glass temperature and thus increase its ductility.

cause a local stress concentration and, finally, fracture of the material. By block copolymerisation, cavity formation can be initiated at the many interfaces within the polymer. A large number of crazes may form, allowing significant plastic deformation before fracture. A similar effect can be achieved by blending (i. e., mixing) different polymers.

This method of influencing crazing is especially important in high-impact polystyrene (HIPS). Polystyrene is copolymerised with butadiene to form small, spherical butadiene (rubber) particles. These induce internal stresses in the material, easing the formation of crazes. If the particles are small, a large number of small crazes form that grow only a short distance. The material has a large strength, but only a small ductility. If the particles are larger, the number of crazes decreases, but they can grow for larger distances, thus decreasing the strength but increasing the ductility [44]. If the material is also copolymerised with acrylonitrile ($CH_2-CH-CN$), it is called acrylonitrile-butadiene styrene (ABS), with an increased strength and ductility and also a larger resistance against solvents.

* 8.8 Environmental effects

Polymers are sensitive to several environmental effects. Polar or nonpolar solvents or irradiation with ultraviolet light can significantly decrease the strength of a polymer.

Polymers may react with different organic or inorganic solvents. Generally, polar solvents may permeate polar polymers and nonpolar solvents nonpolar polymers. For example, PMMA has a limited solubility for alcohol and polyamide (PA, nylon) for water.

If a solvent attacks a polymer, its molecules enter the space between the chain molecules, thus increasing their distance. This increases the volume of the polymer, a phenomenon called *swelling*. The intermolecular bond length is thus increased, weakening the bonds and reducing the glass temperature, strength, and stiffness. This problem is especially severe in polyamides because they can absorb several weight-percent of water. In humid environment, their mechanical properties can thus be severely impaired. On the other hand, this effect can be exploited by increasing the ductility of polymers that would be brittle at service temperature. To achieve this, plasticisers are added. This is the reason why PVC can be used in floorings or plastic bags.

If solvents penetrate a polymer and weaken the intermolecular bonds, they also ease the formation of crazes. On the one hand, the solvent reduces the surface energy, facilitating the formation of free surfaces needed to initiate and propagate crazes. On the other hand, the reduced bond strength reduces the force needed to draw chain molecules out of the bulk material to form fibrils. Crazes formed by this mechanism may, as explained in section 8.4.1, serve to initiate cracks. In this case, the strength of the polymer decreases and it becomes more brittle.

The increase in volume by swelling can also reduce the strength of a component. If the amount of dissolved solvent is different in different parts, the volume increase is non-uniform, and residual stresses may be induced. If, for example, the solvent diffuses out of the surface region, it is under tensile stress and thus sensitive to tensile loads.

One way to protect polymers from solvents is copolymerisation. For example, polystyrene can be attacked by the nonpolar solvent benzene. Copolymerisation with a polar group (for example acrylonitrile) stops the nonpolar solvent from penetrating the material.

Ultraviolet light can also affect polymers because it may break the chemical bonds within the molecular chains. This reduces the chain length and thus the glass temperature. It may also lead to the formation of covalent bonds between neighbouring chain molecules, embrittling the material. This can be avoided by adding light-absorbing or -reflecting particles, for example carbon black or titanium oxide. When irradiated with UV light, polyvinyl chloride dissociates chlorine radicals which may react to form hydrochloric acid (HCl). Stabilisers that can bond the radicals prevent this process.

Elastomers like polybutadiene can also be attacked by oxidation. In this case, additional cross-links between the chain molecules are formed, causing embrittlement.

9

Mechanical behaviour of fibre reinforced composites

In composites, different materials are combined to exploit favourable properties of each. That such combinations may be attractive was already shown in section 6.4.4 for particle strengthening of metals and in section 7.5 for dispersion-strengthened ceramics.

It is rather difficult to exactly define the meaning of the word 'composite'. In the broadest sense, one might consider every material as a composite that comprises two physically distinct phases.[1] However, using this definition would imply that almost every technically used material is a composite, for example almost all steels or precipitation hardened alloys, rendering the definition practically useless. Composites used today are characterised by the following properties:

- A strengthening second phase is embedded in a continuous *matrix.*
- The strengthening second phase and the matrix are initially separate materials and are joined during processing – the second phase is thus not produced by internal processes like precipitation.
- The particles of the second phase have a size of several micrometres at least.
- The strengthening effect of the second phase is at least partially caused by load transfer.
- The volume fraction of the strengthening second phase is at least approximately 10%.

To use these properties as definition is, however, problematic and, furthermore, not future-proof, for further developments (e. g., in nanotechnology) will surely lead to new composites which do not posses some of these properties.

In this chapter, the focus is on *fibre reinforced composites,* or *fibre composites,* for short, in which the particles of the second phase are long fibres, surrounded by a matrix of the other component. One example is *glass-fibre re-*

---

[1] It has to be noted that ceramic composites exist in which both phases are chemically identical, see section 9.1.2.

inforced polymer (GFRP) in which a polymer matrix is strengthened by adding glass fibres. As initially stated, the objective of forming composites is to combine desired properties of the constituents. In the case of glass-fibre reinforced polymers, the glass fibres increase the stiffness and strength, and the surrounding matrix makes the material more ductile and protects the fibres from concentrated loads.

Fibres are frequently used as strengthening component because the load transfer from the matrix is especially effective if the strengthening phase is elongated in the loading direction. We will discuss this in some detail later on. Furthermore, fibres may be advantageous because they are rather thin with a diameter between 1 μm and 25 μm. Defects in the fibre are thus rather small.

9.1 Strengthening methods

Because different materials are joined in composites, there are a large number of possible combinations. Metal, ceramic, or polymer matrix composites can be strengthened with different kinds of particles or fibres. Composites may be classified either by the geometry of the strengthening particles (fibres, fabrics, etc.) or by the matrix material used.

9.1.1 Classifying by particle geometry

If we characterise composites by particle geometry, we can distinguish fibres and particles (in the narrow sense). In fibres, one dimension is larger than the others by at least one order of magnitude, thus they are shaped like long and slender cylinders. In particles, the extension is approximately the same in all directions. Other structures are also possible; the phases may, for example, also be arranged in a sandwich structure or laminate with alternating layers of different materials.

Fibres themselves can be further distinguished by their geometry. If we consider the properties of the composite, we can talk of *long fibres* if the properties of the composite do not change when the fibre length is increased further, whereas a change in length has an influence on the properties when *short fibres* are used.[2] Long fibres with an extension comparable to that of the whole component are frequently called *continuous fibres*. For example, glass fibres (with diameters of a few ten micrometres) are denoted as short fibres (or *chopped fibres*) if their length is less than a millimetre, as long fibres if the length is between one and fifty millimetres, and as continuous, if their length is even larger.

The length of the fibres not only determines the mechanical properties, but is also important for the manufacturing process because long fibres have to be processed different from short fibres. This will be discussed now.

---

[2] This is the topic of section 9.3.2.

Long fibres and fabrics

As we will see later in great detail, strengthening a material with fibres aligned with the direction of the applied tensile stress or maximum principal stress (*uniaxial* alignment) is advantageous. In this configuration, the different elastic properties of fibre and matrix cause stress concentrations at the fibre ends because load has to be transferred from the matrix to the fibre. The longer the fibres are, the more effective is the load transfer, and the smaller are the regions of stress concentrations. Therefore, long fibres are mechanically superior to short ones.

Because the fibres are usually very thin, they are bundled in so-called *rovings,* comprising several thousand single fibres. The diameter of these rovings is thus in the range of millimetres. Alternatively, the fibres can be spun to form yarns.

One disadvantage of the uniaxial structure is that the stiffness and the strength in the transversal direction are markedly inferior (see sections 9.2.2 and 9.3.6). For this reason, different orientations of the fibres may be combined. Mats containing uniaxial fibres are stacked on top of each other to form *laminates,* or the fibres can be woven to *fabrics* that can be laminated again. The orientation of the fibres within the layers can be chosen orthogonal to each other or in more complex ways (for example, with relative orientations of 45°). The properties of such a material can be isotropic within the plane of the fibres, but they are weaker perpendicular to them because there are no fibres in this direction. To make the material isotropic, it is necessary to arrange the fibres in all three directions, but this is rather difficult to achieve with continuous fibres because the fibres have to be woven in three directions.

Long fibres or fabrics have to be positioned correctly within the manufactured component. This can be achieved most easily if the melting point of the matrix material is well below that of the fibres, as in polymer matrix composites. If duromers are used as matrix, the fibres can be laid down in an uncured mixture of resin and hardener. On curing, the resin hardens to form the duromer.

Short fibres

Short fibres can also be directed or at least possess a preferential direction, but frequently they are distributed irregularly in the matrix, rendering the composite isotropic. The main advantage of short fibres is their cheaper production and the easier manufacturing process.

Short-fibre reinforced components can be manufactured by using laminates or mats (chopped-strand mats) containing the fibres. Furthermore, because of the smaller dimension of the fibres, all other manufacturing processes that can be used for the unreinforced matrix material are in principle available. Polymers, for example, can be formed by injection moulding; in metal matrix composites, metal forming processes like rolling can be used.

These processes may not only change the properties of the matrix (as in producing a texture in a metal on rolling), but also on the orientation of the fibres within the matrix. If, for example, a liquid polymer is pressed into a cylindrical mould, the flow velocity is larger in the middle of the cylinder. This difference in the flow velocity causes a preferential orientation not only of the chain molecules, but also of the fibres. Because of the velocity gradient, they are directed perpendicularly to the flow velocity in the middle of the cylinder and parallel to it near the surface [97]. This orientation of the fibres has to be taken into account in designing components because it causes an anisotropy even in short-fibre reinforced composites. Seams, which are created when two partial flows meet, are especially problematic because the flows usually do not mix, resulting in the seam not being bridged by fibres. To avoid these problems, computational fluid dynamics calculations are used to optimise the filling of the mould by ensuring that highly-stressed regions contain fibres with a favourable orientation.

Particles

How particles can be used to change material properties was already discussed in different contexts. Precipitates in metals were covered in section 6.4.4, particle strengthening in ceramics in section 7.5, and copolymers in section 8.7.

Among the particle-reinforced materials are the *cermets* (a word created from 'ceramic' and 'metal') that comprise a metallic matrix containing ceramic particles. Very hard carbides, for example tungsten carbide or titanium carbide, are embedded in a cobalt matrix. These cermets are frequently used as cutting tools. The ceramic particles serve to improve the wear resistance, an effect similar to that already discussed in section 6.4.4 for the case of coarse particles in metals.

Another important particle-reinforced material is concrete. Concrete consists of a cement matrix (a ceramic), containing stone or sand (the so-called aggregate) as reinforcing particles. The aggregate not only serves to moderately increase the stiffness and fracture toughness, but has the additional advantage of being less expensive than the cement. Because of the low fracture toughness, the tensile strength of concrete is rather small (about 4 MPa); the compressive strength is much larger with values of about 30 MPa.[3] For this reason, concrete is mainly stressed in compression in applications. To achieve this, it can be armed with reinforcing steel wires or rods positioned in the mould before pouring the concrete into it (*ferroconcrete*). If these are pre-stressed in tension, they superimpose a compressive load on the concrete after setting, further increasing the strength (*prestressed concrete*).

---

[3] In dealing with concrete, it is rather important to take the volume dependence of the failure stress (equation (7.9)) into account.

Sandwich structures

In sandwich structures, the phases are arranged in layers. This may be useful in plates or shells loaded in bending, with the stiffer outer material being placed well away from the neutral axis and the less stiff material in between, mainly to keep the distance between the load-bearing outer layers constant. The filler material usually has a low density (for example, being a polymer or a foam), making sandwich structures especially suitable for light-weight applications. Furthermore, they often provide good heat insulation. Sandwich structures are mainly used in the aerospace industry, in automotive engineering, and in boat building, for example in light-weight sports boats.

9.1.2 Classifying by matrix systems

All classes of materials can be suitable as matrix material. In this section, we briefly introduce the different matrix systems. A more detailed discussion can be found at the end of the chapter in section 9.4, after the mechanical behaviour of composites has been discussed.

Polymer matrix composites

Polymer matrix composites (PMC) are used to increase the rather low stiffness or strength of polymers by adding stiffer or stronger fibres. If thermoplastics are used as matrix material, short fibres can be added to the granulate material that is subsequently softened by heating and processed, for example in injection moulding. If thermoset resins (duromers) are used, the fibres can be placed into the liquid resin before curing.

There are a large number of possible fibre materials that can be used in reinforced polymers. The most common are glass, carbon, aramid, and polyethylene.

Metal matrix composites

The main advantage of metal matrix composites (MMC) is their increased stiffness and strength compared to the unreinforced material. Long or short fibres can be used for strengthening. Common fibre materials are carbon, silicon carbide, aluminium oxide, boron, and refractory metals like tungsten. Usually, light metals (aluminium, titanium, magnesium) are used as matrix materials. Compared to polymer matrix composites, metal matrix composites have the advantage of larger service temperatures.

One disadvantage is that metal matrix composites have to be processed at markedly larger temperature. This not only increases the requirements on tools used during processing, but also on the fibres which have to withstand these temperatures as well. Metal matrix composites can be manufactured by melting, adding the fibres to the melt. One alternative are powder metallurgic methods in which a metal powder containing the fibres is compacted at temperatures slightly below the melting temperature. The fibre material can also be embedded in layers between layers of the matrix metal. The material is then compressed with high pressure at elevated temperatures to press the metal between the fibres.

Ceramic matrix composites

In ceramic matrix composites (CMC), the fibres mainly serve to increase the fracture toughness. Because one important property of the fibres is their small defect size, it is possible to use the same material for fibre and matrix. The advantage of this is that the elastic properties of fibre and matrix are identical, avoiding the formation of stress concentrations, and that the coefficient of thermal expansion is also the same, so no residual stresses are generated during cooling. Chemical reactions do not occur as well. Possible fibre materials are mainly ceramics.

As in other composites, ceramic matrix composites can be processed with long or short fibres. There are several methods to manufacture long-fibre reinforced ceramics: If the matrix material is a glass, it can encompass the fibres at temperatures above the glass temperature (of about 1000°C) where the matrix behaves like a viscous fluid. Subsequent hot-pressing further compacts the material and removes pores. With this technique, fibre volume fractions of up to 60% are possible. Alternatively, chemical reactions can be used to deposit the matrix around a pre-shaped fibre network, starting with a gaseous or liquid phase [28]. This technique can also be used to coat fibre bundles which can then be processed further by filament winding to form the final component.

Short-fibre reinforced ceramic matrix composites can be produced by sintering at high temperatures, similar to ordinary ceramics (see section 7.1). The fibres are mixed with the powder and the composite is sintered at high temperatures. Fibre volume fractions of 35% can be obtained in this way.

9.2 Elasticity of fibre composites

Young's modulus of a fibre composite is determined by the elastic properties of the constituent materials and also depends on the loading direction. Because the fibres are usually stiffer than the matrix, the modulus is larger in fibre direction than transversally to it.

In this section, we start by considering the simplest case of continuous, uniaxially directed fibres loaded precisely in parallel or transversely to the fibre direction. Afterwards, we will discuss the case of arbitrary load orientation.

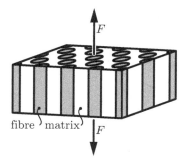
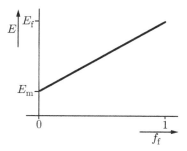

(a) Arrangement of fibres and matrix   (b) Dependence of Young's modulus
on the fibre volume fraction

Fig. 9.1. Parallel connection of fibre and matrix

9.2.1 Loading in parallel to the fibres

If the fibres are oriented parallel to the loading direction, their arrangement is called *parallel connection,* analogous to the connection of springs (figure 9.1(a)). Similar to springs connected in parallel, the deformation in the matrix (subscript 'm') and in the fibre (subscript 'f') must be the same, but the stress may differ:

$$\varepsilon_f = \varepsilon_m, \quad \sigma_f \neq \sigma_m. \tag{9.1}$$

By inspecting the geometry and defining the volume fractions f_f and f_m, with $f_f + f_m = 1$, we find the *(isostrain) rule of mixtures*

$$\sigma = \sigma_f f_f + \sigma_m (1 - f_f). \tag{9.2}$$

Using this together with equation (9.1), we find that Young's modulus of the composite is

$$E_{\parallel} = E_m f_m + E_f f_f = E_m \left[1 + f_f \left(\frac{E_f}{E_m} - 1 \right) \right]. \tag{9.3}$$

Figure 9.1(b) shows how Young's modulus depends on the volume fraction of the fibres. Equations(9.2) and (9.3) are derived in exercise 28.

9.2.2 Loading perpendicular to the fibres

If the fibres are oriented perpendicular to the loading axis, this is called *serial connection* in analogy to springs connected this way. To simplify matters, we assume that the fibres are plates extending over the whole cross section of the material and oriented perpendicularly to the applied load (figure 9.2(a)).

In the serial connection, the balance of forces must hold at each fibre-matrix interface:

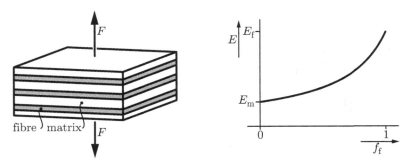

(a) Arrangement of fibres and matrix   (b) Dependence of Young's modulus
                                             on the fibre volume fraction

Fig. 9.2. In-series connection of fibre and matrix

$$\sigma_f = \sigma_m \,. \tag{9.4}$$

Again in complete analogy to springs, both materials can deform in different
ways:

$$\varepsilon_f \neq \varepsilon_m \,.$$

The following (isostress) rule of mixtures applies:

$$\varepsilon = \varepsilon_f f_f + \varepsilon_m (1 - f_f) \,. \tag{9.5}$$

If we use Hooke's law and the condition (9.4), we find, after some rearrangement, Young's modulus

$$E_\perp = \frac{E_m}{1 + f_f \left(\frac{E_m}{E_f} - 1 \right)} \,. \tag{9.6}$$

Figure 9.2(b) shows how Young's modulus depends on the volume fraction of
the fibres or plates. Compared to figure 9.1(b), it is apparent that the increase
in stiffness perpendicular to the fibre direction is much smaller than in the
parallel orientation. The derivation of the (isostress) rule of mixtures and of
Young's modulus for this case is also presented in exercise 28.

*9.2.3 The anisotropy in general

The considerations of the previous sections dealt with the two most extreme
load cases: If loaded in fibre direction, the stiffening effect of the fibres is maximal, if loaded perpendicularly, it is minimal. Under arbitrary loads, it is necessary to calculate the components of the elasticity tensor (see section 2.4.2).
Depending on the fibre arrangement, couplings between normal and shear components can occur. For example, this can be exploited to construct aerofoils
that twist on bending, with normal stresses causing shear strains.

If the fibres are uniaxial, there is still some symmetry in the material, and the number of parameters needed to describe the elastic behaviour is smaller than 21, the value for a triclinic lattice. If the fibres are directed, but their positions in space are irregular or arranged on a hexagonal lattice, the material is transversally isotropic i. e., its properties are the same in all directions perpendicular to the fibre direction. In this case, there are five independent elastic constants (see section 2.4.6). If the fibres are uniaxial and arranged on a rectangular lattice, the material is orthotropic, and the number of independent elastic components is nine (see section 2.4.5).

To determine the elastic constants, empiric equations are frequently used that provide a useful approximation, often even in the case of non-continuous fibres. One example are the so-called *Halpin-Tsai equations* [29].

9.3 Plasticity and fracture of composites

It was already discussed that one of the advantages of composites is the fact that the strengthening phase cannot contain defects larger than its extension. For fibre composites, which are our focus here, the crucial dimension is the fibre diameter. Carbon fibres are used with diameters of less than 5 μm if the objective is to increase the strength as much as possible.

We saw in the previous section that the elastic properties of a composite can be described using a rule of mixtures. This, however, is usually not the case for the plastic and the failure behaviour.

In this section, we start by discussing the behaviour of fibre composites under tensile loads, at first for the simplest case of continuous fibres. Subsequently, we will discuss the load transfer between the matrix and non-continuous fibres and see how this determines the failure properties and the fracture toughness of the material. For this, we also have to consider that fibre properties are statistically distributed. Finally, we will discuss the behaviour under compressive loads, loads perpendicular to fibre direction, and arbitrarily oriented loads.

9.3.1 Tensile loading with continuous fibres

To simplify the discussion, we start by considering the idealised case from figure 9.1(a), with the strengthening fibres being parallel to the load and extending throughout the component. Thus, no effects occurring at the ends of the fibres have to be taken into account, and the fibres are loaded directly by the external load, so no load transfer between fibre and matrix needs to be considered. Finally, we assume that all fibres are identical without any statistical scatter in their strength or diameter.

Failure starts when the stress in matrix or fibre reaches the yield or tensile strength. Similar to the elastic case, we can apply an isostrain rule of mixtures (equation (9.2)) to calculate the stress in the composite:

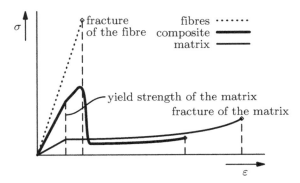

Fig. 9.3. Schematic stress-strain diagram of a fibre-reinforced polymer (after [9])

$$\sigma = \sigma_f f_f + \sigma_m(1 - f_f) \,, \tag{9.7}$$

with σ_f and σ_m being the stresses in fibre and matrix, and f_f being the volume fraction of the fibres. In contrast to equation (9.2), we are now interested in the failure stress of the composite. Therefore, at least one of the stresses in equation (9.7) is a yield or tensile strength.

This rule of mixtures is only approximately valid, even with all simplifying assumptions we have already made. The yield strength of the matrix can be affected by the presence of the fibres in different ways: Adding fibres to a matrix may cause thermal stresses during cooling or other residual stresses, it may render the stress state triaxial if Poisson's ratios of fibre and matrix differ, or it may change the microstructure of a metallic matrix. Young's modulus, being a bulk quantity, is less sensitive to such effects than the yield or tensile strength, making the rule of mixtures more appropriate for the elastic case.

To further discuss equation (9.7), we have to distinguish two cases, depending on whether failure occurs first in matrix or fibre. As the strains in both components are the same because of the parallel connection, the material with the smaller failure strain will fail first (see also exercise 29).

Failure strain in the matrix larger than in the fibre

If the failure strain in the matrix is larger than in the fibre, the fibres fracture before the matrix fails. This is frequently the case in composites with metallic or polymeric matrix. Figure 9.3 shows the resulting stress-strain diagram. It is assumed that the matrix yields plastically before the fibre breaks. The material deforms elastically until the matrix yields. On further increasing the strain, the strengthening fibres fracture, and the stress-strain curve drops to a small stress value that lies below that of the pure matrix material because of the reduced volume. Eventually, failure by fracturing of the matrix occurs. The fracture strain is smaller than in a pure matrix material. This is due to damage initiated by the breaking fibres and the triaxial stress state in a composite.

In component design, the maximum stress in the material is usually the quantity of interest. If the fraction of fibres is so large that the matrix cannot bear a given load after the fibres have fractured (this is the case in figure 9.3), the failure stress is given by the isostrain rule of mixtures, equation (9.7), where σ_f is the failure stress of the fibres and σ_m the stress in the matrix at the failure strain of the fibres. If the volume fraction of the fibres is small, the matrix can still bear the load even after the fibres have fractured. In this case, the failure stress of the composite is $(1 - f_f)\sigma_m$, with σ_m being the failure stress of the matrix. The failure stress is thus reduced, compared to the pure matrix material.[4]

Even if the fibres are long, albeit discontinuous, the stress-strain diagram frequently looks like that shown in figure 9.3. This will be discussed further in section 9.3.4.

Failure strain in the fibre larger than in the matrix

The failure strain of the fibre may also be larger than that of the matrix in some cases, for example in carbon-fibre reinforced duromers or in ceramic matrix composites. After the strain has exceeded the failure strain of the matrix, the complete load has to be borne by the fibres. Similar to the previous case, the maximum stress in the composite depends on the volume fraction of the fibres. If it is sufficiently large, the fibres do not break but can take a load of $f_f \sigma_f$.[5] If the volume fraction is too small, the maximum stress is again determined by the isostrain rule of mixtures, equation (9.7), but now taking σ_m as failure stress of the matrix and σ_f as the stress in the fibre at the failure strain in the matrix. In ceramic matrix composites, the matrix frequently does not fail completely, but forms many small cracks bridged by the fibres. The stress-strain curve for this case will be discussed in section 9.3.3.

9.3.2 Load transfer between matrix and fibre

It was already stated that the considerations of the previous section were simplified. In particular, the assumption of continuous fibres extending throughout the component and loaded directly by the external load is almost never valid. If the fibres are completely embedded within the matrix, the load transfer between matrix and fibre is crucial in determining the strengthening effect. For this reason, we will now discuss this load transfer.

---

[4] As already discussed, the failure stress of the matrix is frequently not the same as it were if no fibres were present, but smaller. In this case, the failure stress at small fibre contents is reduced further. This is not the case for the elastic stiffness.

[5] For this to be true, it has to be assumed that a crack propagating within the matrix does not cause fibre fracture. We will see in section 9.3.3 under which conditions this assumption holds.

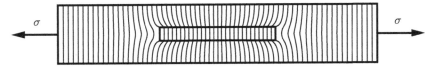

Fig. 9.4. Elastic deformation of the matrix near a fibre under tensile loads. Young's modulus of the fibre has been assumed to be 100 times larger than that of the matrix; the fibre-matrix interface is perfectly bonded and cannot fail. Poisson's ratio of fibre and matrix has been assumed to be the same

If a fibre composite is loaded in tension, the deformation within the material is inhomogeneous. Figure 9.4 plots the elastic deformation of the matrix surrounding a fibre with a Young's modulus that is one hundred times larger. Because the fibre resists the strain more strongly than the matrix, the strain in the matrix increases accordingly. This has two consequences: On the one hand, the strain in the matrix to the left and right of the fibre is larger than besides it. If the total strain is prescribed, the matrix is strained more heavily in this region than the material on average. On the other hand, shear stresses occur near the ends of the fibre at the fibre-matrix interface, increasing the strain in this region. The strain in the matrix of a composite is thus larger than the strain in the fibre and is also larger than the strain in a homogeneous material with the same total deformation. In polymer matrix composites, a rule of thumb is that the maximum strain in the matrix is about twice as high as the global strain of the composite (see also section 9.4.1 and exercise 29).

The load transfer between fibre and matrix is mainly due to friction caused by interface roughness or to adhesion between fibre and matrix on the lateral surface of the fibre. Although the strains at the front and back side of the fibre can be large, only small loads are transferred there because these sides are much smaller. Crucial in determining the properties of the composite is the maximum *interfacial shear stress* τ_i.

The maximum interfacial shear stress is determined by different factors in different composites. In a polymer or ceramic matrix, the adhesion strength between fibre and matrix is the determining factor because it is usually smaller than the yield strength of the polymer or the strength of the ceramic. The adhesion strength is in some cases determined by chemical bonds between fibre and matrix, in others by frictional forces. In a metal matrix composite, the maximum interfacial shear stress is usually determined by the yield strength of the matrix which is smaller than the adhesion strength in most cases. Because the interfacial strength is crucial in determining the strength and toughness of the composite as we will see below, large efforts are usually made to increase or decrease it, depending on the application at hand (one example is discussed in section 9.4.3).

To estimate the stress σ_f within the fibre, we consider an infinitesimal fibre segment with a constant interfacial shear stress $-\tau_i$ acting on its surface (figure 9.5). The forces within the fibre are $-\sigma_f \cdot \pi d^2/4$ at position x and

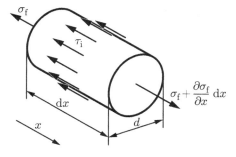

Fig. 9.5. Forces on an infinitesimal fibre segment

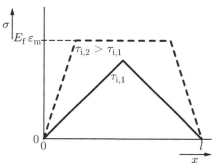

Fig. 9.6. Stress distribution in a fibre for two different values of the interfacial shear stress

$(\sigma_f + \partial\sigma_f/\partial x \cdot dx) \cdot \pi d^2/4$ at position $x + dx$, with d denoting the fibre diameter. The force equilibrium is

$$-\pi\, d\, dx\, \tau_i - \sigma_f \pi \frac{d^2}{4} + \left(\sigma_f + \frac{\partial\sigma_f}{\partial x} dx\right)\pi\frac{d^2}{4} = 0\,.$$

This yields

$$\frac{\partial\sigma_f}{\partial x} = \frac{4}{d}\tau_i\,.$$

The stress changes linearly with position if τ_i is constant.[6]

The maximum stress within the fibre is limited by the strain in the fibre (index 'f') which can never exceed the matrix (index 'm') strain. The maximum possible fibre stress is thus

$$\sigma_{f,max} = E_f\,\varepsilon_m\,.$$

The stress distribution in the fibre is plotted in figure 9.6 for a constant interfacial shear stress τ_i. It has to be noted that in a sufficiently long fibre, there is no interfacial shear stress midway in the fibre.

---

[6] This assumption of a constant interfacial shear stress is a rather good approximation. A more detailed discussion is given in *Chawla* [28].

If we assume that the maximum stress value $E_f \, \varepsilon_m$ is not reached, the maximum stress actually occurring in the fibre is

$$\sigma_{f,\max} = \int_0^{l/2} \frac{4}{d} \tau_i \mathrm{d}x = 2\frac{l}{d} \tau_i \,. \tag{9.8}$$

If we aim at a maximum strengthening effect of the fibres, they have to be sufficiently long to be loaded up to their fracture strength $\sigma_{f,B}$. With the help of equation (9.8), we can thus define the critical fibre length

$$l_c = \frac{d\sigma_{f,B}}{2\tau_i} \,,$$

As can be seen from the equation, the critical fibre length is proportional to the fibre diameter. It is thus not the absolute size of the fibre, but the ratio of its length and diameter (the aspect ratio) that determines its effect.[7] To optimally exploit the fibre strength, they should be longer than the critical length. To achieve this, the maximum interfacial shear strength should be large, as it is the case when chemical bonding between fibre and matrix occurs, and the fibre diameter should be small.

> To estimate the strength of the composite, we can again use the isostrain rule of mixtures, equation (9.7). As the fibres do not extend throughout the component, σ_f is to be taken as the mean stress of the fibres. If the fibre length is equal to the critical length, the maximum stress in the fibre is $\sigma_{f,B}$ and the mean stress is $\sigma_{f,B}/2$.[8] If the fibres are smaller than the critical length, the maximum value of the fibre stress is
>
> $$\sigma_{f,\max} = \frac{2\tau_i l}{d} = \frac{l}{l_c}\sigma_{f,B} \,, \tag{9.9}$$
>
> resulting in a mean fibre stress of $\sigma_{f,\max}/2$. If the fibre length exceeds the critical length, the fibres will break in some places when the load increases, until the fibre fragments have approximately the critical length. These can then bear a maximum stress of $\sigma_{f,B}$. For this reason, it is reasonable to use fibres that exceed the critical length because they can bear a load even after fracture occurs.

9.3.3 Crack propagation in fibre composites

In composites with a brittle matrix i. e., mainly in ceramic matrix composites, the aim is not to increase the strength, but the fracture toughness. The fracture strain of the matrix is usually smaller than that of the fibre, leading to crack propagation in the matrix when the load increases.

---

[7] This is different if creep deformation occurs because in this case there is diffusion of material, see section 11.2.3.

[8] We assume that the fracture strain of the matrix is larger than that of the fibre.

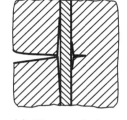

(a) No crack   (b) The crack reaches the   (c) The crack by-passes
                             fibre   the fibre along the inter-
                                                              face

Fig. 9.7. Crack propagation perpendicular to a fibre (after [29]). If the fracture toughness of the interface is sufficiently small, the crack propagates along the interface, causing detachment between fibre and matrix. Thus, the crack can bypass the fibre and propagate further

These cracks propagate within the matrix until they reach a fibre (see figure 9.7). To increase the fracture toughness compared to the pure matrix material, the fibre must not fracture when hit by the crack. This can be achieved if it is not the fibre that fails, but the interface between fibre and matrix, causing a detachment between fibre and matrix as shown in figure 9.7. The interface is perpendicular to the crack and thus parallel to the external tensile load. If we look at the stress state in front of the crack tip (see figure 5.5), we see that there is a tensile component trying to open the interface. If the fracture toughness of the interface is sufficiently small, this tensile stress will cause failure of the interface and a local detachment between fibre and matrix. The crack can propagate along the detached interface and grow further without breaking the fibre. This increase in the crack surface causes an increase in fracture toughness (see also section 7.2.1).

It is crucial for the increase in fracture toughness that the crack is bridged by the fibre after the crack tip has propagated beyond the fibre (see figure 9.8). The fibres can transfer loads and thus hinder further opening of the crack.

The load transfer between the crack surfaces results in a maximum of the stress in the part of the fibre that is situated within the crack. This stress has to be transferred to the matrix on both sides, in a region whose size is approximately that of the critical fibre length.[9] If the stress in the fibre in this region exceeds the fracture stress, the fibre breaks. Fracture usually occurs at a defect, for example a surface defect or a local reduction in diameter. Because of this, fibre fracture occurs not always directly at the crack surface, but at an arbitrary position between the stress maximum near the crack surface and the region where the fibre stress has decreased markedly. The size of this region is

---

[9] In the previous section, we discussed load transfer from matrix to fibre, here we take the opposite point of view, transferring the load from fibre to matrix. Both cases are equivalent.

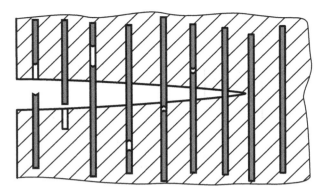

Fig. 9.8. Crack in a fibre composite bridged by fibres

determined by the critical length; the mean distance of the fracture position in the fibre to the crack surface is thus proportional to the critical length.

After the fibre has broken, the fragment still remaining in the matrix that bridges the crack is shorter than the critical length. The stress in the fibre thus decreases. Therefore, the fibre will not break again, but will be *pulled out* of the matrix, doing work against the shear stress τ_i. So far, we have assumed that the fibre is longer than the critical length. If this is not the case, the fibre will not break, but pull-out will occur immediately. Therefore, there is always pull-out, regardless of the fibre length.

The work done on pull-out increases the crack resistance because it impedes crack propagation. Short cracks do not benefit from this because the crack has to grow by some multiple of the fibre distance before a process zone with crack-bridging fibres can form. If the crack has grown a large distance, the fracture toughness approaches a constant value because for each fibre entering the process zone another fibre leaves it. This mechanism is thus one example for the increase in crack-growth resistance on crack propagation, discussed in sections 5.2.5 and 7.2.5.

The crack resistance is the higher, the larger the critical fibre length is. In a ceramic matrix composite, it is thus useful to have a low value of the interfacial shear stress. If fibres are shorter than the critical length, they will not break but will be pulled out on one side of the crack, if they are longer, they will fracture first and be pulled out afterwards.

The energy dissipated on pull-out can be estimated as follows: The force needed to pull out the fibre is

$$F(x) = \tau_i x \pi d$$

if the segment of the fibre remaining in the matrix has a length x (see figure 9.9). From this, we can calculate the energy required to pull out the fibre from the matrix by a distance l' on one side of the crack:

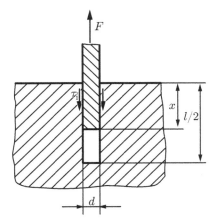

Fig. 9.9. Pull-Out of a fibre from the matrix

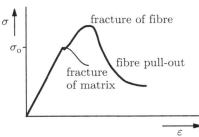

Fig. 9.10. Schematic stress-strain diagram of a fibre-reinforced ceramic (after [29])

$$W_{\mathrm{f},l'} = \int_0^{l'} F\mathrm{d}x = \int_0^{l'} \tau_\mathrm{i} x \pi d \mathrm{d}x = \frac{1}{2}\pi d\tau_\mathrm{i} l'^2 \;.$$

In a simple approximation, we can assume that the pull-out length varies between 0 and half of the critical length l_c. The mean energy dissipation per fibre is thus

$$W_\mathrm{f} = \frac{1}{l_\mathrm{c}/2} \int_0^{l_\mathrm{c}/2} \frac{1}{2}\pi d\tau_\mathrm{i} l'^2 \mathrm{d}l' = \frac{1}{24}\pi d\tau_\mathrm{i} l_\mathrm{c}^2 \;.$$

The fracture toughness of a fibre composite is not determined by the dissipation of one single fibre, but by the total dissipation. To estimate this, we have to take into account how many fibres bridge the crack and can dissipate energy by pull-out. Their number is, if the volume fraction is constant, inversely proportional to the square of the fibre diameter. Using this, the total energy dissipation in the composite is proportional to $\tau_\mathrm{i} l_\mathrm{c}^2/d$.

Figure 9.10 schematically shows the stress-strain diagram of a ceramic matrix composite. First cracks in the matrix occur at a stress of σ_0. The load can be increased beyond that because the bridging fibres can bear larger loads, until they finally fracture.

9.3.4 Statistics of composite failure

So far, we considered one fibre of the composite only, assuming it to be representative of all fibres. However, this implies that all fibres have the same properties.

In reality, fibre properties are statistically distributed. This is true for their geometry (length and diameter), but also, especially in the case of ceramic fibres, for their mechanical properties that are distributed according to Weibull statistics (see section 7.3). Non-ceramic fibres are also usually not identical since they may contain surface defects, for instance. Because of this statistical distribution of their properties, not all fibres fail simultaneously even in a homogeneously loaded composite. Instead, the weakest fibre will fail first. Due to the volume effect (see section 7.3.1), the failure probability of a long fibre is greater than that of a short one.

In the following, we consider the case of long fibres with a length several times larger than the critical length (see equation (9.9)). In this case, the fibre is loaded in tension over most of its length, for load transfer occurs only near its end points (see figure 9.6). The fibre will thus fail by fracture.

If the load on the composite is increased, the weakest fibre will break and will thus not transfer any tensile stresses at the position of failure. This fracture, however, will not unload the whole fibre. If it is much longer that the critical length, the load will be transferred by interfacial shear stresses from the matrix to both fibre fragments. At some distance from this region, both fibre fragments bear the same load as before. Near to the fracture position, the material is weakened and the load is transferred to the surrounding material.

If the fracture toughness of the matrix is low, this increase in stress can cause local failure of the matrix, initiating a crack that propagates from the site of fibre fracture. Because fibre properties are statistically distributed, the crack will usually not cause the next fibre it encounters to fracture and will be stopped there. The increased load is thus distributed to the surrounding fibres.

If the load is increased further, the failure behaviour depends mainly on the fracture toughness of the matrix and the properties of the interface. If the matrix is brittle and the fracture toughness of the interface is large, the stress concentration in front of the crack tip is transferred to the fibre, causing it to break. In this case, the crack propagates on load increase, starting from the site of first fibre fracture. If, on the other hand, the stress concentration in front of the crack tip is not sufficient to cause fibre fracture, another weak fibre somewhere else in the material will fail first, at a position that is completely independent of the first failure position. Thus, fibres will fracture at arbitrary positions in the material, and the load on the material will increase homogeneously, with a decrease in stiffness due to the damaged regions. In this case, the material will fail by a growing number of breaking fibres, eventually failing completely. Typically, the stress-strain curve for a material with a matrix with sufficiently large fracture toughness is similar to that shown in figure 9.3, with the only difference that there is no distinct kink in the curve because the fibres do not fail simultaneously.

Because fibre composites frequently fail in this statistical manner by accumulating local damage, the methods of fracture mechanics are often not too useful. If, on the other hand, a sufficiently long crack in a fibre composite forms, it may propagate. In this case, the fracture toughness K_{Ic} of composites with ductile matrix is often smaller than in the pure matrix material because the fibres cause the stress state to be triaxial (see section 3.5.3). This happens in some polymer matrix composites, but mostly in metal matrix composites in which the fracture toughness may be halved compared to the matrix material [62].

9.3.5 Failure under compressive loads

If a fibre composite is loaded in compression in fibre direction, the deformation mechanism is completely different from the failure behaviour discussed so far. In many fibre composites, the compressive strength is smaller than the tensile strength, a fact that has to be taken into account when designing with these materials. Because the fibres are long compared to their diameter, they may buckle. The buckling load of a cylinder with Young's modulus E loaded in compression is – assuming Euler's case 2 of buckling [18] – determined by

$$\sigma_{\mathrm{b}} = \frac{\pi^2 E}{16} \left(\frac{d}{l}\right)^2 , \tag{9.10}$$

with d and l denoting diameter and length of the fibre [29]. Even in short-fibre reinforced composites with typical fibre lengths of a few millimetres, we usually find $l/d > 100$. If we consider the example of a glass fibre with Young's modulus of $80\,\mathrm{GPa}$ and $l/d = 100$, we find in the ideal case of a perfectly straight fibre a buckling strength of only $5\,\mathrm{MPa}$. Without the presence of the matrix, the compressive strength of the material would thus be vanishingly small.

Buckling of the fibres is impeded by the matrix material that has to deform also when the fibres buckle. A single fibre does not form a single large buckle, but buckles in a sine-shaped wave pattern, keeping the deformation of the matrix smaller. In a fibre composite, the fibres are usually so close to each other that neighbouring fibres cannot deform independently. There are two different deformations patterns, sketched in figure 9.11: Neighbouring fibres may deform either in phase or out of phase.

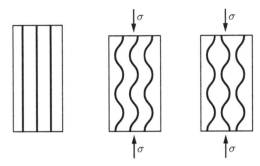

Fig. 9.11. Deformation of a fibre composite under compressive stress. The fibres can bend in an in-phase or out-of-phase pattern

The stress required to form these patterns can be calculated using an energy balance: The energy to compress the material without buckling is compared to that needed for the buckling modes. At small stresses, a homogeneous compression needs less energy, but starting from a certain critical stress value, it is easier to let the fibres buckle than to homogeneously compress the material further. This critical stress is the compressive stress of the material. It is different for the two deformation patterns.

In the out-of-phase deformation mode, the matrix is loaded in tension and compression, in the in-phase mode, it is sheared. Because of this, the modes are sometimes called extension mode and shear mode. Except at small volume fractions of the fibre, the strength of the composite is smaller in in-phase deformation which is thus the mode of interest. If a purely elastic deformation of the matrix is assumed, the calculated strength values for the composite are very large, but the observed values are usually much smaller. In metal and polymer matrix composites, the matrix deforms plastically in the in-phase mode. If we make the simplifying assumption that the matrix is perfectly plastic with a yield strength of $\sigma_{m,F}$, the compressive strength is [122]

$$R_{c,\text{in phase}} = \sqrt{\frac{f_f \sigma_{m,F} E_f}{3(1 - f_f)}} \,. \tag{9.11}$$

This equation is valid only within certain limits. If the volume fraction of the fibres approaches one, the calculated strength becomes infinite, which is obviously not realistic. If the deformation of the matrix is not determined by plastic deformation alone, its Young's modulus also plays a role. Further effects that are not considered in the equation and which may reduce the compressive strength are the fibre orientation, the limited interfacial strength between fibre and matrix, and the possibility that the fibres deform and fail not by buckling, but by kinking. The compressive stress calculated with the given equation is independent of the fibre diameter and the fibre length. In reality, longer and thicker fibres are advantageous because it is easier to align them during processing of the material.

9.3.6 Matrix-dominated failure and arbitrary loads

If a composite with unidirectional fibres is loaded in tension or compression perpendicular to the fibre direction or in axial shear in fibre direction, it can fail without failure of the fibres by fracture, buckling, or kinking. These cases are therefore called *matrix-dominated failure.*

In tensile load perpendicular to the fibres, the strengthening effect of the fibres is small. If their elastic stiffness is larger than that of the matrix, the fibres constrain the transversal contraction of the matrix and cause a triaxial stress state. This may, in a metal matrix composite, for example, shift the yield strength to higher loads. If the matrix is brittle, the triaxiality may facilitate crack formation. If the volume fraction of the fibres is large, the matrix between the fibres has to deform more strongly. The exact arrangement of the fibres plays an important role here, for it determines the geometrically necessary deformation of the matrix.

Under compressive loads perpendicular to the fibre direction, the matrix may shear on planes parallel to the fibres. In this case, the fibres are irrelevant for the compressive strength. Shearing on planes cut by the fibres is not possible because the fibres impede this. If shear occurs in the direction of the fibres, either the matrix itself can shear between the fibres or there may be shearing along the interface. The strengthening effect of the fibres is small in the latter case as well. If the interface is weak, the strength of the composite may even be smaller than that of the pure matrix material [122].

To design components made of fibre composites, for example using the finite element method [15, 63], it is useful to know yield or failure criteria for the composite as a whole that can be evaluated for arbitrary stress states. Several such criteria have been suggested, but all of them are of limited applicability [29, 72, 122].

9.4 Examples of composites

9.4.1 Polymer matrix composites

Polymers are well-suited as matrix materials due to their low density and their low processing temperatures. Accordingly, composites with a polymer matrix are of extreme technical importance. They are indispensable in aerospace industry and many other areas, for example in sports equipment. Polymer matrix composites can be used with long and short fibres. We will start this section by discussing long-fibre polymer matrix composites and then study short-fibred ones.

Long-fibre reinforced polymer matrix composites

Because the strength and elastic stiffness of the fibres used in polymer matrix composites is frequently more than a hundred times larger than that of the

Table 9.1. Density and mechanical parameters (Young's modulus, tensile strength, fracture strain) of some important fibre materials [29, 41, 100, 117, 131, 141]

material	$\varrho/(\mathrm{g/cm^3})$	E/GPa	$R_\mathrm{m}/\mathrm{MPa}$	$\varepsilon_\mathrm{B}/-$
glass fibre	2.5 . . . 2.6	69 . . . 85	1 500 . . . 4 800	1.8 . . . 5.3
aramid fibre	1.4 . . . 1.5	65 . . . 147	2 400 . . . 3 600	1.5 . . . 4.0
polyethylene fibre	0.97	62 . . . 175	2 200 . . . 3 500	2.7 . . . 4.4
carbon fibre	1.75 . . . 2.2	140 . . . 820	1 400 . . . 7 000	0.2 . . . 2.4
silicon carbide fibre	2.4 . . . 3.5	180 . . . 430	2 000 . . . 3 700	1.0 . . . 1.5
aluminium oxide fibre	3.3 . . . 3.95	300 . . . 380	1 400 . . . 2 000	0.4 . . . 1.5

polymer matrix, the mechanical properties of polymer matrix composites are mainly determined by the fibre properties. For this reason, the highest possible fibre volume fractions are aimed at, with maximum values in aerospace industry of about 60 %. Nevertheless, the mechanical behaviour of the matrix is also important because it determines load transfer to the fibres and it must not fail if the strength of the fibres is to be exploited fully. Accordingly, we will start this section by discussing the mechanical behaviour of fibres and derive the requirements on the matrix material from this. Finally, the composite properties are discussed.

The fibres

Table 9.1 contains a survey of some mechanical parameters of commonly used fibre materials. Because glass fibres can have a very high strength of up to 4800 MPa and can also be manufactured inexpensively, it is easy to understand why they are widespread. Their Young's modulus is rather low, with values comparable to that of aluminium. It can be increased somewhat by changing the composition of the glass. However, the Si-O bond is less strong than a C-C bond, and the density of bonds in an amorphous material is always smaller than in a crystalline one. This explains why glass fibres cannot be as stiff as carbon fibres. Accordingly, glass fibres are a reasonable choice if the strength of the composite is the main design variable, but they are less useful for applications requiring a high stiffness.

Carbon fibres are characterised by a high stiffness and strength. However, both parameters cannot be maximised simultaneously. Figure 9.12 plots the tensile strength and Young's modulus of several carbon fibres. In high-strength fibres, Young's modulus does not exceed 400 GPa, in high-stiffness fibres, the tensile strength is reduced.

This variation in the mechanical properties is due to the fibre microstructure. There are two different structures (so-called 'allotropes') of carbon: The diamond structure, shown in figure 1.13, only forms at high temperatures and pressures and is in fact metastable at room temperature. The stable conformation of carbon is *graphite*. In graphite, the carbon atoms are ordered in a

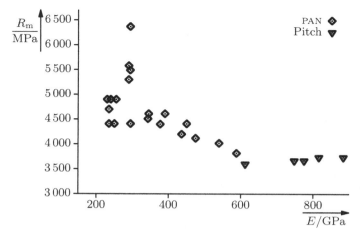

Fig. 9.12. Mechanical properties of technically used carbon fibres from different suppliers [56, 100, 134, 141]. The two types of fibre differ in their manufacturing process

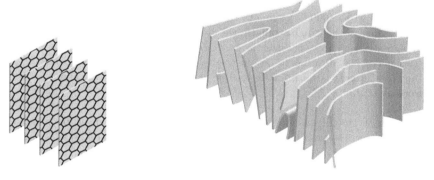

(a) Basal planes in graphite

(b) Arrangement of the basal planes in high-strength carbon fibre

Fig. 9.13. The basal planes of graphite are arranged in parallel to the fibre axis in carbon fibres. In high-strength fibres, the different regions are connected, rendering slip of the planes past each other more difficult (after [29, 97])

hexagonal lattice. The bonds within the hexagonal planes are strong, those between the planes are much weaker (see figure 9.13(a)). The sheets or layer planes can easily slide apart, explaining why it is possible to draw pictures with charcoal sticks.

The microstructure of the high-stiffness carbon fibres is similar to that sketched in figure 9.13(a), with the sheets arranged almost perfectly along the fibre axis. Because the covalent C-C bonds within the sheets are extremely strong, a large Young's modulus in fibre direction results. A strong fibre texture is thus key to the large elastic stiffness.

However, the problem with this microstructure is that the basal planes are only weakly bonded to each other because the bond strength between them is small. Accordingly, the stiffness transversally to the fibre direction is very low (about 6 GPa). Furthermore, this reduces the fibre strength and the interfacial strength between fibre and matrix. To achieve maximal strength, a microstructure is used where the sheets are interwoven, with cross-links between the sheets hampering shearing (see figure 9.13(b)). Because the sheets are oriented obliquely to the fibre axis in this configuration, the stiffness is reduced. Carbon fibres thus have to be optimised either for strength or for stiffness.

> These two microstructures are produced in two different processes. One process starts with polymer fibres, usually made of polyacrylonitrile (PAN). The other process uses pitch produced during refinement of mineral oil. Accordingly, the fibres are called PAN fibres, with high strength, and pitch fibres, with high elastic stiffness (figure 9.12). Although carbon fibres can be rather cheap at 25 €/kg, high-performance fibres can cost as much as 1000 €/kg due to the involved production process.
>
> Because of their high strength, the energy absorption until fracture of high-strength fibres is rather large. For example, a metal with a yield strength of 700 MPa has to be plastically deformed by 10 % to achieve the same energy absorption as a fibre with $R_m = 7000$ MPa and a fracture strain of 2 %.[10]

The strength of the fibres is also determined by their diameter because a thinner fibre contains smaller defects. To achieve a strength of 2000 MPa, a diameter of 10 µm is required, which has to be reduced to 5 µm for a strength of $R_m = 6000$ MPa.

Reducing the fibre diameter has some disadvantages as well. It eases buckling or kinking of the fibres, so that the shear or compressive strength of the composite does not increase as much as the tensile strength does or may even decrease. This limits the applicability of thin fibres.

A further important point is that the fracture strain of high-strength carbon fibres is about 2 % although they deform only elastically. Considering that the strains in the polymer matrix locally exceeds that of the fibre (see figure 9.4), we see that the fracture strain in the matrix has to be rather large. To avoid crack formation in the matrix, its fracture strain should be about twice that of the fibre i.e., 4 % to 5 %. Currently available duromers do not

---

[10] To arrive at this number, it has to be kept in mind that a perfectly plastic material can absorb twice as much energy as a linear-elastic material at identical maximum stress and strain.

meet this requirement, reducing the permissible strain. Thus, the full strength of the fibres can often not be exploited (see also exercise 29).

Polymers can also be strengthened using polymer fibres. As already explained in section 8.4, high strength polymer fibres can be produced by drawing the chain molecules in fibre direction (see figure 8.20 and section 8.5.2). Commonly used fibres are based on aramid or polyethylene. As the density of carbon bonds can never be as high as that in carbon fibres because of the side groups, it is easily understood that the mechanical properties of polyethylene fibres are inferior to that of carbon fibres.

Polymer fibres are viscoelastic even at room temperature. Strength and stiffness are time- and temperature-dependent, a fact that has to be taken into account in the design process. In glass fibres, this is the case only at temperatures of about 200°C, well beyond the service temperature of polymer matrix composites. Carbon fibres are even more stable. Time-dependent behaviour causes a hysteresis between applied load and observed stress that is especially important under cyclic loading (see section 10.4).

The matrix

Although most of the mechanical load is borne by the fibres, there are still several requirements for the mechanical properties of the matrix. Its fracture strain should be sufficiently large to avoid premature damage of the composite by crack formation in the matrix. Its elastic stiffness should be as large as possible to achieve a sufficient support of the fibres under compressive loads and to avoid buckling or kinking of the fibres. Finally, its mechanical behaviour should remain unchanged under different environmental conditions (humidity, temperature, irradiation). Unfortunately, these requirements are partially contradictory. The fracture strain of a duromer matrix, for example, can be increased by decreasing the cross-linking density. This, however, reduces the elastic stiffness. Large fracture strains can also be achieved by using thermoplastic matrices which are considered for aerospace applications for this reason. However, they are less temperature-resistant than duromers and are more difficult to manufacture because they cannot be produced by curing a resin and thus have to be processed at higher temperatures.

Depending on the application, different matrix materials are used. Among the duromers, most common are polyester and epoxy resins. Thermoplastic matrix materials are polyethylene (PE) and polypropylene (PP), but the use of thermoplastics with aromatic rings on the chain and thus with increased temperature stability also grows. One example is polyetheretherketone (PEEK), characterised by high toughness and a glass temperature of about 150°C.

Composite properties

It was already stressed that the properties of fibre and matrix have to be carefully adjusted to obtain optimal properties of the component under mechanical loads. Under tensile loads, the fracture strain of the matrix has to be

Table 9.2. Increase of Young's modulus and tensile strength of a duromer matrix (polyester resin) by addition of glass fibres with a volume fraction of 65% to 70% [77]

type of fibre	E/GPa	$R_\mathrm{m}/\mathrm{MPa}$
none	3.5	90
short fibres, irregular	20	190
short fibre, oriented at $\pm 7°$	35	520
continuous fibres, uniaxial	38	1 300

sufficient for the chosen fibre material. Although cracks in the matrix do not reduce the strength of the component significantly, they can cause consequential damage by penetration of water or other media. In applications with high safety requirements, for example in aerospace industry, the permitted total strain of the composite is limited to a value well below the fracture strain of the fibres for this reason. Because duromer matrix composites are viscoelastic and have no plastic regime, this reduces the permitted stress accordingly. If, for example, the permitted strain is limited to half of the fracture strain, only 50% of the fracture strength can be exploited. This limitation is a crucial reason for the high interest in matrix materials with large fracture strain and temperature stability.

Humidity also has a strong influence on the composite's mechanical behaviour because it changes the properties of the matrix as already discussed in section 8.8. The strength of the matrix decreases whereas its failure strain increases with increasing water content. Some residual humidity can therefore be advantageous in composites with a duromer matrix. Glass or carbon fibres do not absorb any water. If the polymer matrix swells, large residual stresses can be generated. This can also happen in polymer fibres. Aramid fibres, for example, do absorb water, but due to their anisotropic microstructure, they swell mainly in radial direction, also causing large residual stresses.

Short-fibre reinforced polymer matrix composites

The strength and stiffness that can be obtained in short-fibre reinforced polymer matrix composites are well below that of long-fibre reinforced materials. Depending on the chosen processing route, the fibres can be oriented in loading direction or irregularly (see section 9.1.1).

The influence of the fibre direction on the mechanical properties can be seen from table 9.2 for the example of a glass-fibre reinforced duromer matrix. Young's modulus is strongly increased even when irregularly oriented fibres are added. Directing the fibres further increases the stiffness. Using continuous instead of directed short fibres has no significant effect.

Relations are different concerning the tensile strength: Although irregularly oriented short fibres significantly increase the tensile strength, their effect is much smaller than that of directed fibres. The strength further increases by more than a factor of two when continuous fibres are used because the length of the short fibres is below the critical length.[11] Even if the fibres are larger than the critical length, it is experimentally observed that a further increase in fibre length increases the tensile strength [122] because local weak points, caused by irregularities in the fibre distribution, determine the tensile strength.

Mechanically, it is thus best to use fibres that are as long as possible. This, however, is limited by processing technology. For example, long fibres may break or clog the nozzles in injection moulding. Processing technology also limits the volume fraction of short fibres, usually to values that are smaller than in long-fibre reinforced composites.

The same materials can be used as in long-fibre reinforced polymer matrix composites. Short-fibre reinforced polymers are useful in many applications where unreinforced polymers are not sufficient. The design of injection moulded components made of short-fibre reinforced polymers is complicated by the fact that the orientation of the fibre is determined by the fluid flow (see section 9.1.1) and can be irregular within the material.

9.4.2 Metal matrix composites

Metals are especially attractive as matrix material in a composite. As the fracture strain of the matrix is larger than that of common fibre materials, the fibre strength can be fully exploited, and the local strain concentration near the interface (see section 9.3.2) is irrelevant for the composite strength. Since the adhesion between fibre and matrix is frequently strong in metal matrix composites, the maximum interfacial shear stress is usually limited by the metal's yield strength and is correspondingly large. The critical fibre length is thus small and even short fibres result in a high strengthening effect. The large Young's modulus and yield strength of the matrix also lead to a high compressive strength because bending or kinking of the strengthening fibres is avoided (see section 9.3.5). Metal matrix composites can be used at higher temperatures than polymer matrix composites because the temperature stability of the matrix is larger.

The fibres determine the mechanical properties of the composite not only by load transfer, but also by additional effects: Strengthening particles or fibres can pin grain boundaries during processing of the material and thus reduce grain size. This increases the strength by grain boundary strengthening (see section 6.4.2) at low temperatures. The dislocation density can also be increased by adding fibres: If the composite is cooled from the required high processing temperatures, differences in the coefficient of thermal expansion can cause plastic deformation in the vicinity of the fibre. This increases the strength, but also causes residual stresses which may reduce the strength.

---

[11] The maximum interfacial shear stress in polymer matrix composites is determined by the adhesion between fibre and matrix, not by the yield strength of the matrix.

A further increase in dislocation density occurs during plastic deformation because plastic deformation is usually limited to the matrix, leading to a formation of dislocation loops around the fibres (see also section 6.4.4). The Orowan mechanism (see section 6.3.1 and figure 6.45), which would impede dislocation movement, is not relevant, though, because the fibre diameter and distance are too large.

Fibre materials in metal matrix composites are limited to those with a sufficiently high melting temperature because they have to withstand high processing temperatures. Possible materials are carbon, ceramics (for example aluminium oxide or silicon carbide), and high-melting point metals like boron or tungsten. Suitable matrix materials are mainly the light metals aluminium, titanium, and magnesium.

Aluminium is the most frequently used matrix material due to its rather low melting point (depending on the alloy, about 600°C to 660°C) which eases the processing, but also because of its high ductility. In applications, it is not only the strengthening, but also the increase in stiffness that is attractive since Young's modulus of aluminium is rather low (approximately 70 GPa). Adding Al_2O_3 long fibres with a volume fraction of 50% increases its value to 200 GPa [121]; by using carbon fibres, a stiffness of 400 GPa can be achieved [54].

As expected, long-fibre reinforced materials have the best mechanical properties, but are very expensive to produce. The strength values that can be achieved are impressive. For example, an aluminium matrix composite strengthened with continuous silicon carbide fibres can have a room temperature tensile strength of more than 1400 MPa, which even at a temperature of 425°C decreases only to 1050 MPa [49]. If titanium is used as matrix material instead, the strength at room temperature does not increase much because it is determined by the fibre material. However, due to the high melting point of titanium, the material can be used at higher temperatures and the tensile strength at 600°C is still about 1000 MPa [49].

Due to their high specific strength and stiffness, long-fibre reinforced aluminium matrix composites are attractive in aerospace applications. The high-gain antenna boom of the Hubble Space Telescope, for example, is made from a carbon-fibre reinforced aluminium matrix composite [114]. Aluminium oxide reinforced aluminium matrix composites are also suitable for push rods in motorcycle engines and for electrically conductive and mechanically loaded connectors on power poles [1].

Short-fibre reinforced metal matrix composites are significantly less expensive than long-fibre reinforced materials and can thus be used in automotive engineering or in sports equipment. For example, short-fibre reinforced aluminium-silicon carbide composites can be used as pistons in diesel engines at elevated temperatures [49]. Golf clubs and bicycle components can also be manufactured from aluminium matrix composites. Frequently, whiskers (see section 6.2.8) are used as short fibres because of their high strength and favourable aspect ratio.

The stiffness and strength of metals can be increased not only by adding fibres, but also using particles. In contrast to fibres, load is transferred also at the front and back end of the particle, not only by shear stresses. In an aluminium-silicon carbide composite, for example, the tensile strength can be as high as 700 MPa.

Metal matrix composites can be interesting due to other properties as well: The coefficient of thermal expansion of a metal can be strongly reduced by adding carbon fibres and may even become negative.[12] This is important if the component may not distort on thermal loading or when the material has to be joined to a ceramic because the coefficient of thermal expansion of ceramics is usually much smaller than that of metals (see section 2.6). The thermal properties are also of interest in copper-carbon composites because copper has a large thermal conductivity, but is mechanically rather weak. Carbon is especially suited as fibre material not only due to its stiffness and strength, but also because of its high thermal conductivity that may even exceed that of copper.[13]

9.4.3 Ceramic matrix composites

As we saw in chapter 7, ceramics have the attractive properties of high temperature resistance, high strength and stiffness, low density, and high resistance against many aggressive media. Their main disadvantage is their low fracture toughness and the resulting sensitivity to small defects. The main objective in strengthening ceramics with fibres is thus to increase the fracture toughness. It can take values of up to 30 MPa$\sqrt{\mathrm{m}}$ [25, 149], approximately ten times larger than in most unreinforced ceramics. Furthermore, using a fibre composite can also increase the Weibull modulus to about 30, reducing the scatter of strength and thus easing component design.

Suitable fibre materials in ceramic matrix composites are ceramics (for example aluminium oxide or silicon carbide), carbon, and high-melting point metals like boron or tungsten (see table 9.1). In short-fibre reinforced ceramics, whiskers are commonly used because longer irregular short fibres may decrease the tensile strength, though they increase the fracture toughness [25]. The most frequently used matrix materials are aluminium oxide, silicon carbide, or silicon nitride.

Because the increase in fracture toughness is the main objective of using ceramic matrix composites, a pull-out of the fibres must be favoured instead of fibre fracture (see section 9.3.3). The strength of the interface between fibre and matrix thus must not be too large to avoid fibre fracture. On the other hand, it must be strong enough to enable load transfer to the fibre and to

---

[12] This is possible because carbon fibres have a negative coefficient of thermal expansion in fibre direction. The coefficient in the transversal direction is positive.

[13] This is due to the electrons in the basal planes of the graphite, which are highly mobile, similar to those in a metallic bond.

ensure a sufficient energy dissipation during pull-out of the fibres. Chemical bonding between fibre and matrix is therefore usually not desired because it would produce a high-strength interface. Fibre and matrix material thus have to be adjusted to ensure that no chemical reactions occur even at the rather high processing temperatures required.

To design the interfacial properties, the fibres can be coated before the composite is produced by applying thin coatings with a thickness between 0.1 μm and 1 μm. A graphite layer of 1 μm thickness on a fibre based on silicon carbide (called Nicalon), for example, can reduce the interfacial shear strength from 400 MPa to 100 MPa [28].

Furthermore, care has to be taken to ensure a smooth surface of the fibre. Even without chemical bonding between fibre and matrix, a rough surface may impede the pull-out of the fibre by mechanical clamping in the matrix [28].

The coefficient of thermal expansion of fibre and matrix should also not be too different to avoid large thermal stresses during cooling from the processing temperature. Especially problematic is the case of the coefficient of thermal expansion of the matrix being larger than that of the fibre, for the matrix will then shrink onto the fibre and mechanically clamp it, making pull-out difficult. If, on the other hand, the coefficient of thermal expansion of the fibre is larger, the matrix will be under compressive stress in axial fibre direction. This can be advantageous because it impedes the propagation of cracks, as long as the stresses in the fibre do not become too large. To avoid local thermal stresses, a coating interlayer between fibre and matrix may be helpful.

In ceramic matrix composites, the fracture strain of the matrix is usually smaller than that of the fibre, resulting in the matrix to fail first. The stress-strain diagram (figure 9.10) is more similar to that of a material with an apparent yield point (figure 3.5(b)) than to that of a standard ceramic. To design with the composite, the fracture strength of the matrix can therefore safely be used to determine the maximum permissible stress in the component because no catastrophic failure will ensue if the load is exceeded. The composite thus has a higher failure tolerance.

Due to the excellent high-temperature properties of ceramics, ceramic matrix composites are mainly used in aerospace industry and in power engineering. For example, components for gas turbines, rocket engines, or heat shields (e. g., in the Space Shuttle) can be made of ceramic matrix composites. They may also be used in brake discs in aeroplanes or in upmarket cars. One example are the brake discs of the Boeing 767, manufactured from a carbon-carbon composite. Compared to a conventional brake disc, the mass could be reduced by almost 40 % [28].

If market volume is taken as a measure, the most important application of ceramic matrix composites are cutting tools made of SiC-whisker reinforced aluminium oxide for cutting of hard-to-machine materials, especially nickel-base superalloys and hardened steels [25]. Compared to tungsten carbide reinforced hard metals, their wear and temperature resistance is larger. In machining steels, one problem is that carbon may diffuse from the silicon carbide into the steel, causing eventual failure of the tool.

* 9.4.4 Biological composites

Composites are frequently used by organisms in nature to meet the requirements of the environment. In this section, we will discuss three naturally occurring composites.

Different from most man-made materials, biological materials are often characterised by their water content. The mechanical properties of wood or bone in the natural i. e., humid, state are vastly different from that of the dried materials. This requires some effort in testing biological materials because it is difficult to control the water content in the laboratory with sufficient precision.

* Wood

Wood is made of plant cells elongated in the axial direction of the tree or branch. The mechanical properties of wood are determined by the cell walls which are a composite of a natural polymer matrix with cellulose fibres [9, 144]. *Cellulose* is a polysaccharide, a chain molecule with sugar molecules as monomers.[14] The cellulose molecules have a degree of polymerisation of about 10^4 and are arranged in microfibrils with a diameter of 10 nm to 20 nm, with a high crystallinity. The bonds between the cellulose molecules are hydrogen bonds and are very strong due to the ordered structure in the crystalline regions. Up to now, Young's modulus of the crystalline regions can only be estimated theoretically, taking a value of about 250 GPa, whereas the modulus of the amorphous regions is about 50 GPa. The surrounding matrix comprises an amorphous phenylpropanol duromer, called *lignin,* hemicellulose, a short-chained cellulose variant, water, oils, and salts. The volume fraction of cellulose in wood is about 45 %, the lignin and hemicellulose content is about 20 % each.

The cellulose fibres are situated in the cell walls of long, tube-shaped cells, directed in the axial direction of the tree. Within the cell walls, they are arranged in different layers (see figure 9.14). The outer, primary cell wall, contains irregularly arranged fibres. The next layer, the secondary cell wall, consists of three layers. The cellulose fibres in the outer and inner layer are oriented transversally to the cell direction (and the main loading direction), in the medial layer of the secondary cell wall, they are arranged helically, slightly inclined to the longitudinal direction. This helical arrangement of the fibres in the medial cell wall increases the strength because, under tensile loads, the fibres are straightened and have to slide against each other. This is similar to the carbon fibres discussed above, where the non-perfect alignment in loading direction also serves to increase the strength (see section 9.4.1, page

---

[14] There is another polysaccharide, chitin, that is used as 'engineering material' in nature. Most biological polymers, however, are proteins.

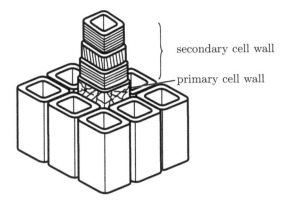

secondary cell wall

primary cell wall

Fig. 9.14. Arrangement of cellulose fibres in the cell wall of wood cells. The diameter of the cells lies between 20 µm and 40 µm, their length between 2 mm and 4 mm. Simplified illustration after [9, 144]

316). Nevertheless, it is much easier to split wood parallel to the fibre than transversally because the crack can run between the cells.

As all fibres, cellulose fibres can bear higher loads in tension than in compression because the fibres can buckle or kink under compressive loads. The arrangement of the cellulose fibres in the outer and inner layer of the secondary cell wall ensures that they are loaded in tension if the wood as a whole is loaded in compression and thus increase the compressive strength. The compressive strength of wood, however, is about 30 MPa, approximately one third of its tensile strength.

The elastic stiffness of wood is much smaller than the theoretical stiffness of single cellulose fibres. This is due to the orientation of the fibres which differs from the loading direction as explained above, but also to the volume fraction of the cell walls which comprise only about 25 % of the total volume. Young's modulus of wood is thus only about 10 GPa in longitudinal direction. Although this is a rather low value, wood is an attractive material for lightweight applications because its density is rather small (with values between $0.2\,\mathrm{g/cm^3}$ and $1.4\,\mathrm{g/cm^3}$). The anisotropy of wood can be avoided by using plywood or flake boards.

> Trees can react to external stresses by adapting their growth. If a tree is loaded asymmetrically (in bending), for example by wind loads or due to growth on inclined ground, it will form so-called reaction wood. In softwoods (as found in conifers), the reaction wood forms on the side that is under compressive loads, in hardwoods on the tensile side. This reaction wood creates residual (compressive or tensile) stresses that tend to straighten the tree [96].
>
> Even in a straightly grown tree, the wood is pre-stressed: In the centre of the tree, stresses are compressive, near the bark, they are

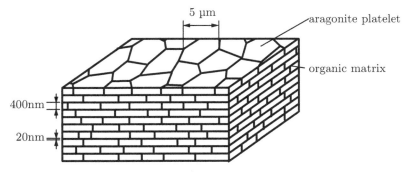

Fig. 9.15. Structure of mother-of-pearl (nacre). Flat aragonite platelets are stacked in a staggered way. The organic matrix lies between the platelets (after [145])

tensile. This has the advantage that these stresses are superimposed to external stresses under bending (for example due to wind loads). The residual stresses thus increase the tensile and decrease the compressive stresses. Because the compressive strength of wood is smaller than the tensile strength, this results in a higher load capacity of the tree.

∗ Nacre

Bivalves, snails, and cephalopods, biologically united as molluscs, often protect themselves with hard shells. These shells are a composite, comprising an organic matrix with included ceramic particles, with a particle volume fraction of 95 % or even more [144].

Because of the high ceramic volume fraction, the mechanical properties of the shells are mainly determined by those of the ceramic. The ceramic component is aragonite, a rhombic crystal modification of calcium carbonate $CaCO_3$, forming prismatic crystals. Young's modulus of aragonite is approximately 100 GPa, its fracture toughness is rather low, with a value of about $0.5\,\mathrm{MPa}\sqrt{\mathrm{m}}$.

There are different shell microstructures in different species. In this section, we only discuss the so-called *nacre* or *mother-of-pearl structure,* found, for example, in the pearl oyster. In nacre, the ceramic takes the shape of polygonal aragonite platelets with a diameter of approximately 5 μm (see figure 9.15). The thickness of the aragonite platelets is only 400 nm. The matrix in between the platelets is organic and is very thin, with a typical thickness of only 20 nm.

The mechanical properties of nacre are highly anisotropic due to the layered structure. If Young's modulus of a shell is measured in the plane of the platelets using a three-point bending test, the result is about 50 GPa. Of much higher interest is the fracture toughness, for it can be as high as $10\,\mathrm{MPa}\sqrt{\mathrm{m}}$, twenty times larger than that of the ceramic component, if the direction of crack propagation is perpendicular to the platelets (vertical direction in the figure).

This high fracture toughness is caused by several mechanisms: Single aragonite platelets are thinner than the critical crack length of aragonite. At a stress of about 150 MPa, the tensile strength, the critical crack length is about 3.5 μm, according to equation (5.2). Therefore, they cannot contain critical cracks. The low fracture toughness of the organic matrix causes a crack to be deflected on reaching a platelet and to propagate around them.[15] Additionally, there may be pull-out of the platelets. Nano-asperities on the platelets cause additional dissipation during sliding of the platelets. In total, the work needed to create fresh surface in nacre is about 1600 J/m² if the crack propagates perpendicularly to the platelets; if it propagates in parallel to the platelets, it is only 100 J/m², but still larger than in pure aragonite, where the value is about 2 J/m², according to equation (5.17).

If we compare the increase in fracture toughness that has been achieved in nacre to those obtained in technical ceramics (see table 7.4), it is rather obvious that it would be highly desirable to technically exploit the same strengthening mechanisms. This is one reason for the strong scientific interest in nacre. The main aim of these studies is to create artificial materials with similar properties. Such materials, which mimic the properties of biological materials, are called *biomimetic materials*.

* Bone

Bone is a biological material of special importance. On the one hand, bones are the characteristic trait of vertebrates which almost exclusively occupy all ecological niches for large animals. Thus, it is of biological interest to understand why having bones is evolutionary advantageous. Even more important is that understanding bone structure enables us to treat or heal bone illnesses or injuries. For these reasons, the structure and mechanical behaviour of bones have been intensely studied [36].

Bone has a complex hierarchical structure on several different length scales. The main components of bone are a ceramic, (modified) hydroxyapatite, and a polymer, the protein collagen. Furthermore, bone contains other proteins, protein-sugar compounds, and, as all biological materials do, water.

Collagen is a protein containing about 1100 amino acids in an exactly defined sequence. This sequence is determined by the genetic code within the DNA. If we consider that there are 20 different amino acids used in common proteins, we see that the number of possible proteins is huge and that an exact control of the amino acid sequence is extremely important to ensure the correct spatial structure of the macromolecule. Collagen molecules form a helical structure, a long, almost straight helix. Three of these helices are

---

[15] This is similar to the crack propagation mechanism in sintered silicon nitride, see page 249.

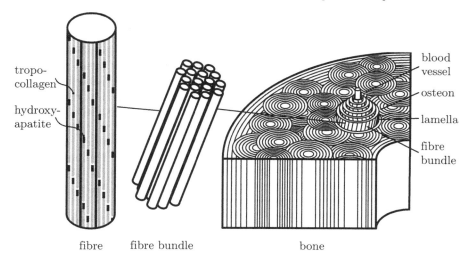

Fig. 9.16. Hierarchical structure of adult human bone. Tropocollagen molecules are arranged in a so-called quarter-stagger structure, with platelets of hydroxy apatite in between. The fibres formed by this structure unite to fibre bundles which in turn form lamellae. The major part of the bone consists of osteons made of ring-shaped lamellae. Near the bone's surface, the lamellae are parallel to the surface. The orientation of the fibre bundles within the lamellae depends on the mechanical loads on the bone; in tensile regions, they are aligned in the loading direction as shown in the figure, in compressive regions, the fibre bundles of some lamellae are perpendicular to the loading direction

intertwined to form a larger component, the tropocollagen molecule, with a length of 296 nm.

The tropocollagen molecules themselves are aligned in parallel in bones and tendons, being shifted by 67 nm in adjacent layers. Within each layer, there are gaps between the molecules that serve as nucleation sites for the crystallisation of the ceramic hydroxyapatite crystals (see figure 9.16).

Hydroxyapatite has the chemical composition $Ca_{10}(PO_4)_6(OH)_2$. In the body, its composition slightly differs from this formula (with the resulting material frequently called being dahllite), for some calcium ions are replaced by other ions, and fluorine ions replace some of the $(OH)^-$ ions.[16] Because of their small size, it is rather difficult to determine the exact shape of the hydroxyapatite crystals in bone, which may also be different in different bones. Typically, they are platelet-shaped, with a thickness of only 5 nm and an edge length between 20 nm and 100 nm. These platelets are situated between the tropocollagen molecules (see figure 9.16).

---

[16] These fluorine ions reduce the solubility of hydroxyapatite in acidic media. To protect our teeth, which have a microstructure very similar to that of bone, tooth paste contains fluorides that improve the acid resistance of the tooth enamel.

This composite of tropocollagen and hydroxapatite forms fibres that unite to form fibre bundles. The fibre bundles are the building blocks of the next hierarchical layer. Depending on the bone type, the fibre bundles may be arranged irregularly, uniaxially, or in a lamellar structure, the latter structure being the one most common in adult humans. Within the lamellar bone, the fibres are arranged in parallel in layers; the fibre bundles in adjacent layers are rotated relative to each other, similar to the fibre layers in a laminate (see section 9.1.1).

In adult humans, these lamellae form tube-shaped structures, called osteons or Haversian systems. A single osteon has a diameter of about 200 µm and a length of a few millimetres or centimetres. In long bones, like limb bones, they are parallel to the bone axis. In the centre of each osteon, there is a blood vessel that supplies the cells within the bone with nutrients.

How the fibres in the lamellae of an osteon are arranged depends on the mechanical load applied to the bones. Long bones are mainly loaded in bending.[17] On the tensile side of the bone, the fibres are oriented in longitudinal direction, on the compressive side, they are arranged either in circumferential direction or alternating between longitudinal and circumferential direction. The arrangement of the lamellae, like that of the osteons, is thus optimised to the external load.

Young's modulus of bone depends on the volume fraction of the hydroxyapatite and on the osteon structure. In the stiffest direction, it is between 12 GPa and 25 GPa. If the strains exceed values of about 0.5 %, bone starts to deform by microcracking. The fracture strain is usually 2 %, but in some specialised bones that are loaded in impact (for example, in the antlers of deer), it may be as high as 10 %. The strength of normal bone is approximately 150 MPa in tension and 250 MPa in compression. The peak loads under normal loading (walking, running, climbing a staircase) are approximately two to four times smaller than this value.

To ensure a favourable orientation of the fibres, bone is permanently rebuilt and adapted to the actual loads. Specialised cells within the bone, the osteocytes, measure the loads and initiate the rebuilding. The old bone is removed by acid-excreting cells (osteoclasts) and is then rebuild by other cells (osteoblasts), forming new osteons. The rebuilding of the bone not only ensures its adaptation to changing load patterns, but it also serves to heal microcracks that may have been formed during excess loading. As long as living bone is not overloaded, it is therefore completely fatigue resistant.

If the load on a bone is changed compared to the load patterns previously encountered, bone material is added or removed. New bone is formed, for example, when a new sports training is begun; bone is removed if it is not loaded anymore, for instance due to long-time illness or to the insertion of implants. Because Young's modulus of bone is markedly smaller than that

---

[17] This is the reason why long bones are hollow and are filled with bone marrow – in mammals – or with air sacs – in birds –, for weight can be saved this way.

of all common implant materials, load is transferred to the implant and the bone is thus partially unloaded. This can cause bone removal, leading to a loosening of the implant. Therefore, large efforts are invested in developing implant materials with a low Young's modulus. Titanium alloys are the most promising candidates because titanium not only has a small Young's modulus, but is also highly biocompatible, usually not causing adverse reactions in the body.

10

Fatigue

So far, we only considered static and monotonically changing loads. In real-world service, components are frequently bearing *cyclic loads,* with the load being time-dependent, but repetitive. Examples are revolving bending loads on rotating shafts, (resonance) vibrations in machines, and starting and stopping processes, for example in turbines.

The ongoing repetition of identical or similar loads strongly reduces the loads the material can bear. Furthermore, failure is not preceded by large plastic deformation even in ductile materials, rendering it more difficult to detect component damage than under static loads – the danger of catastrophic failure is thus rather large. An example of this is the turbine shaft shown in figure 10.1 that did not show any signs of damage caused by crack propagation under cyclic loads until it fractured catastrophically. For these reasons, it is important to consider the *fatigue behaviour* of materials i.e., their behaviour under cyclic loads.

10.1 Types of loads

Cyclic loads can occur in several ways: The load may be determined by forces e.g., centrifugal forces, or by displacements or prescribed strains e.g., thermal strains. Furthermore, frequency or amplitude of the load may differ. Finally, the number of cycles to failure or the number of cycles the component will be exposed to are important.

As an example, consider the engine of a car. All of its rotating parts, for example the crankshaft or the piston rods, are cyclically loaded and will face a large number of cycles during service. If the car drives a distance of 150 000 km on the highway at a speed of 100 km/h and a rotation speed of $3000 \, \text{min}^{-1}$, the total number of cycles is 2.7×10^8. To ensure survival of the engine, it will be designed for an infinite number of cycles, the so-called *fatigue limit.*

Some additional fatigue loads will occur in the engine when it is started since the component walls in the combustion chamber (e.g., cylinder block,

Fig. 10.1. Fatigue failure of a steam turbine shaft made of 28 NiCrMoV 85. The fragment shown has a mass of about 24 t. The crack started at a material defect within the shaft [3]

pistons) will initially be heated at their surface only, causing differential thermal strains between their cold centre and the hot surface that have to be compensated by mechanical (elastic or plastic) strains. This causes thermal stresses. During shut-down, the process is reversed. The effect of this can be significant: If we assume a mean travelled distance of 50 km, the number of cycles in the example is only 3000, with each cycle corresponding to one starting process. It would cause useless oversizing to design the engine for an infinite number of starting cycles, for the number during its life time is rather limited. For this load case, the motor is only designed to survive a finite number of cycles and thus may be loaded beyond the fatigue limit.[1]

As already stated, service loads often have a complex time-dependence. One example is the time-dependent load on car chassis parts during driving on rough roads (figure 10.2). It would be rather expensive to simulate all possible load cases in laboratory experiments. Usually, investigations are restricted to representative cases, for example the sinusoidal and triangular load curves shown in figure 10.3. These curves can be characterised by the *minimum stress* $\sigma_{\min}$, the *maximum stress* $\sigma_{\max}$, and the *mean stress*

$$\sigma_{\mathrm{m}} = \frac{\sigma_{\max} + \sigma_{\min}}{2}, \qquad (10.1)$$

the *stress amplitude* σ_{a}[2]

---

[1] The first gear of a gear box in a car is another example. As it is used only for a small amount of time, it is also designed for finite life only.

[2] Sometimes, the term 'alternating stress' is used for σ_{a}, while, in the strict sense, this is occupied for a certain load case (see below).

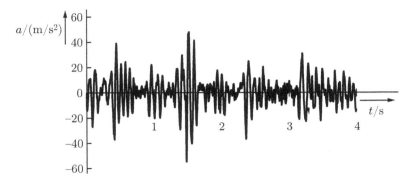

Fig. 10.2. Measured acceleration of the lower control arm in the chassis of a car driving along a rough road. The load of the component is caused by this acceleration

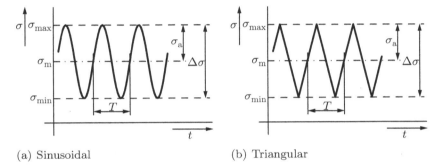

(a) Sinusoidal   (b) Triangular

Fig. 10.3. Typical load curves

$$\sigma_\mathrm{a} = \frac{\sigma_\mathrm{max} - \sigma_\mathrm{min}}{2},$$

the *stress range* $\Delta\sigma = \sigma_\mathrm{max} - \sigma_\mathrm{min}$, and the *period* T. One period corresponds to one *cycle* or *alternation of load*. To reach a number of stress cycles N, a time of $t = NT$ is needed.[3]

If the stress changes its sign during a cycle, entering the tensile and the compressive regime, it is denoted as *reversed* (sometimes also *alternating*) stress. If the load is completely tensile or completely compressive throughout the cycle, it is characterised as *fluctuating* or *pulsating stress*. To characterise the type of loading, an additional parameter is frequently used, the *stress ratio R*, often simply called the *R ratio*. It is defined as[4]

---

[3] Sometimes, the number of cycles N is called 'stress reversals'. This, however, is erroneous because there are two stress reversals during each cycle.

[4] Sometimes, this definition is changed to $R = |\sigma|_\mathrm{min}/|\sigma|_\mathrm{max}$, rendering it impossible to distinguish tensile and compressive pulsating loads.

Table 10.1. Important R ratios

pulsating in compression	$\sigma > 0$ $\sigma < 0$	$\sigma_{max} < 0$	$R > 1$
zero-to-compression	$\sigma > 0$ $\sigma < 0$	$\sigma_{max} = 0$	$R = -\infty$
reversed	$\sigma > 0$ $\sigma < 0$	$\sigma_m < 0$	$-\infty < R < -1$
fully reversed	$\sigma > 0$ $\sigma < 0$	$\sigma_m = 0$	$R = -1$
reversed	$\sigma > 0$ $\sigma < 0$	$\sigma_m > 0$	$-1 < R < 0$
zero-to-tension	$\sigma > 0$ $\sigma < 0$	$\sigma_{min} = 0$	$R = 0$
pulsating in tension	$\sigma > 0$ $\sigma < 0$	$\sigma_{min} > 0$	$0 < R < 1$
static	$\sigma > 0$ $\sigma < 0$	$\sigma_{min} = \sigma_{max}$	$R = 1$

$$R = \frac{\sigma_{min}}{\sigma_{max}}. \tag{10.2}$$

Occasionally, the so-called A *ratio* $A = \sigma_a/\sigma_m$ is used which results in $A = \infty$ for $R = -1$, for instance.

If the load is not prescribed by applied stresses, but by other parameters, for example strains, all parameters are changed accordingly and the type of loading is characterised by adding a subscript to the R ratio e. g., the strain ratio $R_\varepsilon = \varepsilon_{min}/\varepsilon_{max}$.

According to equation (10.2), reversed loads correspond to negative R ratios, pulsating loads to positive values[5]. Table 10.1 summarises the most common R ratios. Most important are the cases of *fully reversed cycling* (with $\sigma_m = 0$ or $R = -1$), of *zero-to-tension cycling* (with $\sigma_{min} = 0$ or $R = 0$), and of *zero-to-compression cycling* (with $\sigma_{max} = 0$ or $R = -\infty$). Therefore, these cases are frequently used in tables.

Figure 10.4 shows the dependence of the R ratio on the mean stress σ_m at constant stress amplitude σ_a. In general, the R ratio increases with increasing mean stress, with the exception of the transition between reversed and compressive pulsating loads. This has to be kept in mind in all considerations involving the R ratio.

---

[5] There is one exception because a pulsating compressive load with $\sigma_{max} = 0$ yields an R ratio of $R = -\infty$.

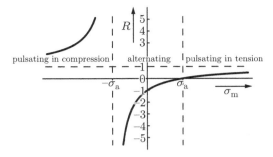

Fig. 10.4. Dependence of the R ratio on the mean stress σ_m

(a) Fatigue fracture of a shaft of a waggon by unilateral bend-pulsating stress

(b) Bilateral bend fatigue fracture (diameter of the axle journal 70 mm) [90]

Fig. 10.5. Examples of fatigue fracture. The smooth regions of the surfaces, containing *beach marks,* are caused by fatigue; the rough regions are the surfaces of final fracture. These regions will be explained in the following sections

10.2 Fatigue failure of metals

In this section, we discuss the mechanisms of fatigue damage and failure in metals.

The fracture surface of a metal that has failed in cyclic loading has a characteristic appearance. Figure 10.5 provides two examples of such *fatigue fractures.* In almost all cases, a smooth, macroscopically weakly deformed, region and a rough region can be discerned. This is due to the different stages of damage evolution under cyclic loads [90]:

- *crack initiation*[6],

---

[6] Frequently, the crack initiation stage is further divided into the two steps *crack formation* and *micro-crack growth.* This is not done here.

- crack propagation under cyclic load,
- final catastrophic fracture of the component.

In the following sections, we will discuss these stages in turn.

10.2.1 Crack initiation

In most cases, a fatigue fracture is starting at a highly loaded position of the component. The high load may be caused by an overload, generating some initial damage in the component that is hard to detect, but may cause ultimate failure.

Local stress concentrations are often caused by notches (see chapter 4). Notches may be part of the design (e. g., at bearings or at undercuts), may be caused during manufacture (e. g., tool marks caused by metal cutting), or may be due to imperfections in the material (e. g., casting pores, brittle precipitates).

In section 5.2.3, we saw that microscopic defects or cracks are usually irrelevant under static loads because they are smaller than the critical crack length from equation (5.27). If loads are cyclic, much smaller defects, like casting pores or inclusions, can initiate fatigue cracks. The fatigue strength of a material is thus much more sensitive to the manufacturing process and material defects than the static strength.

Metal working (e. g., rolling or forging) is one way to remove defects during manufacturing, closing cavities and changing the inhomogeneous casting microstructure to a fine-grained structure formed by recrystallisation (see section 6.4.2). For this reason, components that have been manufactured this way usually have a higher fatigue limit than cast products.

Metal working processes, however, cannot always be employed. A complex shape of the component or the additional costs may require to produce the net-shape by casting. Furthermore, high-strength materials are often not suitable for metal working. For example, turbine blades of stationary gas turbines are often produced by investment casting to enable the use of materials with high strength at service temperature which, however, cannot be forged. These blades are cyclically loaded because of fluctuations in the gas flow. The blades of the last stage have a slender shape so that high stresses are generated by these cyclic loads. To remove casting pores, they are often compacted after casting by hot isostatic pressing (HIP) at temperatures of about 1200°C and pressures between 100 MPa and 200 MPa.

Even if the component is initially defect-free, it is not guaranteed that no cracks will form. Cracks are initiated by a roughening of the surface of the component under cyclic loads (see figure 10.6) caused by plastic deformation. This deformation is due to dislocation movement at stresses below the yield strength $R_{p0.2}$, which is insignificant under static loads.[7] However, dur-

---

[7] According to section 3.2, the yield strength $R_{p0.2}$ is defined as the stress corresponding to a plastic strain of 0.2%. For this, some amount of dislocation movement is necessary.

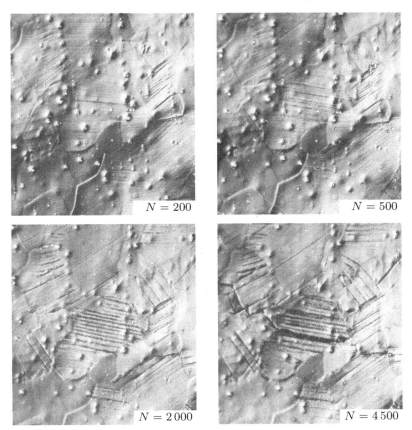

$N = 200$

$N = 500$

$N = 2\,000$

$N = 4\,500$

Fig. 10.6. Formation of slip bands in AlMg 3 in a strain-controlled fatigue experiment ($R_\varepsilon = -1$, $\varepsilon_\mathrm{a} = 0.5\%$, grain size 50 μm). Optical micrograph (after [148])

ing cyclic loading, these small plastic deformations accumulate and initiate a crack as discussed below. The dislocation movement causes a hysteresis in the stress-strain diagram (see figure 10.7). The enclosed area equals the dissipated energy per cycle and unit volume (see section 3.2). If a component is deformed cyclically, the density and position of the dislocations and thus its strength change. This is called *cyclic hardening* or *softening* and will be discussed in more detail in section 10.6.5.

Figure 10.8 illustrates how dislocation movement on slip planes can roughen a surface by forming extrusions and intrusions. Steps on the surface formed by slip of the material (figure 10.8(b)) are not completely removed upon load reversal since a dislocation that has moved in one direction will not necessarily revert to its original position. Instead, another dislocation may move and cause the plastic deformation, leading to a roughening of the initially smooth surface. Furthermore, a thin oxide layer of a few nanometre

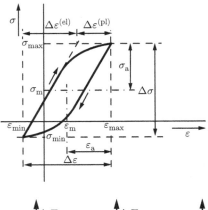

Fig. 10.7. Schematic plot of a hysteresis in the stress-strain response under cyclic loading

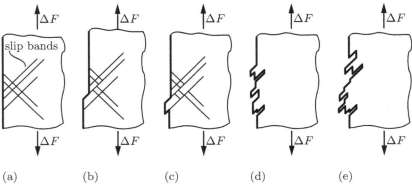

(a)   (b)   (c)   (d)   (e)

Fig. 10.8. Formation of intrusions and extrusions due to cyclic loading. During the cycles, small irreversible plastic deformations occur on parallel slip bands that cause accumulation of extrusions and intrusions on the surface. The number of cycles increases from the left to the right

thickness forms on a freshly created surface in air. Thus, the material cannot re-weld even if the deformation is completely reversible, and a sharp crack is initiated.

Altogether, a large number of surface notches form that serve as starting points for cracks. These cracks propagate into the material, growing along the crystallographic slip planes. They grow preferentially in grains with slip planes that are oriented at approximately 45° to the maximum principal stress. The formation of fatigue slip bands is shown in figure 10.6 in an aluminium alloy (AlMg 3) under strain-controlled cyclic loading. Figure 10.9 shows extrusions and intrusions, the initiating points for cracks, at the surface of a ferritic steel.

According to the definitions from section 5.1.1, the newly formed micro-cracks are initially not loaded in mode I, but in mode II or III. This stage of fatigue crack propagation is called *propagation stage I*. During this stage,

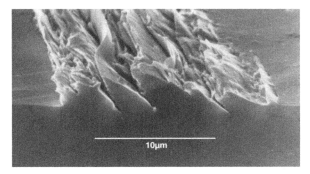

10µm

Fig. 10.9. Scanning electron microscope micrograph of extrusions in a ferritic steel [73]

crack growth is slow [19, 90]. If the crack reaches a grain boundary, it has to propagate into a neighbouring grain, usually with less favourably oriented slip systems. This may slow down or even completely stop the crack. Thus, many initially created cracks are stopped, and only those that accidentally start at the most favourable conditions grow further.

With increasing crack growth, the stress at the crack tip increases until less well-oriented slip systems can be activated. This enables the crack to change its orientation to mode I, perpendicular to the maximum principal stress. The crack becomes a macrocrack. This stage of the process is called *propagation stage II* and will be discussed in the next section. The transition between stage I and stage II usually occurs at a crack length of approximately 0.05 mm to 2 mm [130]. Usually, the transition length is smaller in high-strength materials than in those with low strength. Because the stress concentration is largest at the largest crack, this crack grows fastest, resulting in one crack leaving the others behind and dominating the fatigue process. This crack finally causes failure of the component. Figure 10.10 illustrates the transition between stage I and stage II.

The initiation cracks needs a long time in smooth specimens loaded with stresses well below the yield strength, for the amount of dislocation movement is very small in this case, so that the greatest part of the components life time is spent during this initial stage. If the surface is hardened, for example by shot peening (work hardening) or nitriding of steel, dislocation movement near the surface is impeded further. This increases fatigue life. This argument is only valid if the propagating crack starts at the surface, not at inner defects. If the crack initiation occurs at defects, like tool marks or pores, the fatigue life reduces accordingly.

Microscopic defects do not have a very strong influence on the total life if the stress amplitude is large and the number of cycles to failure is small (usually smaller than 10^4 cycles). In this case, initial cracks form comparatively

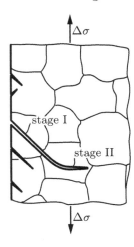

Fig. 10.10. Stages I and II of crack propagation (after [8, 19, 113])

quickly by dislocation movement. The increase in crack propagation rate is smaller, resulting in a larger part of the life time of the component spent in propagation stage II. This stage is discussed in the next section.

10.2.2 Crack propagation (stage II)

As we saw in the previous section, a fatigue crack reorients during crack propagation to be perpendicular to the maximum principal stress as soon as a certain crack length is exceeded. The crack is now loaded in mode I (see figure 10.11 (a)). Because of the stress concentration, a plastic zone forms in the direct vicinity of the crack tip (figure 10.11 (b) and (c)), causing a small crack tip opening and similarly small crack propagation.[8] Upon unloading, the elastic strain in the bulk material is removed, and the crack closes. However, the crack is widened at the crack tip due to the localised plastic deformation, resulting in compressive residual stresses. These stresses cause an opposite plastic deformation and close the crack even at the crack tip (figure 10.11 (d)). Upon unloading, the crack surfaces touch before the external load is reduced to zero (figure 10.11 (e)). This is due to the plastic deformation which causes a roughening of the crack surfaces. Furthermore, the crack often follows the crystallographic planes within the grains, resulting in roughness on the scale of the grain size. The compressive stresses are uncritical since they are transferred at the crack surfaces (figure 10.11 (f)). During a single cycle, there is thus some crack propagation because the crack does not return to its initial position.

The tensile load in the next cycle re-opens the crack. Because of the compressive residual stress in the unloaded material, some minimum threshold

---

[8] This was already discussed in section 5.3.3 in the context of elastic-plastic fracture mechanics, see figure 5.22 (2).

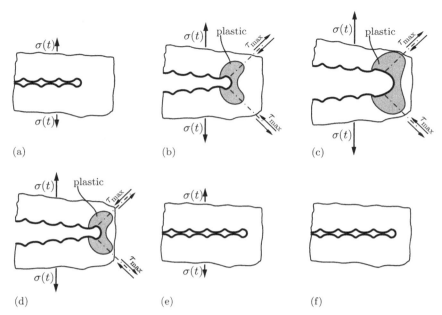

Fig. 10.11. Model of crack propagation under cyclic loading (after [35, 76]). Upon loading ((b) and (c)), the region near the crack tip yields. The crack propagates stably, blunting the crack tip. Upon unloading (d), the elongated crack is compressed. Because of the deformation, the crack surfaces touch before the external load is zero (e). After complete unloading, residual stresses remain, compressing the crack (f)

stress is required. External loads that are smaller than the residual stress do not open the crack and thus cannot cause crack propagation. If the stress is exceeded, the crack again propagates by a certain increment, resulting in a crack length that grows with every cycle and thus in considerable crack growth. Because the crack does not propagate further under static loads, the crack propagation is *stable*.

Frequently, the crack propagation can be seen on the fatigue crack surface as so-called *fatigue striations* (figure 10.12), with a distance that depends on the load and is usually between 0.1 µm and 1 µm [90]. This distance is an approximate measure of the crack propagation per cycle.[9] Macroscopically, the striations are invisible and the fatigue crack surface appears rather smooth. The formation of striations is due to the plastic deformation near the crack tip shown in figure 10.11.

Even with the naked eye, so-called *beach marks* (sometimes also called *arrest lines*) can frequently be discerned on the fracture surface (see figure 10.5 on page 337). They are due to changes in the load during component service

---

[9] Each striation is generated in exactly one cycle, but not every cycle does generate a striation.

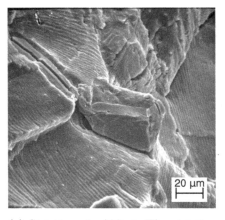 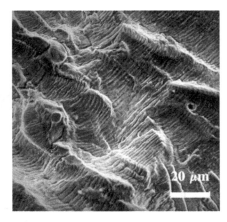

(a) Striations in AlCu 5. The striations   (b) Striations in copper
are deflected at the Al$_2$Cu particle [90]

Fig. 10.12. Scanning electron microscope micrograph of striations on two fatigue fracture surfaces

e. g., a change in the revolution speed, a short-time overload, or a machine downtime during a weekend. These changes cause local differences in the surface roughness or the surface oxidation that are visible as beach marks. The arrangement of the beach marks often makes it easy to determine the position of the initial crack and the crack propagation direction.

Crack propagation in a specimen is also determined by the grain size. The larger the grains are, the rougher the crack surfaces will be. Therefore, the surfaces will touch earlier during unloading, causing larger compressive stresses. To open the crack again, a larger stress is thus needed. The crack propagation rate is thus smaller in coarse-grained materials than in fine-grained ones. This is different in specimens without initial cracks. In this case, the fatigue life of the fine-grained specimen is larger because its strength is higher due to grain boundary strengthening, making crack initiation at the surface more difficult.

10.2.3 Final fracture

Final fracture occurs when the crack has grown to a size where the stress intensity factor (see equation (5.2)) equals the fracture toughness K_{Ic} of the material. The crack propagates unstably throughout the specimen, causing catastrophic fracture. The appearance of this part of the fracture surface differs from that created by fatigue crack growth. Macroscopically, the surface is jagged; microscopically, a dimple fracture surface, a brittle cleavage fracture, or a mixture of both types results, depending on the material's ductility (see section 3.5).

Even if there is ductile failure, no significant plastic deformation can be seen macroscopically since plastic deformation concentrates near the crack tip where the stresses are highest [90].

10.3 Fatigue of ceramics

Contrary to metals, ceramics do not yield plastically when loaded. Therefore, cracks cannot grow under cyclic loading by localised plastic deformation at the crack tip. Furthermore, no intrusions and extrusions can form on the surface of an initially smooth specimen by dislocation movement. Because of this, many ceramics do not exhibit any cyclic effects i. e., there is no difference between their behaviour under static and cyclic loads. All loads that they can bear once, they can bear infinitely many times. For example, this is the case in fine-grained ceramics with a single phase i. e., many technical ceramics.

> If subcritical crack growth occurs in a ceramic (see section 5.2.6), the crack propagates subcritically during each cycle. This can cause apparent fatigue behaviour, although it is not the number of cycles but the accumulated loading time that is relevant. This behaviour is frequently called, rather misleadingly, 'static fatigue'.
>
> To compare the behaviour of such a ceramic under cyclic loads with that under static loads, the accumulated loading time of cyclic loading can be converted to an *effective loading time* t_{eff}, the time that would cause the same damage under static loads. This is discussed in detail in *Munz / Fett* [104].

Nevertheless, some ceramics can fail by cyclic crack propagation. Under cyclic loads, cracks in the material are opened and closed. If energy is dissipated in this process, this energy can be used to propagate the crack. If we consider the stress-strain diagram for cyclic loading, fatigue can only occur if there is a hysteresis in the diagram. As we saw in section 7.2.5, many of the mechanisms that increase the crack-growth resistance of a ceramic cause such a hysteresis (see figures 7.5 and 7.7). This increase in fracture toughness under static loads thus makes the material sensitive to fatigue, and the strength under cyclic loads drops below the static strength.

One example for this are ceramics exhibiting crack bridging (see section 7.2.2). The crack surfaces rub on each other and dissipate energy during opening and closing of the crack. The repeated opening and closing of the crack surfaces under cyclic loads causes wear of the surfaces, reducing the crack bridging effect. This can cause cyclic crack propagation. Crack deflection at grain boundaries can also cause similar bridging effects [69].

Fatigue is also observed in transformation-toughened ceramics, like partially stabilised zirconium oxide (see sections 7.2.4 and 7.5.4), where phase transformations occur near the crack tip. This effect is attributed to the formation of microcracks in the vicinity of the crack tip [66].

10.4 Fatigue of polymers

Similar to metals, polymers can deform plastically and thus can fail in a similar way under cyclic loading. However, the microscopic mechanisms are not the same as in metals.

> If we load a polymer cyclically with a non-zero mean stress, the viscoelastic and viscoplastic deformation causes an increase of the strain. This effect can cause failure under cyclic loads, but this is not true fatigue, for it is not the number of cycles, but the total loading time that determines failure.

10.4.1 Thermal fatigue

Polymers deform viscoelastically. Under cyclic loads, the stress-strain curve upon unloading is not the same as upon loading. Therefore, there is a hysteresis between stress and strain, causing energy dissipation during the deformation, thus producing heat. This hysteresis is discussed in detail in exercise 26.

The heat generated under cyclic loads cannot easily dissipate into the surroundings because the thermal conductivity of polymers is small. If heat generation exceeds heat dissipation, the temperature increases in each cycle until the temperature-dependent strength of the material is exceeded and the material fails. If the stress is reduced, the heat generation reduces as well, and the number of cycles to failure increases. For this reason, this phenomenon is called *thermal fatigue*. If the stress is reduced further, an equilibrium between heat generation and heat dissipation will be established without exceeding the strength.

Thermoplastic polymers fail by plastic yielding under thermal fatigue because the yield strength decreases with increasing temperature. Elastomers and duromers can also fail by thermal fatigue due to the reduction of Young's modulus with temperature which causes a continuously growing deformation.

Thermal fatigue is observed mainly in stress-controlled loading because the strain amplitude increases in this case due to the reduction of stiffness with increasing temperature. The heat generated per cycle thus increases with time. If the loading is strain-controlled, thermal fatigue is usually not problematic because the stress decreases in this case.

If thermal fatigue occurs, the load frequency can strongly influence fatigue life. On the one hand, longer cycles provide more time to dissipate the heat, on the other hand, Young's modulus and the size of the hysteresis in the stress-strain diagram are dependent on the frequency. This frequency dependence will be discussed in section 10.6.2. If sufficiently long unloading occurs between the cycles, the generated heat can be dissipated, and there is no thermal fatigue. The geometry of the component is also important, for it determines the heat dissipation.

10.4.2 Mechanical fatigue

If the temperature increase during cyclic loading is sufficiently small so that no thermal fatigue occurs, the polymer fails by mechanical fatigue. Similar to metals, we can distinguish the stages of crack initiation and crack propagation.

Cracks can form in a polymer in different ways, depending on the dominating deformation mechanism [130]. Fracture of single chain molecules or sliding of chain molecules can weaken the material locally and thus serve as initiating points for cracks. In many polymers (for example, polystyrene, polycarbonate, PMMA), crazing (see section 8.4.1) plays an important role, as it does in plastic deformation. Under cyclic loads, crazes may form and grow even if the loads are comparatively small, until they have grown sufficiently to act as microcracks. If a crack propagates, the stress concentration near the crack tip can initiate further crazes (see figure 8.14) that coalesce with the crack. Because crazing depends on the hydrostatic stress state, the mean stress is especially important in this case. Alternatively, shear bands may form and be starting points for microcracks. Similar to metals, fatigue striations (see figure 10.11) may be generated in polymers due to plastic deformation at the crack tip [19]. If there are phase boundaries within the polymer, for example, between the amorphous and crystalline regions of a semi-crystalline polymer or between chemically different phases in a copolymer, these boundaries may initiate cracks due to the stress concentration they cause. Nevertheless, the fatigue strength of semi-crystalline polymers is in most cases superior to that of amorphous polymers [19].

Whether a polymer fails by mechanical or thermal fatigue is determined by many factors e. g., the load frequency, the stress level, the temperature, and the geometry of the component. In general, polymers with weak viscoelastic effects, which produce only a small amount of heat in each cycle, fail by mechanical fatigue. Among these are polystyrene and many duromers. Thermal fatigue is important in materials with a large hysteresis in the stress-strain diagram, for example polyethylene, polypropylene, and polyamide. Finally, both effects can interact because the increase in temperature changes not only the static, but also the dynamic material properties. In this case, the material heats up initially, and crack propagation sets in afterwards. This phenomenon can be observed in PMMA, PET and polycarbonate.

10.5 Fatigue of fibre composites

In fibre composites, the presence of the fibres can change the fatigue strength of the matrix in several ways. On the one hand, global effects due to load transfer and the corresponding change in the stress and strain fields can play a role, on the other hand, local effects can occur, especially at the fibre-matrix interface.

In most cases, the fatigue strength of fibre composites is larger than that of the matrix alone. If Young's modulus of the fibre is larger, partially unloading the matrix, and if the load is stress-controlled, the reduced stress in the matrix material may increase fatigue life. This is not the case if the load is strain-controlled because in this case the increased stiffness of the material causes an increase in the stress, and fatigue life may be reduced.[10] If loaded perpendicularly to the fibre direction, fatigue life of the composite can be reduced compared to the matrix material alone because strains in the matrix increase and the stress state is triaxial.

The shear stress at the interface between fibre and matrix locally increases the strain in the matrix (see section 9.3.2). This increased strain may cause local damage in the matrix and initiate cracks. This is especially important for short fibres because shear stresses occur only near the fibre ends in long fibres.

In polymer matrix composites, the fatigue strength is usually larger than that of the matrix material, provided the fibres are aligned in load direction or are irregularly oriented. Since the stiffness is larger than in the unreinforced material, the strain is reduced (if the load is stress-controlled). In a polymer matrix, this also serves to reduce heat generation and thus further increases fatigue life [107]. Carbon fibres, with their high thermal conductivity, also reduce thermal fatigue because they can dissipate localised temperature peaks. In metal matrix composites, fatigue life is also significantly larger than in the matrix material if the fibres are aligned to the loads (see also page 365).

Damage in polymer matrix composites and metal matrix composites usually starts with local detachment between fibre and matrix at weak points, for example, by fracture of very thin fibres [122]. This may happen already after only a few cycles. Subsequently, microcracks form, starting from these damaged regions, by fatigue of the matrix material. Thus, the material is increasingly damaged, a fact that can be observed by a reduction in stiffness. The component finally fractures when the number of microcracks has become so large that they coalesce. This failure behaviour is similar to the stochastic failure discussed in section 9.3.4. In metal matrix composites, this behaviour is usually observed if the fibres are irregular or arranged in laminates.

If there are tensile and compressive components in the cyclic load (i.e., if $R < 0$), fatigue life can be significantly reduced compared to pulsating tensile loading. The reason is that those fibres whose matrix interface has failed are sensitive to failure by buckling [131].

Fatigue can also occur in ceramic matrix composites. If a crack propagates under tensile loading, it will be bridged by the fibres and partially unloaded (see section 9.3.3). This can cause stable crack growth under static loading, with the crack not propagating further unless the load is raised (see

---

[10] This would not be true in the idealised case if the fibres were continuous and loaded directly as in figure 9.1(a). In real-world composites, even continuous fibres are always loaded by load-transfer from the matrix.

section 5.2.5). Under cyclic loading, the weak interface between fibre and matrix allows movement between them. Friction occurring in this process can cause damage, reduce the unloading effect of the fibre, and thus enable the crack to propagate [29]. Failure usually occurs not by propagation of a single crack through the material, but by accumulation of damage [28].

10.6 Phenomenological description of the fatigue strength

In the previous sections, we discussed the failure mechanisms of the different material classes. Using the example of metals, we discussed in some detail how cracks may be initiated in a component and how they propagate, until final fracture occurs. To safely design components, we need tools to describe and assert their life time. Two different approaches can be used:

In many cases, macroscopic cracks are known or supposed to be present in the component. Turbine shafts in gas turbines used for power generation, for example, are tested non-destructively using ultrasonic testing. If no cracks are detected, it is assumed that the largest crack present in the shaft has a length equal to the detection limit (usually a few millimetres). In this case, the methods of fracture mechanics (see chapter 5) can be used to describe crack propagation. The component is designed to ensure that these (hypothetical) cracks do not cause impermissible crack propagation.

However, even if no cracks are initially present in the component, it is not exempt from fatigue failure. Microcracks may, for example, form at the surface of the component, propagate, and cause ultimate failure as described in section 10.2 for the case of metals. Thus, we also need methods to assert the life time of uncracked components.

10.6.1 Fatigue crack growth

If there is a macroscopic crack in a material that may grow upon cyclic loading, the fatigue life is determined by its propagation rate. This crack is sufficiently long to be described using the methods of continuum fracture mechanics described in chapter 5. Because fatigue crack propagation occurs at much smaller stresses than propagation under static loads, the plastic zone near the crack tip is comparatively small. The conditions to apply linear-elastic fracture mechanics are thus fulfilled even in ductile materials and small specimens. Thus, the stress intensity factor K is sufficient to describe the stress field.

> For a crack to be considered as macroscopic or 'long', its length must be large compared to the length scale of microstructural features, especially the grain size. In this case, local changes in the crack resistance e. g., due to different grain orientations, are averaged out. Furthermore, the plastic zone must be restricted to the vicinity of the crack tip

(see sections 5.2 and 5.3). This is not the case in microcracks (also called 'short cracks') because they are completely embedded into the plastic field [113]. This difference can also be seen in the fact that macrocracks grow perpendicular to the maximum principal stress (in mode I), whereas microcracks often do not (see section 10.2.1).

According to equation (5.2), the stress intensity factor K_I depends on the external stress σ and the crack length a. Transferring this to the case of cyclic loads with a stress range $\Delta\sigma$, we arrive at the *cyclic stress intensity factor*

$$\Delta K = \Delta\sigma\sqrt{\pi a}\,Y\,,\tag{10.3}$$

using the definition $\Delta K = K_{\max} - K_{\min}$, where $K_{\max}$ and $K_{\min}$ are the maximum and minimum stress intensity factor in the cycle. The R ratio for the stress intensity factor is[11]

$$R = \frac{K_{\min}}{K_{\max}}\,.$$

The crack propagation is described using the *crack-growth rate* $\mathrm{d}a/\mathrm{d}N$, defined as the crack growth $\mathrm{d}a$ per cycle.[12]

Fatigue-crack-growth threshold

As we saw in section 10.2.2, a threshold stress is required to open the crack under cyclic loading because compressive stresses occur near the crack tip that have to be overcome. Accordingly, there is a threshold value of the stress intensity factor below which there will be no crack propagation (see figure 10.13). If we denote this value of the stress intensity factor as K_{op} (subscript 'op' for 'opening'), the crack can only grow if $K_{\max} \geq K_{\mathrm{op}}$. The value of K_{op} depends not only on the material, but also on the cyclic loading. As it is determined by the deformation near the crack tip during crack opening and closing, it depends on the maximum stress and the R ratio.

It is common to describe cyclic loads not using $K_{\min}$ and $K_{\max}$, but rather with the cyclic stress intensity factor ΔK and the R ratio. The crack propagates if the cyclic stress intensity factor exceeds a certain value, the *fatigue-crack-growth threshold* ΔK_{th}. If we rewrite the criterion for crack propagation, $K_{\max} = K_{\mathrm{op}}$, for the cyclic stress intensity factor, we find the following relation for the fatigue-crack-growth threshold:

$$\Delta K_{\mathrm{th}} = K_{\mathrm{op}} - K_{\min} = 2(K_{\mathrm{op}} - K_{\mathrm{m}}) = (1 - R)K_{\mathrm{op}}\,.\tag{10.4}$$

---

[11] In principle, the R ratio for K should be denoted R_K. However, because $R_K = K_{\min}/K_{\max} = \sigma_{\min}/\sigma_{\max} = R$ holds, we can simply write R instead.

[12] Because the crack growth is not continuous during the cycle, the crack-growth rate is defined as $\mathrm{d}a/\mathrm{d}N = \lim_{\Delta N \to 1} \Delta a/\Delta N$ [113]. Mathematically, it is thus not a differential quotient. Nevertheless, it is common to write it in this way.

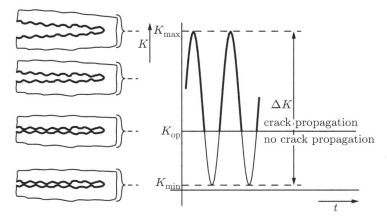

Fig. 10.13. At loads smaller than K_{op}, the crack does not open. Beyond this limit, the crack opens

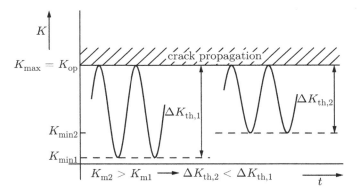

Fig. 10.14. Illustration of the decrease of ΔK_{th} with increasing mean stress intensity factor K_m

Here, K_m is the mean stress intensity factor, analogous to equation (10.1). The crack starts to grow when $\Delta K = \Delta K_{th}$. According to this equation, ΔK_{th} decreases with increasing mean stress intensity factor, leading to crack growth at decreasing load amplitudes. As illustrated in figure 10.14, this is simply due to the description of crack growth using ΔK; the criterion $K_{max} = K_{op}$ itself remains unchanged.

Equation (10.4) seems to imply that ΔK_{th} depends linearly on the R ratio. However, it has to be taken into account that K_{op} itself also depends on R. For practical applications, it would be useful to know the exact R-dependence, for this would allow to make measurements at one R ratio only and to extrapolate from there. Several, sometimes contradictory, approaches can be found in the literature, for example in *Schott* [130]:

$$\Delta K_{\mathrm{th}}(R) = \begin{cases} (1-R)^\gamma \Delta K_{\mathrm{th}}|_{R=0} & \text{for } R < R_{\mathrm{t}}, \\ \text{const} & \text{for } R \geq R_{\mathrm{t}}. \end{cases} \qquad (10.5)$$

with $R_{\mathrm{t}} = 0.5\ldots0.7$. In low- to medium-strength ferritic steels, $\gamma \approx 1$, in high-strength martensitic steels, $\gamma \to 0$.

Because the stress intensity factor K_{op} needed to open the crack depends on the deformation near the crack tip, it also depends on Young's modulus, for the crack opening in a linear-elastic material is the smaller, the higher Young's modulus is (see equation (5.3)). Accordingly, *Schwalbe* [133] provides the following approximation for the fatigue-crack-growth threshold in metals:

$$\Delta K_{\mathrm{th}}(R) = (2.75 \pm 0.75) \times 10^{-5} E (1-R)^{0.31} \sqrt{\mathrm{m}} \quad \text{for } R < 1. \qquad (10.6)$$

Equation (10.6) also shows the dependence of the fatigue-crack-growth threshold on the R ratio, which, however, is not in agreement with equation (10.5) above.

Although equations like these exist, it should be kept in mind that K_{op} and thus ΔK_{th} depends on many other material parameters e. g., the grain size. Accordingly, large differences in the exact values can be found even within a certain material class. Nevertheless, the equations are useful in estimating the order of magnitude of ΔK_{th}. If we take steel as an example (with $E = 210\,000\,\mathrm{MPa}$), we find $\Delta K_{\mathrm{th}} = 5.8\,\mathrm{MPa}\sqrt{\mathrm{m}}$ for $R = 0$. This is more than one order of magnitude smaller than the static fracture toughness K_{Ic} of ductile steels and thus illustrates how dangerous even small cracks can be under cyclic loads.

Crack propagation

If the cyclic stress intensity factor ΔK exceeds the fatigue-crack-growth threshold ΔK_{th} (i. e., if $K_{\max} > K_{\mathrm{op}}$), the crack grows in every cycle. The crack-growth rate $\mathrm{d}a/\mathrm{d}N$ is determined by those parts of ΔK that exceed K_{op} i. e., (for the case $K_{\min} < K_{\mathrm{op}}$) by the effective cyclic stress intensity factor $\Delta K_{\mathrm{eff}} = K_{\max} - K_{\mathrm{op}}$. Because K_{op} is usually unknown, $\mathrm{d}a/\mathrm{d}N$ cannot be plotted against ΔK_{eff}. Instead, its dependence on ΔK and R is used. As figure 10.15 illustrates, ΔK_{eff} increases with ΔK and with the mean stress intensity factor K_{m} i. e., the R ratio.

During crack propagation, the cyclic stress intensity factor ΔK increases due to the increase of the crack length. Therefore, the crack-growth rate $\mathrm{d}a/\mathrm{d}N$ also increases even if the cyclic load of the component is constant. If the maximum stress intensity factor $K_{\max}$ approaches the fracture toughness, the crack accelerates rapidly and eventually becomes unstable after a few more cycles.[13] Final fracture of the component ensues. Similar to the fatigue-crack-growth threshold, the transition to unstable crack propagation is determined

---

[13] Because of the preceding cyclic crack propagation, the crack may not become unstable exactly when the stress intensity factor equals K_{Ic} (cf., for example,

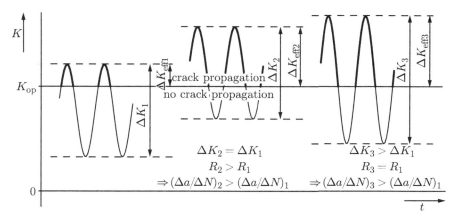

Fig. 10.15. Illustration of the increase of the crack growth per cycle, $\mathrm{d}a/\mathrm{d}N$, with increasing R ratio (corresponding to an increasing mean stress intensity factor K_m) or cyclic stress intensity factor ΔK. The fraction of the cycle with opened crack increases

by the maximum stress intensity factor K_max ($K_\mathrm{max} = K_\mathrm{Ic}$). According to equation (10.4), the critical stress intensity range ΔK_Ic can be determined:

$$\Delta K_\mathrm{Ic} = 2(K_\mathrm{Ic} - K_\mathrm{m}) = (1 - R)K_\mathrm{Ic}. \tag{10.7}$$

Again, an increase of K_m or the R ratio decreases the allowed cyclic stress intensity factor ΔK_Ic.

If we plot the crack-growth rate $\mathrm{d}a/\mathrm{d}N$ versus the cyclic stress intensity factor ΔK for a constant R ratio in a double-logarithmic plot, we get a crack-growth curve or $\mathrm{d}a/\mathrm{d}N$ curve (figure 10.16). A marked increase of the crack-growth rate is apparent in region III where the maximum stress intensity factor K_max approaches K_Ic ($\Delta K \to \Delta K_\mathrm{Ic}$). The crack slows down in region I when K_max approaches K_op from above ($\Delta K \to \Delta K_\mathrm{th}$). In between, there is a region marked 'II' where the dependence between $\log(\mathrm{d}a/\mathrm{d}N)$ and $\log(\Delta K)$ is almost linear. Accordingly, the crack-growth rate follows the so-called *Paris law*

$$\frac{\mathrm{d}a}{\mathrm{d}N} = C\Delta K^n = C^* \left(\frac{\Delta K}{K_\mathrm{Ic}}\right)^n \tag{10.8}$$

in this region. Here, C is a constant depending on the material and the R ratio. Similar to the subcritical crack growth of ceramics in equation (7.1), the unit of the constant C depend on the exponent n, whereas C^* has the units of a length. In metals, the exponent n is usually in the region $2 \leq n \leq 7$ [35], but

___

section 5.2.5). However, this is irrelevant under cyclic loads because this effect can only alter the life time of the component by a few cycles. For simplicity, we will use K_Ic in the following.

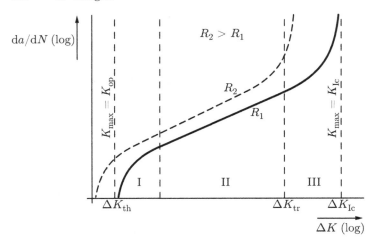

Fig. 10.16. Crack-growth curve, plotting da/dN versus the cyclic stress intensity factor ΔK in a double-logarithmic plot. There are three characteristic regions as shown for the curve with the R ratio R_1

in brittle materials it can be as large as 50 [120]. For ferritic-pearlitic steels, *Landgraf* [87] states the following upper limit for the da/dN curve at $R = 0$:

$$\frac{da}{dN} = 6.9 \times 10^{-9} \frac{\text{mm}}{\text{cycle}} \times \left(\frac{\Delta K}{\text{MPa}\sqrt{\text{m}}} \right)^3 .$$

If we load a crack with a constant stress range $\Delta\sigma$ with $\Delta K > \Delta K_{\text{th}}$, the crack grows. According to equation (10.3), ΔK increases, and the loading point in the da/dN curve in figure 10.16 moves to the right. The crack-growth rate increases in each cycle until ΔK_{Ic} is reached, and final fracture destroys the component.

If we consider a specific material and increase the mean stress intensity factor K_{m} (thus usually also increasing the R ratio, see section 10.1), K_{max} increases as well. The cyclic stress intensity factor ΔK that the component can bear decreases, shifting the curve to the left, as shown by the dashed line in figure 10.16. This is a direct consequence of what we discussed above concerning the mean-stress dependence of ΔK_{th}, ΔK_{Ic}, and da/dN. In equation (10.8), this shift of the curve is accounted for by the R-dependence of the factor C. There are a large number of, sometimes contradictory, approaches to describe the dependence of C on the R ratio and the da/dN curve in all three regions (see, for example, *Broek* [23], *Radaj* [113], and *Schott* [130]).

Some exemplary da/dN curves are shown in figure 10.17. Only in the case of the steel was the range of the cyclic stress intensity factor sufficiently large to capture all three regions of the curve. The slope of the curve is much larger for ceramics than in the Paris region of metals, resulting in a cyclic stress intensity factor that is almost the same for negligible and rapid crack growth. The reason for this is that the strength of ceramics is at most only slightly

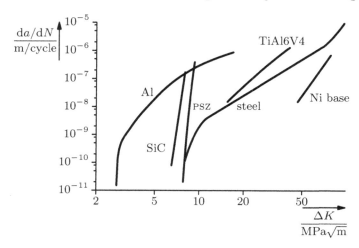

Fig. 10.17. Crack-growth curves of several materials: An aluminium alloy (AlZn 6 CuMgZr), a steel (20 MnMoNi 4-5), a nickel-base superalloy (Waspaloy), a titanium alloy (TiAl 6 V 4), silicon carbide, and a PSZ ceramic [6, 38, 53, 57, 87]

reduced by cyclic loading. The fatigue-crack-growth threshold ΔK_{th} is almost the same in ceramics and metals. Nevertheless, metals do have one advantage if loaded cyclically: Their fatigue strength is larger if they are designed against a small number of cycles since the Paris region can be exploited in the design. Because of the extended Paris region, significant crack growth must take place before the component fails so that regular inspection cycles may detect the growing crack before failure.

In polymers, the da/dN curves are similar to those of metals. Below a certain threshold value ΔK_{th}, there is no crack growth, at larger values, three regions can be distinguished, with region II being described by a Paris law. The exponent n takes a value of about 4 in many polymers [97].

In composites, the fatigue behaviour can frequently not be described adequately by da/dN curves because the material usually fails by accumulating local damage, not by propagation of a single crack. Measuring da/dN curves is thus a rather involved procedure [29]. If a single crack determines the failure behaviour, the da/dN curves can be described with a Paris law. Compared to the matrix material alone, K_{Ic} is often reduced in polymer and metal matrix composites (see section 9.3.4), but ΔK_{th} is increased. Despite the reduced fracture toughness, the fatigue life of a composite may thus be larger than that of the matrix material.

Assessing life times

Using equation (10.3), we can calculate the critical crack length a_{f} at which unstable or accelerated crack growth occurs (transition between regions II and III in figure 10.16). If we require that this crack length must not be exceeded, we can calculate the number of cycles to failure for a given initial crack length $a_0 < a_f$.

To do so, we exploit the equality $\Delta K = \Delta K_{\mathrm{Ic}}$ or $\Delta K = \Delta K_{\mathrm{tr}}$ (for the transition between region II and III), respectively. The number of cycles until the critical crack length is reached can be estimated for the initial crack length a_0 by [8, 40]:

$$N_f(a_0) = \int_0^{N_f} \mathrm{d}N = \int_{a_0}^{a_f} \frac{1}{C} \left(\frac{1}{\Delta K} \right)^n \mathrm{d}a \, . \tag{10.9}$$

Here we assume that we are already in region II at the initial crack length. Inserting ΔK from equation (10.3) and assuming a constant stress range $\Delta\sigma$, we find

$$N_f(a_0) = \frac{1}{C} \left(\frac{1}{\Delta\sigma\sqrt{\pi}} \right)^n \int_{a_0}^{a_f} \frac{1}{(Y\sqrt{a})^n} \mathrm{d}a \, . \tag{10.10}$$

If the geometry factor Y is independent of the crack length – an assumption unfortunately not true in most cases –, we can take Y out of the integral and solve the integral.[14] Otherwise, equation (10.10) must be integrated numerically.

The result is (for $n \neq 2$ and geometry factor Y independent of the crack length)

$$N_f(a_0) = \frac{1}{C} \left(\frac{1}{\Delta\sigma\sqrt{\pi}Y} \right)^n \cdot \frac{2}{2-n} \left(a_f^{\frac{2-n}{2}} - a_0^{\frac{2-n}{2}} \right) \, . \tag{10.11}$$

This equation can be used to estimate the number of cycles to failure for a known length of the largest crack (see exercise 30).

Growth of short cracks

As explained above, the crack-growth rate $\mathrm{d}a/\mathrm{d}N$ depends on the cyclic stress intensity factor $\Delta K = \Delta\sigma\sqrt{\pi a}\,Y$ and on the R ratio. According to this, a short crack loaded with a large stress range will propagate with the same rate as a long crack loaded with a small stress range provided the cyclic stress intensity factor ΔK is the same. In many cases, this simple picture is correct.

However, the statement of the previous paragraph only holds for macrocracks. Microcracks may grow faster than expected from the $\mathrm{d}a/\mathrm{d}N$ curve (figure 10.16), and they may even grow at a cyclic stress intensity factor below ΔK_{th} [21, 113]. On the one hand, this is due to the fact that the crack growth resistance of the material varies on the microscopic scale. A microcrack that is, for example, surrounded by favourably oriented grains may grow rather

---

[14] For the case $n = 2$, we have to integrate $1/a$, leading to $\ln a$. This case is dealt with in exercise 30.

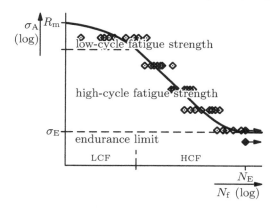

Fig. 10.18. Example of an *S-N* diagram with data points

quickly, whereas another crack is stopped at a grain boundary because neighbouring grains are less favourably oriented. On the other hand, short cracks may remain open even under compressive loads because they are embedded in a plastic deformation field [113]. This explains why crack propagation may take place even below ΔK_{th}.

If a component contains only microcracks or is not cracked at all, $\mathrm{d}a/\mathrm{d}N$ curves cannot be used to assess the life time. In this case, other methods are required that are the subject of the next section.

10.6.2 Stress-cycle diagrams (*S-N* diagrams)

At the beginning of the chapter, we already saw that the complex load-time curves occurring in real life are usually replaced by simplified curves in the laboratory e. g., using sinusoidal loading. Frequently, smooth specimens are used, similar to the tensile specimens discussed in section 3.2. They are loaded cyclically with a fixed period, prescribing the stress amplitude σ_a or the strain amplitude ε_a, and also the R ratio (R or R_ε, respectively). The advantages and disadvantages of these two experimental procedures will be discussed at the end of this section; in the following, we will consider stress-controlled experiments only.

For each fatigue experiment, the number of cycles to failure[15] is measured. If several fatigue experiments are performed and the number of cycles to failure N_f is plotted versus the stress amplitude σ_A or the stress range $\Delta\sigma$, the resulting diagram is called a *stress-cycle* (or *S-N*) *diagram* (sometimes also *stress-life* or *Wöhler diagram,* see figure 10.18). We denote the stress values in the *S-N* diagram with capitalised subscripts. For example, we denote the stress amplitude that causes failure after N_f cycles as σ_A instead of σ_a. The number of cycles can also be specified in the subscript, as in σ_{AN_f}, stating, for example,

---

[15] Failure can be defined as fracture of the specimen or occurrence of a crack.

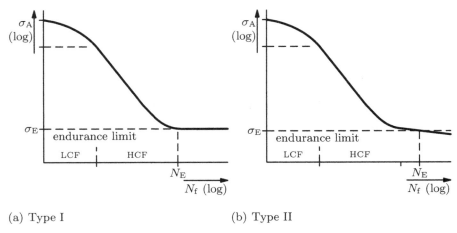

(a) Type I   (b) Type II

Fig. 10.19. The characteristic types of S-N curves

$\sigma_{A(1.5\times10^4)}$ = 130 MPa. The number of cycles to failure is always plotted logarithmically in the S-N diagram; the stress can be plotted logarithmically or linearly.

Some materials exhibit a true *fatigue limit* (sometimes also called the *endurance limit*). In this case, there exists an *limiting number of cycles* N_E, with the S-N curve being almost horizontal at a larger number of cycles. In this case, the S-N diagram is of type I (figure 10.19(a)). A specimen that has survived N_E cycles never fails. The experiment can be stopped and the specimen can be marked accordingly, usually with an arrow in the diagram (sometimes denoted as 'run out', see figure 10.18). Frequently, N_E takes values between 2×10^6 and 10^7, depending on the material. The stress level that corresponds to N_E in the S-N curve is called the *fatigue strength, endurance limit,* or *fatigue limit* σ_E.

In many materials, there is no horizontal part of the S-N curve (type II, figure 10.19(b)). Although the slope of the S-N curve becomes smaller beyond a certain number of cycles, failure can still occur. These materials thus have no true fatigue limit. To ensure safety of the component, a limiting number of cycles of 10^8 is often used, ten times larger than the usual value for materials with a true fatigue limit. To state explicitly that a fatigue strength corresponds only to a certain number of cycles, not to a true fatigue limit, the number of cycles can be added to the subscript, as in $\sigma_{E(10^8)}$.

So far, we have only looked at large numbers of cycles, the so-called *high-cycle fatigue* (HCF) regime. As we already saw in the introduction of the chapter for the example of the car engine (section 10.1), it is sometimes necessary to design against a rather limited number of cycles. If this number is smaller than about 10^4, the term *low-cycle fatigue* (LCF) is used. However, the number of cycles that characterises the transition from low- to high-cycle fatigue is not well-defined [130]. A stress amplitude that causes failure in the LCF regime is called *low-cycle fatigue strength,* an amplitude causing failure in the HCF regime is called *high-cycle fatigue strength.*

As can be seen from figure 10.18, the slope of the *S-N* curve is usually much smaller in the LCF than in the HCF regime so that a small change in the stress amplitude has a large effect on the number of cycles. This phenomenon is restricted to metals and polymers and will be discussed for the case of metals in the next section.

If the maximum stress σ_{max} reaches the strength of the monotonous experiment in the first cycle (the tensile strength R_m for the case of axial loading), the specimen fractures during this cycle. Often, the number of cycles to failure is then taken to be $N_f = 0.5$. The left end of the *S-N* curve is thus determined by $\sigma_{A(0.5)} = 0.5(1 - R)R_m$.

Independent of the material tested, the scatter of the cycles to failure is usually rather large, for even small defects in the material or on the surface can have a strong effect on the life time. Different specimens thus are never identical. For this reason, several experiments have to be performed at each stress level (usually 6 to 10) to allow ascertaining the width of the scatter band. Using statistical methods, limiting curves can be constructed that represent a certain probability of failure (for example, 95%). This is elaborated on in *Forrest* [50], *Radaj* [113] or *Schott* [130].

As the introductory example of a car engine (see section 10.1) shows, real-life fatigue loads can be stress- or strain-controlled. Stress-controlled loads occur if the loads are determined by external forces, strain-controlled loads, for example, if there are temperature changes causing thermal strains. In many cases, LCF loads are strain-controlled and HCF loads stress-controlled. This, however, cannot be used as a rule. For example, loads in a rotating disc are determined by centrifugal forces. Since these are constant during rotation, switching the device on and off corresponds to a single cycle. The load is thus stress-controlled, but the number of cycles is low (LCF). Usually, stress- or force-controlled experiments are easier to perform than strain-controlled experiments and are thus often preferred. This is especially true in the HCF regime.

If we look at an *S-N* curve (figure 10.18), we can see that the number of cycles to failure strongly depends on the stress in the LCF regime. Small scatter in the stress-strain properties of different specimens (due to scatter in the material properties, for example) would cause large changes in the number of cycles to failure measured in the experiment. The scatter band would thus be rather wide. In this regime, strain-controlled experiments are more useful since, with a prescribed strain amplitude, the scatter of the stress amplitude is small. Furthermore, stress-controlled experiments would also cause more rapid failure due to the reduction in the cross section of the specimen caused by crack propagation [113].

To assert the influence of notches and inhomogeneous stress distributions on fatigue life, experiments can also be performed with notched components or specimens, resulting in specific S-N curves. The influence of notches on fatigue life is discussed in more detail in section 10.7.

S-N curves of metals

In a double-logarithmic plot, the S-N curve of many metals is a straight line for a wide range of the number of cycles (see figure 10.18). This line can be described by the *Basquin equation* [14]

$$\sigma_A = \sigma'_f (2N_f)^{-a}. \tag{10.12}$$

The *fatigue strength coefficient* σ'_f is related to the tensile strength. In plain carbon and low-alloy steels, a rule of thumb states $\sigma'_f = 1.5R_m$; in aluminium and titanium alloys, $\sigma'_f = 1.67R_m$ holds approximately [113]. The *fatigue strength exponent* a depends on the material and the specimen geometry; in many materials, it takes values between 0.05 and 0.12 if smooth specimens are used [8, 113].

In plain carbon steels and titanium alloys with body-centred cubic lattice, there is a true fatigue limit with a horizontal S-N curve at a number of cycles beyond 2×10^6 to 10^7 [130] (type I, figure 10.19(a)). This, however, is not true for notched specimens (and thus also for components) or if corrosion or oxidation occur during the experiment.

Face-centred cubic metals and hardened steels do not have a true fatigue limit (S-N curve of type II, figure 10.19(b)). At a number of cycles beyond 10^7, the slope of the S-N curve is rather small and a limiting number of cycles of $N_E = 10^7$ to 10^8 can be used to design safely against fatigue [130].

> Recently, it has been found even in body-centred cubic metals that a specimen can fail in fatigue even beyond the limiting number of cycles (10^7). At a very large number of cycles (more than 10^{10}), the S-N curve may drop again [93, 135]. This is called *ultra-high-cycle fatigue* (UHCF) or *very-high-cycle fatigue*.
>
> In contrast to failure at smaller numbers of cycles, which usually start from the surface, failure in the UHCF regime is caused by microcracks being initiated at microscopic inclusions slightly below the surface of the specimen, visible as so-called *fish eyes* at the surface [135, 139].

S-N curves of metals have a small slope at low numbers ($N_f \lesssim 10^3$) of cycles as well as in the regime $N_f > N_E$. In this region, the yield strength of the material is exceeded, and the strain amplitude increases rapidly with the stress amplitude. A slight increase of the stress causes much larger plastic deformations and thus strongly reduces the life time.

If we plot the strain amplitude $\varepsilon_A(N_f)$ versus N_f in a double-logarithmic plot, we get a *strain-cycle diagram* as shown in figure 10.20. Two linear regimes, with a smooth transition between them, can be discerned. As we will see soon,

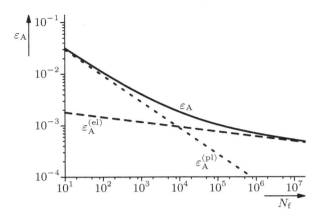

Fig. 10.20. Strain-cycle diagram with $\sigma'_f = 470\,\mathrm{MPa}$, $E = 210\,000\,\mathrm{MPa}$, $a = 0.1$, $\varepsilon'_f = 0.1$, and $b = 0.5$

these linear regimes are related to the elastic and the plastic part of the total strain. The total strain amplitude can be decomposed as

$$\varepsilon_A = \varepsilon_A^{(el)} + \varepsilon_A^{(pl)} . \tag{10.13}$$

Large numbers of cycles (HCF) can only be reached with a small stress amplitude so that the amount of plastic deformation is small. The total strain thus corresponds mainly to the elastic part of the strain. The line can be described using the Basquin equation, re-written with the help of Hooke's law:

$$\varepsilon_A^{(el)} = \frac{\sigma'_f}{E} (2N_f)^{-a} . \tag{10.14}$$

At a small number of cycles (LCF), the stresses are large and the total strain is mainly determined by plastic deformation. In the LCF regime, a good approximation for the relation between plastic strain amplitude and cyclic life is given by the *Coffin-Manson equation* [32, 94, 95].

$$\varepsilon_A^{(pl)} = \varepsilon'_f \cdot (2N_f)^{-b} . \tag{10.15}$$

For the *fatigue ductility coefficient* ε'_f, the true fracture strain in tensile loading can be used as a good approximation. The *fatigue ductility exponent* b depends on the hardening of the material. Typical values for b are in the range of 0.4 to 0.73 [113, 130].

Adding both parts of the strain (equation (10.13)) yields

$$\varepsilon_A = \frac{\sigma'_f}{E} (2N_f)^{-a} + \varepsilon'_f \cdot (2N_f)^{-b} , \tag{10.16}$$

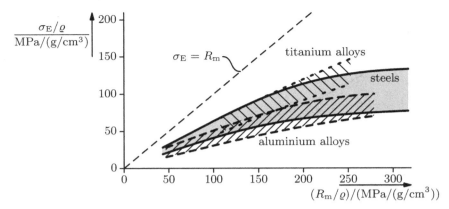

Fig. 10.21. Dependence of the fatigue limit for fully reversed loading ($R = -1$) on the tensile strength in different material classes [113]

the strain-cycle diagram shown in figure 10.20. This equation frequently is called *Coffin-Manson-Basquin equation*.

As we saw in section 10.2.1, the crack initiation – and thus the fatigue strength – of smooth specimens of ductile materials is determined by accumulated plastic deformation which usually occurs at the surface. For this reason, the fatigue limit σ_E for fully reversed loading, $R = -1$, is related to the strength under static loads. The most suitable parameter to quantify this relation is not the yield strength R_p, as might be expected, but the tensile strength R_m or a combination of both, as already used in the failure-assessment diagram in section 5.2.3 [113].

In low-strength materials, the fatigue limit is usually proportional to the static strength. In high-strength metals, the fatigue limit increases only slightly within a material class (figure 10.21). The reason for this is that high-strength materials are very notch-sensitive, and the fatigue limit is thus determined by surface or inner defects. A large number of approximation formulae for the fatigue strength can be found in the literature [90, 113, 130, 147]; some of them are listed in table 10.2, taken from *Radaj* [113].

S-N diagrams of ceramics

As already explained in section 10.3, many ceramics do not exhibit any cyclic effects and can thus bear infinitely many cycles of any load that is smaller than the static strength (e. g., under tension or bending). In these ceramics, the S-N curve is simply a horizontal line at $\sigma_{Max} = R_m$ or $\sigma_A = 0.5(1 - R)R_m$ (for the example of a uniaxial load). For a fully reversed stress, this results in

$$\sigma_E|_{R=-1} \approx R_m \, . \tag{10.17}$$

Table 10.2. Approximate fatigue limit of some metals

material class	fatigue limit σ_E for $R = -1$	
steels	$= (0.35\ldots0.65) \times R_m$	for $R_m < 1\,400\,\mathrm{MPa}$
	$\approx 700\,\mathrm{MPa}$	for $R_m \geq 1\,400\,\mathrm{MPa}$
cast irons	$= (0.3\ldots0.4) \times R_m$	for $R_m < 500\,\mathrm{MPa}$
aluminium alloys	$= (0.3\ldots0.5) \times R_m$	for $R_m < 325\,\mathrm{MPa}$
	$\approx 130\,\mathrm{MPa}$	for $R_m \geq 325\,\mathrm{MPa}$
titanium alloys	$= (0.45\ldots0.65) \times R_m$	for $R_m < 1\,100\,\mathrm{MPa}$
	$\approx 620\,\mathrm{MPa}$	for $R_m \geq 1\,100\,\mathrm{MPa}$

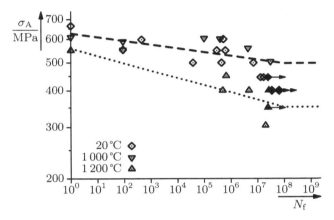

Fig. 10.22. S-N diagram of Si_3N_4 at different temperatures (measured in bending at $R = -1$) [113]. The dashed line is a fit according to the Basquin equation, common to temperatures of 20°C and 1000°C, whereas the dotted line is valid at 1200°C

This can be exploited to test components with the proof test (section 7.4). If the component does not fail during the test, it can be assumed that it will not fail by fatigue in service.

If mechanical fatigue occurs in a ceramic, equation (10.17) does not hold anymore, and an S-N curve is useful. Figure 10.22 shows such a curve for silicon nitride at three different temperatures. As it is usual for ceramics, the slope of the S-N curve is small. Slightly reducing the stress thus can significantly increase the life time. The fatigue limit is only slightly below the static strength. Fatigue occurs in this ceramic because the crack propagates on the glassy phase of the grain boundaries (see section 7.5.2), resulting in crack bridging effects as explained in section 10.3. However, the effect is rather small.

The number of cycles to failure is almost identical for 20°C and 1000°C so that the same fit curve can be used to describe both. Raising the tempera-

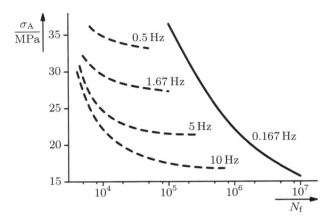

Fig. 10.23. *S-N* diagram of polyoxymethylene (polyacetal) at different loading frequencies (after [102]). The solid line corresponds to mechanical fatigue; at higher frequencies (dashed lines), the material fails by thermal fatigue. If the load is reduced, the thermal fatigue curves join with the curve for mechanical fatigue

ture to 1200 °C has a marked influence on the fatigue strength of the ceramic because creep (see chapter 11) occurs in this case. This small temperature dependence over a wide temperature range is also typical of ceramics.

S-N curves of polymers

We already saw in section 10.4 that the fatigue behaviour of polymers strongly depends on the load frequency because of their viscoelastic properties. If the frequency is sufficiently large, the polymer can fail by thermal fatigue due to the heat generated during deformation. This is shown for the example of a thermoplastic polymer in figure 10.23. At low frequencies, the thermoplastic fails by crack formation and propagation, similar to a metal, at higher frequencies, thermal fatigue occurs (section 10.4.1), and the fatigue strength strongly decreases. The load frequency is for this reason usually limited to 10 Hz.

S-N curves of different polymers are depicted in figure 10.24. In many polymers (e. g., PVC, PP, PA), the *S-N* curve is horizontal at a large number of cycles, corresponding to a curve of type I (see figure 10.19(a)). However, as figure 10.23 shows, this may be due to thermal fatigue, and in this case the horizontal part of the curve meets the curve for true mechanical fatigue at higher numbers of cycles.

S-N curves of polymers have to be used with caution in designing components. The fatigue strength depends much more strongly on the load frequency than in metals because the equilibrium between heat production and dissipation plays a crucial role. To design components, experiments should be as close to real service conditions as possible.

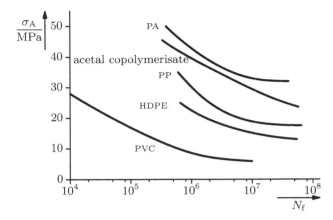

Fig. 10.24. *S-N* curves for fully reversed bending of several polymers at a temperature of 20°C and a loading frequency of 10 Hz (polyamide: 15 Hz, after [41])

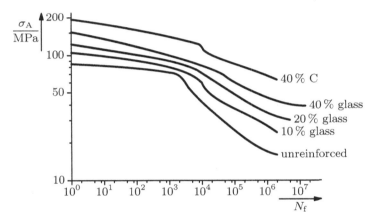

Fig. 10.25. Comparison of the *S-N* curves of unreinforced as well as glass- and carbon-reinforced polysulfone (simplified plot after [29])

S-N curves of fibre composites

According to section 10.5, the fatigue strength of fibre composites is usually higher than that of the matrix material alone. This is shown in figure 10.25, using the *S-N* curves of unreinforced and short-fibre reinforced polysulfone. The increased fatigue strength is apparent from the figure. Long carbon fibres are especially efficient in increasing the fatigue strength of polymer matrix composites, not only because of their high stiffness, but also because of their thermal conductivity. The picture is similar in metal matrix composites. For example, adding 20% silicon carbide fibres to an aluminium matrix doubles the fatigue strength [140].

Because fibre composites usually do not fail by formation and growth of a single crack, but by accumulating damage, their stiffness decreases with increasing number of cycles. This is similar in unreinforced materials since the growing crack reduces their stiffness as well. However, a significant reduction in stiffness is usually observed only shortly before ultimate failure, whereas a damaged composite may have a long life time despite its reduced stiffness.

10.6.3 The role of mean stress

In the S-N diagram (see the previous section), we plot all values at constant R ratio. To quantify the influence of the mean stress or the R ratio for the whole range of numbers of cycles, from the LCF regime to the fatigue limit, an extensive number of experiments are required. If this is done, the result is as should be expected: The curves shift to smaller stress amplitudes with increasing mean stress. In many cases, only the dependence of the fatigue limit is of interest. In this case, the fatigue strength diagrams after Smith and Haigh can be used.

Smith's fatigue strength diagram

To draw a *Smith's fatigue strength diagram,* the stress amplitude at the fatigue limit σ_E is measured for different values of the mean stress σ_m.[16] The maximum and minimum stress, σ_{Max} and σ_{Min}, are plotted in a diagram as shown in figure 10.26. As can be seen from the figure, the stress amplitude σ_E decreases with increasing mean stress as expected. Because plastic flow of the material is not allowed, the diagram is limited horizontally by the yield strength R_p or R_e in the tensile, and by the compressive strength R_c (in ductile metals, this is usually equal to R_p) in the compressive region.

Haigh's fatigue strength diagram

A *Haigh's fatigue strength diagram,* or *Haigh's diagram* for short, corresponds to a Smith's fatigue strength diagram in which the stress amplitude σ_E is plotted versus the mean stress, instead of the maximum and minimum stress. The distance of a data point from the abscissa in Haigh's diagram thus corresponds to the vertical distance of the point from the diagonal in Smith's diagram. Because the stress amplitude is the same above and below the mean stress, only the upper part of the curve is drawn (figure 10.27(a)). To plot the R ratio $R = \sigma_{min}/\sigma_{max} = (\sigma_m - \sigma_a)/(\sigma_m + \sigma_a)$ in the figure, this equation is rewritten as

---

[16] Lowercase subscripts are used for the prescribed quantities, e. g., the mean stress σ_m, while uppercase subscripts are used for the resulting quantities for endurance, e. g., the stress amplitude σ_E. Or, in other words, for a given mean stress σ_m, we search the fatigue limit σ_E.

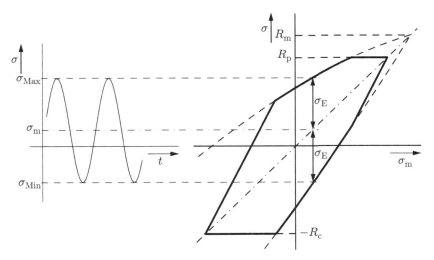

Fig. 10.26. Stress-time diagram and Smith's fatigue strength diagram. As shown, the points in the fatigue strength diagram can be taken directly from the different stress-time diagrams. In metals, the compressive yield strength R_c is usually the same as R_p

$$\sigma_a = \frac{1 - R}{1 + R}\, \sigma_m\,. \tag{10.18}$$

According to this, the relation between σ_a and σ_m corresponds to a straight line through the origin, with a slope that depends on the R ratio, as drawn in figure 10.27(a).

The limits at $+R_p$ and $-R_c$ are not as easy to draw in Haigh's diagram as in Smith's. The relation $\sigma_{Max} = \sigma_m + \sigma_E = R_p$ yields $\sigma_E = R_p - \sigma_m$; from $\sigma_{Min} = \sigma_m - R_c$ we get $\sigma_E = R_c + \sigma_m$. Both limits thus correspond to straight lines with a slope of $\pm 45°$, intersecting the axis at R_p and R_c, respectively. These are shown in figure 10.27(a).

Different approximations can be used to describe the fatigue strength diagram between these limits [130]. Frequently, a linear *Goodman equation* is used, corresponding to a straight line that connects the fatigue strength σ_E at $R = -1$ and the tensile strength R_m at $R = 1$ (figure 10.27(b)). This approximation is valid for positive mean stresses ($-1 \le R \le 1$). Mathematically, it can be described by the equation [8, 76]

$$\sigma_E(\sigma_m) = \left(1 - \frac{\sigma_m}{R_m}\right) \sigma_E|_{R=-1}\,. \tag{10.19}$$

Furthermore, the condition $\sigma_{Max} \le R_p$ must hold to avoid yielding. The curve constructed in this way serves as a reasonable and conservative approximation for ductile materials in its range of validity.

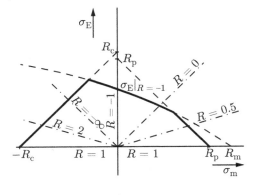

(a) Exact fatigue strength diagram. The straight lines through the origin correspond to constant R ratios

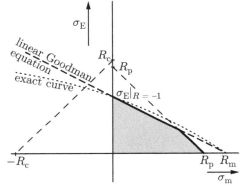

(b) Approximation of the fatigue strength diagram by a linear Goodman equation. The region where the approximation is safe is shown in grey

Fig. 10.27. Haigh's fatigue strength diagram

*10.6.4 Fatigue assessment with variable amplitude loading

The S-N diagram plots the life time of a material at constant stress amplitude and R ratio. It is, however, not possible to assert the life time, using the diagram, if the load amplitude changes. The most obvious way to determine the life time in this case is to simulate the service load history in the laboratory. Unfortunately, this is a rather involved procedure that is not feasible in most cases. It would be helpful if it were possible to estimate the life time directly from the S-N curves. One way to do this is to use *Miner's rule* (also known as *Palmgren-Miner rule*) [99] that will be explained now. The rule is rather simple and thus easy to employ, but it has some disadvantages, discussed at the end of this section.

To use Miner's rule, a partial damage of the component is calculated for each loading step. Assume that the component is loaded with k different stress amplitudes $\sigma_{a,j}$, $j = 1, \ldots, k$ and that the number of cycles for amplitude j is n_j. We can then use the S-N curve to determine the number of cycles to failure for each of the stress amplitudes, $N_{f,j}(\sigma_{a,j})$. Miner's rule now assumes that each step 'uses up' part of the components life time, with a partial damage $D_j = n_j/N_{f,j}$. The component fails when the total damage D equals one:

$$D = \sum_{j=1}^{k} \frac{n_j}{N_{f,j}} = 1 . \tag{10.20}$$

The sequence of the load steps is not taken into account in equation (10.20). It is easy to see that this is not valid in some cases. Assume that we use two stress amplitudes, $\sigma_{a,1}$, being smaller than the fatigue limit σ_E, and $\sigma_{a,2}$, being larger. If we start with a sufficient number of cycles at the higher load $\sigma_{a,2}$, microcracks will form according to section 10.2.1. Afterwards, the smaller stress amplitude $\sigma_{a,1}$ is sufficient to further propagate the crack. Miner's rule, however, associates no damage with the smaller load because the number of cycles the component would live at this load alone is $N_{f,1} = \infty$, corresponding to $D_1 = 0$. Thus, a calculational damage of $D = 1$ does never occur, and the specimen fails although D stays smaller than 1.

If we load another specimen with the reversed sequence, no damage will be caused by the first stress amplitude $\sigma_{a,1}$, and we find $D_1 = 0$ for any n_1. All damage in the material is accumulated during the second loading step, at stress amplitude $\sigma_{a,2}$, causing failure at $n_2 = N_{f,2}$ when $D = D_2 = 1$ holds. As the example shows, the sequence of loading may influence the fatigue life strongly, an effect neglected in Miner's rule.

> Sometimes, cyclic hardening occurs during the first loading step (see section 10.6.5 below). The specimen will then yield less in the next loading steps. This so-called *coaxing* is especially important if the stress amplitude is increased in many loading steps. If a training effect occurs, the specimen may fail at damage values $D > 1$.

To summarise, it can be stated that Miner's rule is useful to provide a first approximation of the life time, but it has to be used with great care, for the calculated life time may not be a conservative estimate.

∗ 10.6.5 Cyclic stress-strain behaviour

We already mentioned in section 10.2.1 that the stress-strain behaviour of metals may change during cyclic loading. Depending on the initial state, different effects may occur.

∗ Cyclic hardening and softening

If we perform, for example, a fatigue experiment at constant strain amplitude ε_a, the stress amplitude σ_a changes during the experiment. Figure 10.28 shows typical cases. If the stress increases during loading, *cyclic hardening* occurs (figure 10.28(a)); if the stress decreases, the phenomenon is called *cyclic*

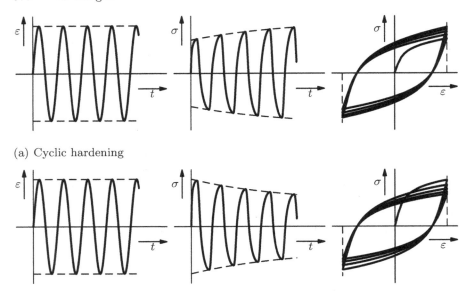

(a) Cyclic hardening

(b) Cyclic softening

Fig. 10.28. Cyclic stress-strain behaviour at the beginning of strain-controlled fatigue experiments (after [130]). The controlled variable is shown on the left, the material answer in the centre

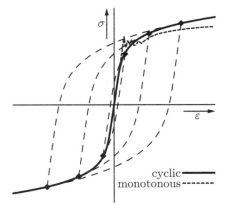

cyclic ——
monotonous ------

Fig. 10.29. Cyclic stress-strain diagram. A static stress-strain curve is shown for comparison

softening. Frequently, the stress amplitude changes only initially and then stabilises to a constant value.

If we perform cyclic experiments at different strain amplitudes and plot the stabilised values of the stress amplitude, we arrive at the *cyclic stress-strain diagram* sketched in figure 10.29. Usually, it does not coincide with the result of a monotonous tensile or compressive test. If cyclic hardening occurs,

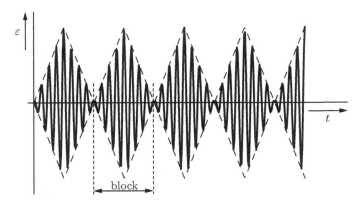

Fig. 10.30. Time-dependence of the strain to determine the cyclic stress-strain curve in an incremental-step test

the curve lies at higher stress values; if there is cyclic softening, the curve lies below the monotonous one.

The cyclic stress-strain curve is frequently approximated by the Ramberg-Osgood law, equation (3.15), using modified parameters K' and n':

$$\varepsilon_{\mathrm{a}} = \frac{\sigma_{\mathrm{a}}}{E} + \left(\frac{\sigma_{\mathrm{a}}}{K'}\right)^{1/n'} .$$

Cyclic hardening and softening are caused by dislocation movement under cyclic loads. New dislocations form, existing dislocations rearrange to reduce the stored energy, and dislocations may also annihilate. If the dislocation density is initially small, multiplication of dislocations usually causes cyclic hardening. If the dislocation density is large, for example because of work hardening of the material prior to cyclic loading, the material may soften by dislocation rearrangement and annihilation.

The microstructure of the material may also change under cyclic loads. In precipitation-hardened alloys with underaged precipitates (see sections 6.3.1 and 6.4.4), the precipitates may be destroyed by repeated cutting, reducing the strength of the material. In ferritic steels, the dislocations may detach from their surrounding carbon atoms so that no apparent yield strength exists in the cyclic stress-strain diagram, and the cyclic curve lies below the static one in this strain range (see section 6.4.3). Cyclic softening results.

To reduce the experimental efforts in measuring cyclic stress-strain curves, the *incremental-step test* can be used. In this test, the strain amplitude is varied block-wise between zero and a maximal value as sketched in figure 10.30. After the block has been repeated several times, the material behaviour does not change anymore and a stationary state is arrived at. If the stress is measured at each of the strain maxima, the

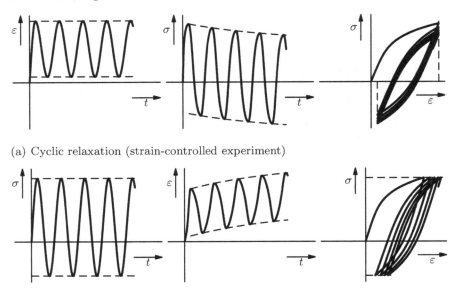

(a) Cyclic relaxation (strain-controlled experiment)

(b) Cyclic ratchetting (stress-controlled experiment)

Fig. 10.31. Cyclic stress-strain behaviour at the beginning of fatigue experiments with non-zero mean strain or stress, respectively (after [130]). The controlled variable is shown on the left, the material answer in the centre

cyclic stress-strain curve can be obtained. In this way, the whole curve can be measured using only one specimen.

The cyclic stress-strain curve can be used, for example, to perform finite element simulations of cyclic loadings. To simulate the complete experiment in the computer, it would be necessary to obtain information on the hardening of the material (isotropic and kinematic hardening) and to determine a material model that correctly describes it. This would be an extremely complicated procedure. Furthermore, the entire number of cycles would have to be calculated, which would require an immense amount of computing time. Instead, the flow curve, taken by the finite element software to be monotonous, can be replaced by the cyclic stress-strain curve. A single, monotonous loading of the component is then simulated. Stresses and strains calculated in this way correspond well with those in the cyclically loaded component.

∗ Cyclic relaxation and ratchetting

If a strain-controlled fatigue experiment is performed at a non-zero mean strain, cyclic relaxation may occur in addition to cyclic hardening or softening, with the mean stress decreasing over time (figure 10.31(a)). If, on the other hand, the experiment is stress-controlled at a non-zero mean stress, the hysteresis loop frequently shifts along the strain axis as shown in figure 10.31(b). This phenomenon is called *ratchetting*.

Again, dislocation movement is responsible for this. In a simple model, we can imagine (for the case $\sigma_m > 0$) that there are obstacles to dislocation movement in the material (for example, precipitates) which are strong enough that they cannot be overcome by the maximum shear stress occurring at σ_{max}. Under cyclic loading, the dislocations move backwards and forwards between the obstacles. Since the deformation starts elastically upon load reversal before dislocation movement is activated, there is a hysteresis in the stress-strain diagram (see also figures 3.30(b) and 10.7). Occasionally, thermal activation (see section 6.3.2) or other external processes may enable a dislocation to overcome one of the obstacles. Another mechanism to overcome obstacles is provided by dislocation pile-up which locally increases the stress. Thus, a dislocation may occasionally surmount one of the obstacles and cause an additional plastic strain. Because the absolute value of σ_{max} is larger than that of σ_{min}, the shear stress and thus the force on the dislocations is larger at σ_{max}, resulting in a higher probability of overcoming the obstacle. This causes the shift of the hysteresis loop shown in figure 10.31(b). The argument is the same if $\sigma_m < 0$, but in this case, the maximum shear stress occurs at σ_{min}. Altogether, plastic strain increases slightly in each cycle, causing ratchetting.

* 10.6.6 Kitagawa diagram

In sections 10.6.1 and 10.6.2, we discussed two different ways to design with cyclically loaded materials. Using fracture mechanics, we found in section 10.6.1 that a crack cannot propagate further if the cyclic stress intensity factor ΔK is smaller than a limiting value ΔK_{th} which depends on the material and the load. This argument, however, was only valid if the methods of linear-elastic fracture mechanics can be applied. This is the case for macrocracks, cracks with a length larger than the typical length scale of microstructural features (e. g., grains) in the material. Furthermore, the plastic zone near the crack tip has to be small compared to the size of the crack (see also sections 5.2 and 5.3). As already stated in section 10.2.1, this explains why smooth specimens may fail by formation and growth of microcracks, although the threshold value ΔK_{th} is not exceeded initially.

This fact can be visualised using the so-called *Kitagawa diagram*. In this diagram, the stress range $\Delta\sigma = \sigma_{max} - \sigma_{min}$ is plotted versus the crack length a on a double-logarithmic scale (figure 10.32), and a line is drawn to separate the regions of finite and infinite life times. If the crack length exceeds a critical value a^*, the fatigue strength is determined by linear-elastic fracture mechanics:

$$\Delta\sigma_{th} = \frac{\Delta K_{th}}{\sqrt{\pi a}\, Y} \, . \tag{10.21}$$

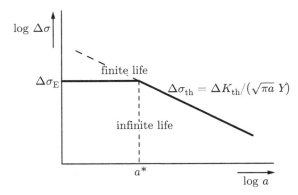

Fig. 10.32. Kitagawa diagram to determine the influence of defects on the fatigue limit. The stress range $\Delta\sigma = 2\sigma_A$ is plotted versus the crack length a using a double-logarithmic scale. $Y = 1 = \text{const}$ was assumed

If the geometry factor Y does not depend on the crack length, a straight line with slope -0.5 results in the double-logarithmic plot. This line, however, cannot be extended to arbitrarily small crack lengths, for it is limited by the fatigue limit $\Delta\sigma_E = 2\sigma_E$ measured on smooth specimens. As already stated, this limit is due to the fact that the stress level is sufficient to initiate microcracks that can propagate through the specimen. When these cracks have become large enough to be treated as macrocracks, the stress intensity factor ΔK is larger than ΔK_{th}, and the crack will not stop, causing failure after a finite number of cycles.

The *critical crack length* a^* is determined by

$$a^* = \frac{1}{\pi}\left(\frac{\Delta K_{th}}{\Delta\sigma_E Y}\right)^2. \tag{10.22}$$

In metals, it is usually a few hundredth to tenth of a millimetre [113]. If $a < a^*$, linear-elastic fracture mechanics is not valid anymore, rendering the stress intensity factor useless. As shown in figure 10.32, the maximum allowed stress range does not depend on the crack length in this case.

If the geometry factor depends on the crack length, the right part of the curve in the Kitagawa diagram deviates from a straight line. Because the critical crack length a^* is usually small, this length dependence is only of minor importance at a^*.

The transition from the fatigue limit measured on smooth specimens to the fracture-mechanically controlled part of the curve is not as abrupt in reality as sketched in the idealised curve of figure 10.32, but is smoothed. This is shown in figure 10.33. The figure also illustrates that even small defects with a size below the already small value a^* can significantly reduce the fatigue strength. In the diagram, for example, the stress range for infinite life $\Delta\sigma$ reduces to 82 % of the fatigue limit $\Delta\sigma_E$ even at $a/a^* = 0.5$. This again serves

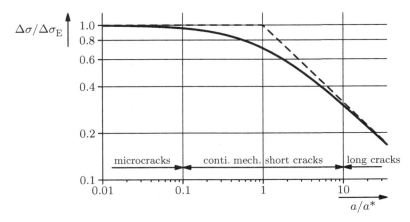

Fig. 10.33. More precise approximation of a Kitagawa diagram (after [113])

to illustrate how strongly the fatigue strength of a material depends on the material quality (surface roughness, porosity, etc.), compared to the static case.

* 10.7 Fatigue of notched specimens

As explained in chapter 4, notches cause a stress concentration in a component. Thus, it should be expected that notches also affect the fatigue strength of a component. The stress concentration at the notch root is again described by the stress concentration factor K_t according to equation (4.1):

$$\sigma_{a,max} = K_t \sigma_{a,nss} \, .$$

If we assume that the maximum stress amplitude in the component must not exceed the fatigue limit of a smooth specimen, σ_E, we should expect that the maximum nominal stress amplitude for a notched specimen is

$$\sigma_{E,nss,expected} = \frac{\sigma_E}{K_t} \, , \tag{10.23}$$

where the stress amplitude $\sigma_{E,nss}$ is calculated at the notched cross section. Experimentally, it is found that a notched specimen can bear larger loads in fatigue than predicted by equation (10.23). Thus, we define the *fatigue notch factor* (or *fatigue strength reduction factor*) K_f as quotient of the fatigue limit of a smooth specimen σ_E and that of the notched specimen $\sigma_{E,nss}$

$$K_f = \frac{\sigma_E}{\sigma_{E,nss}} \, . \tag{10.24}$$

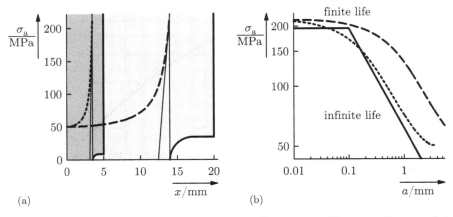

(a)   (b)

Fig. 10.34. Illustration of the notch support effect using a Kitagawa diagram. Subfigure (a) shows two geometries with stress concentration factor $K_t = 3$. The stress fields of the geometries differ in the gradient at the notch root. If a crack starts there, it is in the upper (finite life time) part of the Kitagawa diagram (subfigure (b)). If the gradient is large, the crack is unloaded and stops; if the gradient is small, the crack continues to grow. In the figure, $Y = \text{const}$ was assumed

The fatigue notch factor K_f is limited by $1 \leq K_f \leq K_t$. The difference between K_f and K_t depends on the material, the notch geometry, and the load case. Since the fatigue limit of the notched specimen is larger than expected from the maximal stress at the notch root, it is frequently said that the notch has a supporting effect. This is a rather misleading term because the fatigue limit of the notched specimen is never larger than that of the un-notched one at the same net-section stress.

To explain the unexpectedly high fatigue limit of a notched specimen, the crack propagation in stage I is crucial in a ductile material. As we saw in section 10.6.6, it is not the stress intensity factor which is important at this stage, but the stress at the position of the crack. Since fatigue cracks usually start at the surface of the specimen at the notch root, a growing crack enters a region with smaller stress and can thus be partially unloaded.[17] Figure 10.34 shows the Kitagawa diagram for this case. With increasing crack length, the stress at the crack tip reduces and the crack may enter the part of the diagram where the life time is infinite and the crack may thus stop.

Whether a growing crack can be stopped in this way depends on how rapidly the stress decreases at the notch root. To quantify this decrease, we define the *relative stress gradient* at the notch root

---

[17] As we will see below, this is not true anymore as soon as linear elastic fracture mechanics can be applied. In this case, the increase of the stress intensity factor due to the growing crack is larger than the decrease of the stress in the notch root, see figure 10.38.

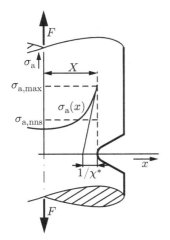

Fig. 10.35. Geometrical interpretation of the relative stress gradient χ^*

Table 10.3. Relative stress gradient of some geometries [76]

geometry	χ^*/mm^{-1}	
	tension-compression	bending
	$\dfrac{2}{\varrho}$	$\dfrac{2}{b}+\dfrac{2}{\varrho}$
	$\dfrac{2}{\varrho}$	$\dfrac{2}{d}+\dfrac{2}{\varrho}$
	$\dfrac{2}{\varrho}$	$\dfrac{4}{D+d}+\dfrac{2}{\varrho}$

$$\chi^* = \frac{1}{\sigma_{a,max}} \left. \frac{d\sigma_a}{dx} \right|_{x=X} , \tag{10.25}$$

where x and X are defined in figure 10.35, and where the maximum stress amplitude at the notch root is $\sigma_{a,max}$. The unit of the relative stress gradient is mm^{-1}. The relative stress gradient corresponds to the inverse of the distance between the notch root and the intersection of the tangent at $\sigma_a(x)|_{x=X}$ with the coordinate axis as in figures 10.34 and 10.35. It only depends on the geometry and can be found in approximation equations or tables [43,76]. Some example values are given in table 10.3.

Here we want to stress the difference between the relative stress gradient χ^* and the stress concentration factor K_t which both depend on geometry only. The stress concentration factor quantifies the concentration of the stress at

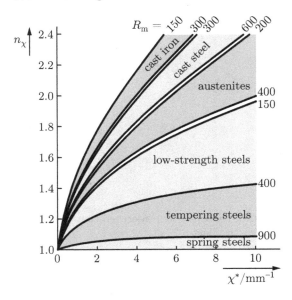

Fig. 10.36. Dependence of the notch support factor on the relative stress gradient for several iron-base materials (after [147])

the notch root. The relative stress gradient quantifies how quickly the stress decreases from this maximum value. As shown in figure 10.34, the relative gradient can be different even if the stress concentration factor is the same. The size of the component is especially important: If we scale the geometry isometrically, the stress concentration factor remains unchanged, whereas the relative stress gradient changes because it has the units of an inverse length. Even at the same stress concentration factor, the larger component is thus more notch-sensitive than the smaller.

If the relative stress gradient is small, the stress decreases only slowly with increasing distance from the notch root. To have the crack enter the region of the Kitagawa diagram where the life time is infinite, the maximum stress at the notch root must be smaller. Therefore, the fatigue limit $\sigma_{E,\mathrm{nss}}$ decreases with decreasing relative stress gradient. This is also confirmed by figure 10.36: In this figure, the *notch support factor*

$$n_\chi = \frac{K_t}{K_f} = \frac{\sigma_{E,\mathrm{nss}}}{\sigma_E/K_t} = \frac{K_t\sigma_{E,\mathrm{nss}}}{\sigma_E} \tag{10.26}$$

is plotted versus the relative stress gradient. The notch support factor n_χ does not compare the strength for static and cycling loading, but the observed fatigue limit $\sigma_{E,\mathrm{nss}}$ with the expected fatigue limit if the fatigue behaviour were determined by the elastic stress concentration at the notch root (σ_E/K_t). It is not a material parameter, but depends also on the geometry via the quantity

χ^*. The following inequality holds: $1 \leq n_\chi \leq K_t$. At $n_\chi = 1$, the notch exerts its full influence and the fatigue limit decreases as predicted by the stress concentration factor. This is the case in large components and for comparable values of K_t i.e., when χ^* is small. If $n_\chi = K_t$, $K_f = 1$ or, equivalently, $\sigma_{E,nss} = \sigma_E$ holds. In this case, the component is not weakened by the presence of the notch. If an n_χ value larger than K_t results from the diagram 10.36, the notch has no influence and n_χ is actually equal to K_t.

> The fact that a stress gradient at the surface increases the strength of a component can also be seen in bending. In bending, the stress decreases linearly with the distance from the surface. The gradient increases with decreasing component size. For this reason, components can bear higher loads at the surface if loaded in bending than in tension, and thin components have a higher fatigue limit than thick ones [76].

Figure 10.36 also shows a correlation between n_χ and the material strength. If we consider a material class, the notch support factor usually decreases with increasing tensile strength R_m. The notch support factor of a ferritic spring steel, for example, is smaller than that of a low-strength steel (see figure 10.36). This can be explained as follows: Materials with lower strength (and higher ductility) have a smaller fatigue limit σ_E but a comparable threshold value ΔK_{th} for crack propagation. According to equation (10.22), the critical crack length a^* is larger, making it possible for cracks starting at a notch root with a smaller relative stress gradient to enter the region of the Kitagawa diagram where the material survives. The cracks can thus be stopped. This may not be possible in the high-strength material because of the smaller critical crack length a^*, and the notch support factor becomes smaller. Figure 10.37 illustrates this by surveying the yield strength R_p (static test) and the fatigue limit (under fully reversed stress) of un-notched (σ_E) and notched specimens ($\sigma_{E,nss}$). The fatigue limit of a notched specimen depends not only on the material, but also on the notch geometry. In the example, it remains almost constant with increasing yield strength, but it may even decrease for some geometries. Since high-strength materials usually have to serve at higher loads, the notch sensitivity increases even if the fatigue limit of the notched specimen remains constant.

It may seem surprising that the notch support factor of brittle materials, like cast iron, is rather large (see figure 10.36). This is due to the fact that cracks in cast iron start at the graphite particles which act as inner defects and are statistically distributed. It is rather improbable that the crack that determines failure behaviour (the largest crack) is situated exactly at the notch root where the stress concentration becomes important. This is analogous to the dependence of the failure probability on the material volume according to the Weibull statistics (see section 7.3). This notch support in brittle materials also occurs under static loads.

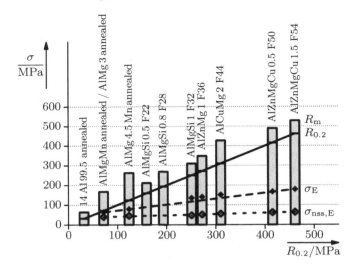

Fig. 10.37. Plot of the fatigue limit $(R = -1)$ versus the yield strength of some aluminium alloys (after [5])

* Notches in pre-cracked components

So far, in considering the effect of notches, we have assumed that the notch root contains no initial crack and that the component is designed against its fatigue limit. Therefore, we compared the fatigue properties with those measured for smooth, crack-free specimens. At finite life, cracks may form at the notch root and enter stage II of crack propagation. They can then be described using the stress intensity factor. In this case, the remaining life time is of interest, which cannot be calculated using the considerations made so far. To assess the life time, we consider the two limiting cases of the crack length a being small and large.

If the crack length is small, the crack 'feels' the stress concentration at the notch root almost completely. We can thus imagine the crack to be situated in an un-notched component loaded at the stress in the notch root, $K_t \Delta\sigma_{nss}$, ignoring the small decrease of the stress when moving away from the notch root to achieve a conservative design. In other words, the stress intensity factor for a notched specimen with stress concentration factor K_t and external load $\Delta\sigma_{nss}$ is

$$\Delta K = K_t \Delta\sigma_{nss} \sqrt{\pi a} \cdot Y_{short} \,. \tag{10.27}$$

Y_{short} is the geometry factor for a surface crack in a semi-infinite geometry (e.g., $Y = 1.1215$ for a crack in a plane geometry, see table 5.1). It is used here because the crack starts at the notch surface and is small compared to the component's dimensions.

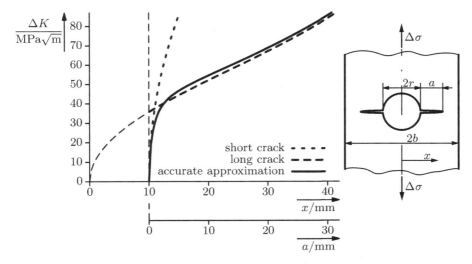

Fig. 10.38. Dependence of the stress intensity factor on the crack length a in a flat tensile specimen of width $2b = 160\,\mathrm{mm}$ with a central hole with diameter $2r = 20\,\mathrm{mm}$ at a load with a nominal stress amplitude of $200\,\mathrm{MPa}$. It is assumed that the crack grows symmetrically on both sides of the hole

If the crack is large, it has moved so far away from the notch stress field that it 'feels' only the far-field stress $\Delta\sigma$.[18] The notch is now in the region unloaded by the crack, and the fact that the notch opening is much wider than that of the crack is irrelevant for the mechanical behaviour. Thus we have to add the notch depth to the crack length, resulting in

$$\Delta K = \Delta\sigma\sqrt{\pi(a+t)}\,Y_{\text{long}}\,. \tag{10.28}$$

We already exploited this fact when we considered CT specimens in section 5.2.7 where we also measured the crack length starting at the loading point, not at the beginning of the crack (see figure 5.14(b)).

> It is also possible to estimate the cyclic stress intensity factor for in-termediate values of the crack length. This is explained in detail in *Radaj* [113] and *Dankert* [37]. Figure 10.38 shows the approximations for short and long cracks and an improved approximation for arbitrary crack length for an example, following *Dankert*. For short cracks of length up to $a_{\text{short}} = 0.5\,\mathrm{mm}$, equation (10.27) provides a good approximation. For cracks larger than about $a_{\text{long}} = 3\,\mathrm{mm}$, the more precise approximation exceeds the simple approximation formula for long cracks and approaches it gradually. This is due to the fact that

---

[18] The stresses $\Delta\sigma_{\text{nss}}$ from equation (10.27) and $\Delta\sigma$ from equation (10.28) differ because we use the cross section at the notch in the first and the total cross section in the second case.

the specimen becomes more compliant due to the hole in the middle of the crack, similar to a specimen with a surface crack being more compliant than one with an inner crack (see section 5.2.2). For very long cracks, the influence of the hole on the crack is negligible.

As figure 10.38 shows, the stress intensity factor always increases with increasing crack length, the stress gradient in the notch root notwithstanding. A crack in stage II, loaded in mode I, cannot be stopped by the decrease of the stress at the notch root. This can only happen for short cracks in stage I that cannot be described using fracture mechanics.

Using equations (10.27) and (10.28), the cyclic stress intensity factor can be estimated and it can be checked, according to section 10.6.1, whether the crack propagates ($\Delta K \geq \Delta K_{\text{th}}$) or how large the crack-growth rate $\mathrm{d}a/\mathrm{d}N$ is.

11

Creep

Creep is the time-dependent, plastic deformation of a material.[1] According to the definition from section 2.1, creep processes are *viscoplastic* processes. The time-dependent plastic deformation of polymers has already been covered in chapter 8. Here we discuss creep of metals and ceramics.

11.1 Phenomenology of creep

If a metallic or ceramic component is stressed at elevated temperature i. e., at a *homologous temperature* T/T_m (T_m: absolute melting temperature) of at least 0.3 to 0.4, the strain of the component increases with time at constant load. A typical, schematic plot of the strain is shown in figure 11.1 which also shows the strain rate $\dot{\varepsilon}$ versus the time and the strain. Initially, the component reacts with a time-independent strain ε_0 which consists of an elastic and a plastic part. The strain increases further with time, with the strain rate changing strongly, usually decreasing continuously. This region (region I) of the creep curve is called *primary creep* or *transient creep*. Following this is a region (region II) of *steady-state creep* or *secondary creep* with approximately constant strain rate.

This shape of the creep curve occurs only in materials that do not change their microstructure during the creep process. This is the case in simple alloys, but not in many technical alloys (see section 11.2.1 for more about this). A constant strain rate is also only observed if the stress in the component is kept constant. Since the cross section of the component decreases under tensile load during the deformation, the force on the component has to be reduced over time. In service, this is usually not the case so that no region of constant strain

---

[1] In the context of polymers, the time-dependent *elastic* deformation (retardation and relaxation, see section 8.2.1) is frequently denoted as creep as well. To avoid confusion, this is not done in this book, excepting the standard term 'creep modulus'.

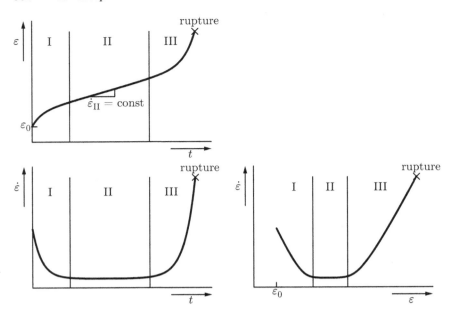

(a) Plot of the strain and strain rate   (b) Plot of the strain rate versus strain
versus time

Fig. 11.1. Stages of creep at constant stress

rate is observed. In these cases, the minimum creep rate is used instead of the
constant creep rate during secondary creep to quantify the creep behaviour.

After most of the life time has passed, the creep rate strongly increases
until final fracture ensues. In this region of *tertiary creep* (region III), massive
inner damage occurs in the material as we will discuss in detail in section 11.3.
This strongly reduces the load-bearing cross section and thus explains the
strong increase of $\dot{\varepsilon}$.

The primary part and the secondary part of the creep curve are frequently
described using empirical laws e. g., the *Garofalo equation* [40]:

$$\varepsilon = \varepsilon_0 + \varepsilon_t \left(1 - e^{-rt}\right) + \dot{\varepsilon}_{II} t \,. \tag{11.1}$$

In this equation, ε_t describes the additional strain during transient creep, $1/r$
quantifies the transition time between regions I and II, and $\dot{\varepsilon}_{II}$ is the constant
creep rate during secondary creep. Figure 11.2 illustrates how the total strain
is decomposed into these parts.

Experimentally, it is frequently observed that, during secondary creep, the
strain rate $\dot{\varepsilon}_{II}$ depends with a power law on the stress σ and exponentially on
the temperature T:

$$\dot{\varepsilon}_{II} = B\sigma^n \exp\left(-\frac{Q}{RT}\right) \,, \tag{11.2}$$

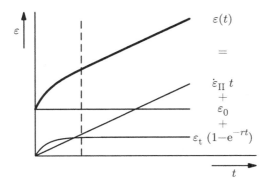

Fig. 11.2. Decomposing the total strain into its terms according to equation (11.1) (after [40])

where B is a constant[2], n is the *creep exponent,* and Q is an activation energy characterising the creep process. This creep law is called *power-law creep* or *Norton creep.* According to this law, creep is a thermally activated process. In crystalline materials, it is found that the activation energy Q is roughly equal to the activation energy for self-diffusion[3] of the material. This suggests that diffusion is important in creep and also explains why the onset of creep depends on the homologous temperature T/T_m: High-melting materials have a large value of the binding energy and thus need a large amount of energy to create and move vacancies. The activation energy for self-diffusion is thus large, and the exponential term in equation (11.2) can reach the size of that in a low-melting material only at higher temperatures. As shown in table 11.1, a rule-of-thumb is that the maximum service temperature of mechanically highly stressed metals and ceramics is approximately $T/T_m = 0.5$.[4] An important exception to the rule are nickel-base superalloys. Although their melting temperature is lower than that of steels, their maximum service temperature is much higher. This class of materials is thus of special importance in high-temperature applications. Modern aero engines, for example, would be unthinkable without them. The reasons for this exceptional behaviour are discussed in section 11.4.

The creep resistance of materials can be visualised using *creep diagrams* that plot the stress until fracture or until a certain plastic deformation is reached versus the time at a certain temperature. An example is shown

---

[2] The unit of B depends on the exponent and is $\mathrm{s}^{-1}\mathrm{MPa}^{-n}$. To avoid this awkward unit, the stress can be normalised, for example by dividing it by Young's modulus, resulting in a unit of s^{-1}.

[3] Self-diffusion is the diffusion of atoms in a matrix of the same atoms.

[4] The maximum service temperature also depends on the application. Materials in rocket engines that are used only for a few minutes can serve at higher temperature than materials used in power plants where the service time may exceed ten years.

Table 11.1. Approximate maximum service temperature T_{max} of several technical materials under high mechanical loads compared to the melting temperature T_m. Values of T_m refer to the pure material, not to the alloy which may start to melt at temperatures considerably lower than T_m

material	T_m/K	T_{max}/K	T_{max}/T_m
aluminium alloys	933	450	0.48
magnesium alloys	923	450	0.49
ferritic steels	1 811	875	0.48
titanium alloys	1 943	875	0.45
nickel-base superalloys	1 728	1 300	0.75
Al_2O_3	2 323	1 200	0.52
SiC	3 110	1 650	0.53

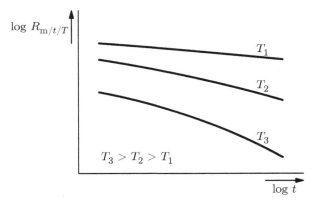

Fig. 11.3. Schematic creep diagram. At different temperatures, the stress that causes fracture after a certain time is plotted. Each point on the curve corresponds to one experiment. $R_{m/t/T}$ is the stress in a specimen that fails after a time t at temperature T. For example $R_{m/100\,000/550}$ is the failure stress after 10^5 h at 550°C

schematically in figure 11.3. Each point in the diagram represents one experiment. The diagram again illustrates that the life time at high temperatures is in principle finite because the creep strain accumulates over time.

> If a component is to be made of a new, temperature-resistant material, a particular problem has to be solved: Imagine a turbine shaft of a new steam turbine that is to be manufactured from a new ferritic steel promising higher steam temperatures and thus a higher efficiency. With a required life time of the shaft of 200 000 h, it is not sensible to wait for creep data measured experimentally at comparable times, for this would require about twenty years. To avoid this, data at shorter testing times, which are partially measured at temperatures beyond the service temperature, are extrapolated to larger service times. This is frequently done using the so-called *Larson-Miller parameter,* moti-

vated by the following considerations: The exponential dependence of the creep rate on temperature shows that thermal activation plays an important role (see appendix C.1). Assume that a certain number of microscopic processes are required to obtain a certain strain or damage of the material, without considering the actual mechanisms behind these processes. The probability of such a process depends on the temperature via the *Boltzmann factor* $\exp(-Q/kT)$. If the time until a specific total strain or a specific damage is accumulated is denoted by t_f, the quantity

$$\theta = t_f \exp\left(-\frac{Q}{kT}\right) \tag{11.3}$$

is a constant according to this assumption. The energy Q depends on the stress because the external stress may have an effect on the microscopic processes by supplying additional energy, analogous to the processes during thermally activated passing of obstacles by dislocations (section 6.3.2) or during relaxation processes in polymers (section 8.2.2). The microscopic explanation is postponed to section 11.2.2.

If we take the logarithm of equation (11.3), we find

$$\frac{Q}{k} = T \ln \frac{t_f}{\theta}. \tag{11.4}$$

The left-hand side of this equation is an unknown function of the stress, called the *Larson-Miller parameter* P. The quantities t_f and θ are divided by their unit h to render them unit-free. The equation can be re-written as

$$P = T\left(\ln \frac{t_f}{h} + C\right). \tag{11.5}$$

The constant $C = -\ln\theta$ is called the *Larson-Miller constant,* usually taking values between 35 and 60 when the time is measured in hours and the temperature in kelvin. In practice, a value of $C = 46$ is frequently used. The exact value of the constant depends on the material and has to be measured experimentally.

Frequently, the decadic logarithm is used instead of the natural logarithm in describing the time t_f. In this case, a pre-factor $\ln 10$ has to be factored out:

$$P = T\left(\ln 10 \lg \frac{t_f}{h} + C\right) = T \ln 10 \left(\lg \frac{t_f}{h} + \frac{C}{\ln 10}\right).$$

This factor is usually absorbed into the (now modified) Larson-Miller parameter, resulting in the equation

$$P' = T\left(\lg \frac{t_f}{h} + C'\right). \tag{11.6}$$

Since the Larson-Miller parameter is only used qualitatively, this scaling is irrelevant. The modified Larson-Miller constant takes a value of $C' = 20$ if $C = 46$. If we plot the stress versus P and choose the appropriate value of C, curves measured at different temperatures can in most cases be mapped onto a single master curve with sufficient precision. Using the example of the steam turbine shaft and assuming a parameter of $C' = 20$ and a service temperature of 600°C, testing times of 10 000 h at 650°C are sufficient to get some clues on the service strength at 600°C and 200 000 h. Obviously, the predictions will become less precise if the ratio of extrapolation becomes larger. One possible source of error are phases in the material that are unstable at the higher temperature. If these phases embrittle the material, extrapolating to longer times at lower temperatures can severely overestimate the life time. For this reason, a new material will only be used if data have been acquired at testing times of several ten thousand hours. In parallel, creep experiments with estimated fracture times of 100 000 h and more will be started to provide a time buffer in the case of unexpected material behaviour.

A further example of using the Larson-Miller parameter can be found in exercise 32.

11.2 Creep mechanisms

Depending on the temperature and the stress, different microscopic processes are important in determining creep behaviour. These will be discussed in this section. We will see that different processes are important at different temperature and stress values; a fact that can be visualised using so-called deformation mechanism maps.

11.2.1 Stages of creep

Before we discuss the different microscopic creep mechanisms in detail, we want to explain the difference between primary and secondary creep in this section.

As in time-independent plastic deformation, dislocations play an important role in the time-dependent plastic deformation of metals. At the onset of creep deformation, the number of dislocations in the material usually increases, causing hardening that can be experimentally observed by the reduction in the creep rate at constant stress. However, the dislocation density cannot increase arbitrarily since recovery occurs simultaneously (see section 6.2.8), with dislocations annihilating by climb. This process becomes the easier, the closer the dislocations are. Accordingly, after some transition time, an equilibrium between the generation of additional dislocation segments by plasticity and the annihilation of dislocations by recovery will be found. This equilibrium causes the creep rate to become constant in the secondary stage.

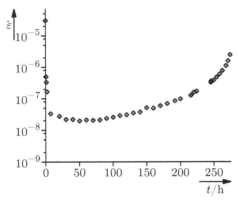

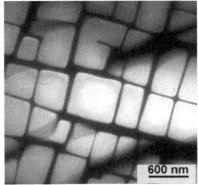

(a) Experimental results for the creep rate versus time

(b) Micrograph. The bright particles are the γ' precipitates (see also page 404)

Fig. 11.4. A nickel-base superalloy changes its microstructure during loading. There is no stationary value of the creep rate analoguous to figure 11.1(a)

In some materials, the microstructure can change at elevated temperatures. For example, the particles in precipitation hardened alloys usually coarsen over time. In this case, there will be no stationary region with constant creep resistance of the material. Instead, the creep rate continuously increases after a minimum has been reached. This is shown for the example of a nickel-base superalloy with a high fraction of cuboid precipitates in figure 11.4.

11.2.2 Dislocation creep

Analogously to the deformation at low temperatures, creep deformation can also occur by dislocation movement in metals. However, there is one crucial difference. If an edge dislocation encounters an obstacle e. g., a precipitate, it needs a certain minimal stress to overcome the obstacle at low temperatures; otherwise it will be stopped. At elevated temperatures, the dislocation can evade the obstacle by adding or emitting vacancies (see figure 6.31 on page 197). Using this mechanism, called *climb*, the dislocation can leave its original slip plane as we already saw in section 6.3.4.

In this case, the strain rate is determined by the rate of emission or absorption of vacancies. Figure 11.5 shows the example of two edge dislocations pinned at two obstacles. Dislocation 1 has to absorb vacancies to climb; dislocation 2 needs to emit them. Thus, vacancies can be transported from one dislocation to the other, with one dislocation acting as *vacancy source*, the other as *vacancy sink*. The *vacancy current density, j*, determines the rate of deformation. This quantity can be estimated.

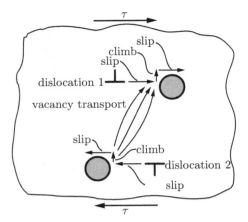

Fig. 11.5. Pile-up of dislocations at obstacles and vacancy diffusion. Dislocation 1 is a vacancy sink, dislocation 2 a vacancy source

If there is no external stress, the vacancy concentration n is determined by the enthalpy required to create a vacancy, Q_V. In this case, the vacancy concentration is (see appendix C.1)

$$n = \exp\left(-\frac{Q_V}{kT}\right). \tag{11.7}$$

To study the effect of an external stress, we can use the following consideration: If there is an external stress, dislocation 2 will move a bit on emitting a vacancy. The external stress τ does the work τV^* according to section 6.3.2, where V^* is the so-called *activation volume*. The total energy needed to create a vacancy is reduced by this amount. If we assume that the vacancy concentration is still in equilibrium, its value at the position of dislocation 2 is

$$n_2 = \exp\left(-\frac{Q_V - \tau V^*}{kT}\right). \tag{11.8}$$

Thus, it is larger than without an externally applied stress.[5]

Near dislocation 1, additional energy is required to create a vacancy. The vacancy concentration is thus reduced because the dislocation tends to absorb vacancies. The concentration becomes

$$n_1 = \exp\left(-\frac{Q_V + \tau V^*}{kT}\right).$$

---

[5] In the discussion of the Larson-Miller parameter in section 11.1, we used a stress-dependent energy in the Boltzmann factor. Here, we found the reason for this dependence.

Thus, there is a vacancy concentration gradient between dislocation 1 and 2. It is determined by the difference of the two densities and by the distance l between the dislocations. In a material containing several obstacles, l is proportional to the mean distance of the obstacles. This gradient causes diffusion of vacancies from dislocation 2 to dislocation 1. In this argument, we assumed that the vacancy concentration at both dislocations can still be described by using the Boltzmann equation which is valid only in thermal equilibrium. This is a valid assumption provided that the energy τV^* is small compared to the enthalpy of vacancy formation Q_V.

According to *Fick's law*, the *vacancy current density* is proportional to the gradient of the vacancy concentration:

$$j = -D_0 \exp\left(-\frac{Q_{ex}}{kT}\right) \frac{\partial n}{\partial x}. \tag{11.9}$$

Here, D_0 is the diffusion constant of vacancy diffusion and Q_{ex} is the activation energy for vacancy migration i. e., the exchange of a vacancy with a neighbouring atom.

The vacancy concentration gradient $\partial n/\partial x$ can be assumed as constant in the case of small fluctuations in concentration:

$$\frac{\partial n}{\partial x} = \frac{n_1 - n_2}{l}. \tag{11.10}$$

Since τV^* is usually small compared to the enthalpy of vacancy formation Q_V, the gradient can be approximated as follows:

$$\frac{\partial n}{\partial x} = \frac{n_1 - n_2}{l} = -\frac{1}{l} \exp\left(-\frac{Q_V}{kT}\right) \left(\exp\left(\frac{\tau V^*}{kT}\right) - \exp\left(\frac{-\tau V^*}{kT}\right)\right)$$

$$\approx -\exp\left(-\frac{Q_V}{kT}\right) \frac{2\tau V^*}{lkT}. \tag{11.11}$$

Here we used the approximation formula $\exp x \approx 1 + x$ for $x \ll 1$.

We finally find the vacancy current density as

$$j = -D_0 \exp\left(-\frac{Q_{ex}}{kT}\right) \frac{n_1 - n_2}{l}$$

$$= D_0 \exp\left(-\frac{Q_{ex}}{kT}\right) \frac{2\tau V^*}{lkT} \exp\left(-\frac{Q_V}{kT}\right).$$

According to this calculation, the vacancy current density is

$$j = \frac{2\tau V^*}{lkT} D_0 \exp\left(-\frac{Q_V + Q_{ex}}{kT}\right), \tag{11.12}$$

where the product of the diffusion constant D_0 and the exponential term is the diffusion coefficient for volume diffusion, D_V. The quantities Q_V and Q_{ex} are the enthalpy of vacancy formation and vacancy migration, τ is the externally applied shear stress, l is the distance between the obstacles, and V^* is the activation volume (see section 6.3.2). The strain rate $\dot{\varepsilon}$ is proportional to the vacancy current density j and to the dislocation density ϱ. Thus, $\dot{\varepsilon} \propto j\varrho$ holds. In the stationary state, the dislocation density ϱ is usually proportional to the square of the stress [51].[6] If we insert the equation for the current density (and use $\tau \propto \sigma$), we finally find

$$
\begin{aligned}
\dot{\varepsilon} &= \frac{A\sigma^3}{kT} D_0 \exp\left(-\frac{Q_{\mathrm{V}} + Q_{\mathrm{ex}}}{kT}\right) \\
&= \frac{A\sigma^3}{kT} D_{\mathrm{V}}(T).
\end{aligned}
\tag{11.13}
$$

A is a material parameter that has to be determined experimentally. The strain rate thus depends exponentially on the temperature and with a power law on the stress. Equation (11.13) can only be used to describe secondary creep because it does not take into account the evolution of the dislocation density during primary creep or the damage processes occurring in tertiary creep.

The exponential dependence of the strain rate on the temperature has been confirmed experimentally. The relation between strain rate and stress is found to follow a power law, as predicted by the equation (see also section 11.1), but in reality the creep exponent typically takes values between 3 and 8. Due to the large variations of the creep exponent in different materials, the value of the factor A can differ by several orders of magnitude. The activation energy in equation (11.13) is frequently stated per mole in the units kJ/mol. In this case, Boltzmann's constant k has to be replaced by the *gas constant* R in the equation as explained in appendix C.1.

In deriving equation (11.13), we assumed that the vacancy diffusion occurs through the undistorted crystal (*volume diffusion*). However, vacancy transport can also occur mainly along lattice defects like dislocation lines (*dislocation pipe diffusion*). In this case, the activation energy for site exchange is smaller due to the lattice distortion. Because of this difference, vacancy transport along dislocation lines dominates at low temperatures, whereas volume diffusion is the faster mechanism at higher temperature. Large stresses also favour transport along dislocations because the dislocation density increases due to the formation of new dislocations (work hardening). We also required in deriving equation (11.13) that the product of external stress and activation volume is small compared to the thermal energy kT. This is not the case at high deformation speeds and stresses. In this case, the relation between strain rate and stress is exponential and the power law is not valid anymore (*power-law breakdown*) [51].

---

[6] A similar relation was already discussed in the context of work hardening, see equation (6.20).

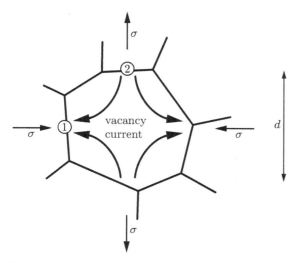

Fig. 11.6. Movement of vacancies in diffusion creep

11.2.3 Diffusion creep

At high temperatures, dislocation creep – i. e., dislocation movement aided by vacancy diffusion – is not the only mechanism contributing to deformation. Vacancy diffusion alone can cause a deformation without any dislocations being involved. In this process, called *diffusion creep*, grain boundaries are sources and sinks of vacancies. It is mainly this mechanism that determines the creep behaviour of ceramics. As illustrated in figure 11.6, vacancies are formed at grain boundaries with a normal vector oriented in the direction of the tensile stress. These vacancies move to grain boundaries with compressive stresses or lower values of the tensile stress. The material itself moves in the opposite direction from regions with compressive to those with tensile stresses.

The derivation of the strain rate in diffusion creep is analogous to that of the previous section. Again, the vacancy current density is calculated and related to the strain rate.

As in section 11.2.2, we start by calculation the vacancy concentration. In equilibrium and without externally applied stress, the vacancy concentration is $n = \exp(-Q_V/kT)$, where Q_V is again the energy for the formation of a vacancy. If we now consider a grain boundary whose normal direction is in the direction of the tensile stress (figure 11.6), the material can elongate in the loading direction if an atom from the crystal lattice is added at the grain boundary, creating an additional vacancy in the crystal. The external stress does some work because the material lengthens (on average) by the quotient of the volume Ω of the vacancy and the cross section of the grain boundary considered. The force equals the stress multiplied with the cross section of the grain boundary. The work done is thus $\sigma\Omega$, independent of the size of the grain boundary. On the other hand, work $\sigma\Omega$ must be done to create a vacancy in a region that is under compressive stress σ.

The vacancy concentration in the regions ① and ② from figure 11.6 is thus

$$n_1 = \exp\left(-\frac{Q_V + \sigma\Omega}{kT}\right) ,$$

$$n_2 = \exp\left(-\frac{Q_V - \sigma\Omega}{kT}\right) .$$

The vacancy concentration gradient can be estimated, according to equation (11.10), as $(n_1 - n_2)/d$, with the grain size d replacing the dislocation distance. Thus, a vacancy current density

$$
\begin{aligned}
j &= -D_0 \exp\left(-\frac{Q_{ex}}{kT}\right)\frac{n_1 - n_2}{d} \\
&= \frac{2\sigma\Omega}{dkT}D_0 \exp\left(-\frac{Q_V + Q_{ex}}{kT}\right)
\end{aligned}
\tag{11.14}
$$

results, with d being the mean diffusion length, approximately equal to the grain size.

The creep rate $\dot\varepsilon$ is proportional to the vacancy current density divided by the size of the grain. Thus, we find

$$
\begin{aligned}
\dot\varepsilon &= A_{NH}\frac{\sigma\Omega}{kT}\frac{D_0}{d^2}\exp\left(-\frac{Q_V + Q_{ex}}{kT}\right) \\
&= A_{NH}\frac{\sigma\Omega}{kT}\frac{D_V}{d^2} .
\end{aligned}
\tag{11.15}
$$

Here, A_{NH} is a material parameter, σ the external stress, Ω the volume of a vacancy, d the grain size, and D_V the diffusion coefficient for self-diffusion through the bulk material. This process is called *Nabarro-Herring creep*. Since the stress dependence is linear, Nabarro-Herring creep is most important at low stresses, whereas dislocation creep is more important at high stresses.

Since the creep rate is inversely proportional to the square of the grain size, creep is favoured if the grains are small. In contrast to time-independent plastic deformation, where small grains are preferred (grain boundary strengthening, see section 6.4.2), large grains are advantageous in materials that creep.

Vacancy diffusion needs not to occur through the bulk material in diffusion creep. Instead, vacancies may move directly along the grain boundaries (see figure 11.7). The activation energy of vacancy diffusion along a grain boundary is smaller than in the bulk because the lattice is distorted.

As before, the vacancy current density j is inversely proportional to the grain size. The derivation of the current density in the bulk material can thus be copied exactly, simply replacing the activation energy with

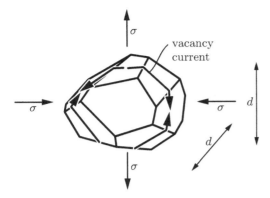

Fig. 11.7. Movement of vacancies along grain boundaries in diffusion creep

the smaller activation energy along the grain boundary. The number of vacancies moving through the grain per unit time is given by the vacancy current density multiplied by the cross section of the region the vacancies are moving through. This is $d\delta$, where δ denotes the thickness of the grain boundary.

In total, the rate of vacancy diffusion is thus $jd\delta$. The growth rate equals this rate, normalised by the cross section of the grain i. e., $jd\delta/d^2 = j\delta/d$. To get the strain rate, we again need to divide by the grain size, yielding the final result $\dot\varepsilon \propto j\delta/d^2 \propto 1/d^3$.

The strain rate for grain boundary diffusion creep is thus

$$\dot\varepsilon = A_{\mathrm{C}} \frac{\sigma\Omega}{kT} \frac{\delta D_{\mathrm{GB}}}{d^3}\,. \tag{11.16}$$

δ is the thickness of the grain boundary, D_{GB} is the diffusion coefficient of self-diffusion along the grain boundary, and A_{C} is another material parameter. This process is called *Coble creep*. Due to the strong dependence on the grain size, Coble creep is most important if the grains are small. Since the activation energy of self-diffusion along the grain boundaries is smaller than in the volume, Coble creep is also dominant compared to Nabarro-Herring creep at low temperatures.

Because the shape of the grains changes in diffusion creep, neighbouring grains have to deform in a compatible manner, analogous to the compatibility of deformation in grain boundary strengthening as discussed in section 6.4.2. This is one cause of *grain boundary sliding*, described in the next section.

Diffusion creep is also important in fibre composites. It was shown in section 9.3.2 that the load transfer to a fibre is determined only by the aspect ratio, the quotient of length and diameter. That the length is considered to be important in technical applications is only due to the fact that the fibre diameter cannot be made arbitrarily small, whereas the length can be as large as desired. However, at high temperatures, the fibres can be unloaded by diffu-sion processes. Atoms of the matrix can move along the fibre-matrix interface and relax the stress between fibre and matrix. Similar to equation (11.16), the absolute size of the fibre now becomes important and only sufficiently long fibres can have a strengthening effect.

11.2.4 Grain boundary sliding

At high temperatures, grains in metals and ceramics can move against each other. This process is called *grain boundary sliding*.

The strain rate of grain boundary sliding cannot be estimated as simply as for the other processes. It is [26]

$$\dot{\varepsilon} = A_{\mathrm{GBS}} \frac{\delta \sigma^n D_{\mathrm{GB}}}{d} \,. \tag{11.17}$$

As usual, A_{GBS} is a material parameter, δ is the thickness of the grain bound-ary, σ the externally applied stress, D_{GB} the diffusion coefficient of grain boundary diffusion, and d the diameter of the grain. The creep exponent n of grain boundary sliding usually takes values between 2 and 3.

In metals, grain boundary sliding usually contributes only slightly to the overall deformation, but it is nevertheless important for two reasons: First, in diffusion creep, grain boundary sliding ensures the compatibility of the grains during the deformation (see also section 6.4.2 and the end of the previous sec-tion) as sketched in figure 11.8. Second, at points where three grain boundaries meet (triple points), movement of the grain boundaries by sliding can cause a large concentration in local stresses and thus induce damage by rupture of the grain boundaries (see also section 11.3). It is thus doubly advantageous to increase the resistance of the grain boundary against sliding: deformation by diffusion creep is impeded, and the danger of early damage is reduced. This will be discussed further in section 11.4.

In ceramics, the strength at high temperatures is often limited by grain boundary sliding. The reason for this is the presence of a glassy phase at the grain boundaries (see also section 7.1). These amorphous regions have a much lower softening temperature than the grains themselves. This 'lubricating film' eases sliding of the grains, without dislocation movement inside the grains be-ing necessary. One important goal in manufacturing ceramic high-temperature materials is thus to reduce the amount of glassy phase as much as possible.

11.2.5 Deformation mechanism maps

The various creep mechanisms discussed so far differ in their temperature de-pendence because the activation energy of the mechanisms is different. Further-more, they differ in their stress-dependence. The creep exponent takes values

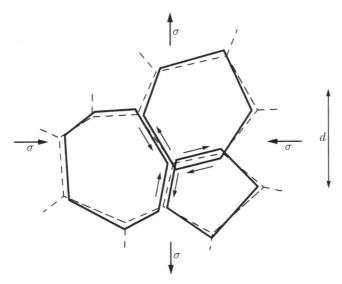

Fig. 11.8. Grain boundary sliding ensures the compatibility of grains which would be violated if only diffusion creep would occur

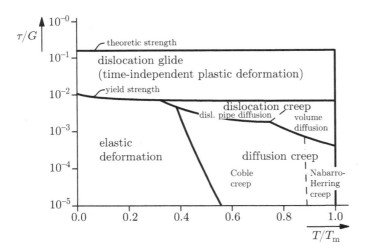

Fig. 11.9. Idealised deformation mechanism map (after [26])

between 1 in diffusion creep and 3 in dislocation creep, with even higher values occurring in reality. Thus, depending on the external conditions, different creep mechanisms dominate the behaviour.

So-called *deformation mechanism maps* allow to read off the dominant mechanism under different conditions. Figure 11.9 shows a schematic deformation mechanism map. In the diagram, the temperature and the external

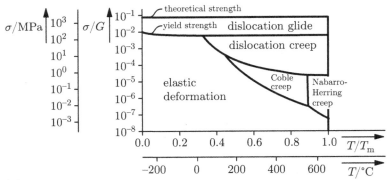

(a) Aluminium

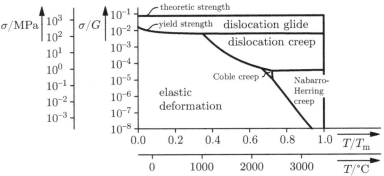

(b) Tungsten

Fig. 11.10. Deformation mechanism maps (after [35]). The grain size is 32 µm in both cases

stress, normalised by the relevant material parameters (melting temperature and shear modulus), are used as axes so that the dominant deformation mechanism can be read off.

At low external stresses and low temperatures, the material deforms elastically. At higher temperatures, diffusion creep starts, being stronger at small stresses than dislocation creep because of its lower creep exponent. Because of the lower activation energy for grain boundary diffusion, this mechanism is more important than bulk diffusion at low temperatures. Since the creep exponent is the same in both cases, the two regions are separated by a vertical line.

If we move on to larger stresses, dislocation creep with its larger creep exponent becomes dominant. Vacancy diffusion along dislocation lines is more important than diffusion through the bulk material at lower temperatures since its activation energy is smaller. Because the creep exponent is larger for diffusion along the dislocations than through the bulk, the separating line is inclined.

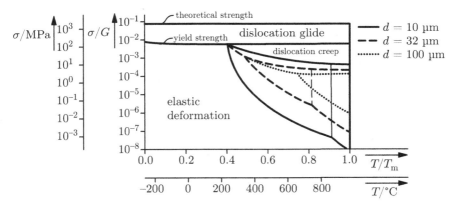

Fig. 11.11. Deformation mechanism map of silver as function of the grain size (after [35])

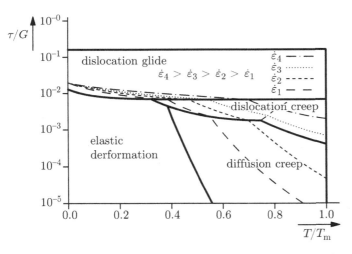

Fig. 11.12. Idealised deformation mechanism map at different strain rates (after [26])

At even higher stresses, time-independent plastic deformation begins. If the stress level reaches about one tenth of the shear modulus, the theoretical strength of the material is reached.

Diagrams like this can be compiled for different materials and material states. Figure 11.10 shows the deformation mechanism maps of aluminium and tungsten at a grain size of 32 μm. Although both maps have the same overall structure as the schematic map from figure 11.9, they nevertheless differ in the size and exact shape of the different regions.

Figure 11.11 shows how the grain size changes the deformation mechanisms, using the example of silver. The regions are shown for three different

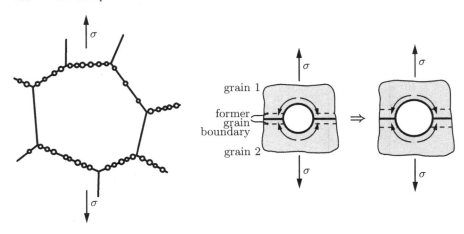

(a) Accumulation of cavities at   (b) Mechanism of formation
the grain boundaries

Fig. 11.13. Schematic illustration of cavern-type pores at grain boundaries

grain sizes, making it easy to see that creep processes are more important at small grain size and start at lower temperatures. This is due to the grain size dependence of diffusion creep.

Since creep processes are time-dependent, the dominant mechanism also depends on the strain rate. This can also be represented in the diagrams as shown in figure 11.12. At high strain rates i.e., high stresses, diffusion creep becomes less important in comparison to dislocation creep.

11.3 Creep fracture

After sufficiently long loading times, creeping materials fail by *creep fracture*. The strain rate, which attained its minimum value during secondary creep, increases again, and *tertiary creep* starts, ending with the final fracture, so-called *creep rupture*. In most cases, creep fracture is distinguished by material failure at the grain boundaries, not inside the grains. In contrast to ductile fracture, creep fracture is thus usually intercrystalline. Transcrystalline fracture usually occurs only at high stresses [40].

That fracture occurs at the grain boundaries indicates a damage process there. It is due to the formation of pores and microcracks. Microscopically, two different types of damage are distinguished in a creeping material. On the one hand, oval cavern-type pores can be formed at grain boundaries which are loaded under tension, on the other hand, wedge-type pores may be induced at triple points where three grain boundaries meet.

Cavern-type pores form by diffusion processes in which the material increases its length in the direction of the tensile stress by moving atoms from

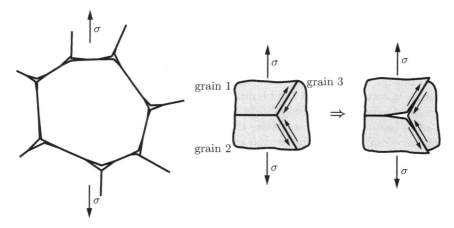

(a) Wedge-type pores at   (b) Mechanism of formation
triple-points of the grains

Fig. 11.14. Schematic illustration of wedge-type pores at triple-points

the region of the forming pore to neighbouring zones (figure 11.13). Completely analogous to the precipitation of particles (see section 6.4.4), a nucleation barrier has to be overcome to form a cavern-type pore, for the energy required to form an inner surface is proportional to the square of the pore diameter, whereas the energy gain depends cubically on the diameter. For small pores, the surface energy dominates. Pores are therefore not formed initially, but only after a long time. Their formation is favoured at high temperatures and large testing times because this raises the probability to overcome the nucleation barrier by thermal activation. Local stress concentrations, for example due to precipitates on the grain boundary or dislocation pile-up, increase the energy gain and thus favour pore formation.

Wedge-type pores are formed by grain boundary sliding at triple points (see figure 11.14). In these regions, there is a large stress concentration (see section 11.2.4) that can cause failure of the grain boundaries if their cleavage strength is exceeded. Accordingly, this type of damage usually occurs at high stresses.

Both damage mechanisms cause a decrease of the effective cross section of the specimen and stress concentrations by notch effects. This is the reason for the rapid increase in the strain rate observed in tertiary creep.

11.4 Increasing the creep resistance

Materials heavily loaded under creep conditions must meet particular requirements.

We can conclude from the discussion of the previous sections on mechanisms that a large activation energy of vacancy diffusion is advantageous because vacancy diffusion is important in almost all of the mechanisms discussed. Vacancy diffusion is weak if the formation of a vacancy is difficult and if the diffusion of any formed vacancy is impeded. The enthalpy of vacancy formation is correlated with the binding forces in the material and thus with the melting temperature. Therefore, the homologous temperature T/T_m can be used as parameter to characterise the creep properties.

Vacancy diffusion occurs by the exchange of a vacancy with its neighbouring atom. The atom has to overcome an energy barrier formed by the surrounding atoms. This barrier is the higher, the closer packed the atoms are; close-packed structures are thus more creep resistant. For example, the diffusion coefficient for self-diffusion of iron is

$$D_\alpha = 2.0 \times 10^{-4} \cdot \exp\left(\frac{-251\,\text{kJ/mol}}{RT}\right)$$

in α iron (body-centred cubic) and

$$D_\gamma = 1.8 \times 10^{-5} \cdot \exp\left(\frac{-270\,\text{kJ/mol}}{RT}\right)$$

in γ iron (face-centred cubic) [51]. At a temperature of 600°C, the diffusion speed is 150 times larger in α than in γ iron.

The grain size also affects the creep properties, particularly so in diffusion creep. The coarser the grain is, the longer the diffusion paths of the vacancies becomes, thus reducing the strain rate. The grain size is also important in creep damage since cavern-type pores form at grain boundaries under tension. Grains elongated in the load direction are also advantageous because, due to their larger area, they reduce the shear stresses that evolve along grain boundaries during diffusion creep. Elongated grains can be achieved, for example, by drawing wires. They are one important reason for the long life-time of tungsten filaments in light bulbs. Gas turbine blades with strongly elongated grains in the loading direction are also highly creep resistant. One additional effect in these blades is that a crack cannot propagate easily even if single grain boundaries fail because the next grain boundary that is perpendicular to the loading direction is usually far away (see figure 11.15). The grains may in fact be as long as the component, completely avoiding transversal grain boundaries (see figure 2.12). Single-crystal alloys are obviously especially suitable.

Similar to time-independent plastic deformation, creep deformation in metals is dominated by dislocation movement, especially at higher stresses. Mechanisms that impede dislocation movement are thus also important in producing creep-resistant materials. However, these mechanisms have to be temperature resistant.

Surveying the strengthening mechanisms discussed in chapter 6, we see that grain boundary strengthening is not suitable in creep applications because we need large grains as explained above. Work hardening can also not

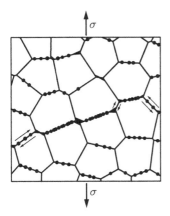

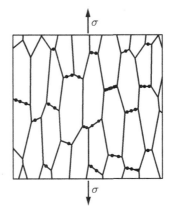

(a) Isotropic microstructure

(b) Directionally solidified micro-structure

Fig. 11.15. Creep damage in an isotropic and a directionally solidified material (after [26]). Crack propagation is strongly impeded in the directionally solidified material. In a technically used directionally solidified alloy, the elongation of the grains is much more pronounced than in this sketch, see figure 2.12

be used since a large initial dislocation density would rapidly reduce by recovery processes to a value that is determined by temperature and strain rate, destroying any initial hardening effect.

A suitable mechanism is solid solution hardening, provided the dissolved elements have a large activation energy for diffusion and are thus diffusing slowly. Accordingly, carbon in steel cannot strengthen the material at high temperatures because the interstitially dissolved carbon atoms diffuse rapidly and move along with the dislocations.[7] High-melting foreign atoms, on the other hand, can contribute significantly to the creep strength of metallic high-temperature materials since their bonds with the matrix atoms are usually strong, causing a limited mobility and thus making it difficult for dislocations to take the atoms with them. Examples are molybdenum, tungsten, and rhenium that are added as solid solution strengtheners to nickel-base superalloys at weight fractions of up to 10 %.

Another possible mechanism is precipitation hardening of metals. Here, however, the following problem may occur: To achieve a fine distribution of the precipitates, coherent particles are needed. These, however, are usually only metastable and transform to an incoherent equilibrium phase at high temperatures. The interfacial energy strongly increases, causing accelerated particle growth and a loss of the hardening effect. This is the crucial reason why precipitation-hardened aluminium alloys cannot be used above $0.5\,T/T_{\mathrm{m}}$

---

[7] Carbon is, nevertheless, an important alloying element in creep-resistant steels, for it can form precipitates in the form of carbides.

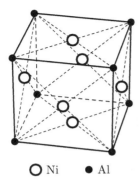

○ Ni   ● Al   **Fig. 11.16.** Unit cell of the γ' phase Ni_3Al

in long-term applications. Nickel-base superalloys play a special role in this context because a coherent equilibrium phase exists which only coarsens slowly (see figure 11.4(b)). This is the so-called γ' phase with stoichiometry Ni_3Al. Its unit cell only differs slightly from that of the face-centred cubic matrix: the nickel atoms in the binary system Ni_3Al are situated almost exclusively on the face-centred positions, whereas the aluminium atoms occupy the corners of the cell (figure 11.16).[8] While the lattice constants of precipitates and matrix in the binary system Ni-Al differ by more than one percent, adding other alloying elements can reduce the lattice mismatch to 0.1% to 0.2%. This reduces the interfacial energy to a minimum and causes a very slow coarsening process. For this reason, these alloys can be used at 75% of their melting temperature, justifying the name 'superalloys'.

The volume fraction of the second phase in modern superalloys can be as high as 70%. At these large values, dislocations are constrained to move in the small channels between the cuboid precipitates, forming highly elongated dislocation loops (figure 11.17). This impedes dislocation movement and causes the high strength of these alloys. The dislocations can cut the particles only at rather high stresses. Thus, stresses of $100\,MPa$ can be applied at temperatures of $1000\,°C$ and service times of several thousand hours without material failure.

> Dispersion-strengthened materials (see section 6.4.4) also have a high creep resistance. If a power law according to equation (11.2) is used to describe the relation between strain rate and stress, extremely high creep exponents with values between 20 and 200 result. The strain rate rapidly reduces to very small values with decreasing stress, resulting in a high creep strength.
>
> This unusual creep property is caused by the interaction of the dislocations with the dispersoids. The dispersoids are small enough to be

---

[8] In multi-component alloys, titanium and tantalum may take some of the sites usually occupied by aluminium. Therefore, the notation $Ni_3(Al, Ti, Ta)$ is frequently used.

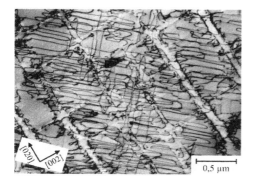

Fig. 11.17. Elongated dislocation loops in a nickel-base superalloy (transmission electron microscopic picture) [127]

easily passed by climbing. Nevertheless, they strongly impede dislocation movement because the line tension of the dislocation is reduced at the interface to the dispersoid, resulting in an attractive interaction. To detach the dislocation from the dispersoid, a high stress is required. If the external stress is not sufficient, part of the required energy has to be provided by thermal activation. Because the required energy depends on the external stress, which thus enters the Boltzmann factor in the exponent, the dependence on the stress is unusually strong and can be approximately described by a power law with large creep exponent.

To enable this mechanism, the line energy of the dislocation at the particle interface must be reduced.[9] This is also the case if the crystal contains a cavity instead of a particle. Tungsten alloys, for example, can be strengthened by adding potassium particles which evaporate at high temperatures and thus form gas-filled cavities.

Another example of how the different strengthening mechanisms interact is shown in table 11.2. The table shows the *creep rupture strength* at a loading time of $100\,000\,\mathrm{h}$ for different steels. Whereas the plain carbon steel C 35 has no significant strength at temperatures above 450°C, adding only 1 % chromium and 0.4 % molybdenum in the steel 13 CrMo 4-4 significantly increases the creep resistance. Raising the chromium or molybdenum content further and adding vanadium strengthens the effect. Strengthening is, on the one hand, due to carbide particles whose stability increases from Fe_3C to $Cr_{23}C_6$ to VC_X.[10] On the other hand, molybdenum and chromium not bound in the carbides cause solid solution strengthening. The service temperature can be

---

[9] In contrast to the obstacles discussed in section 6.3, it is now important that the interaction is attractive, not repulsive, because a repulsive obstacle can be easily avoided by climbing.

[10] Vanadium carbide does not precipitate in a unique chemical composition and is thus denoted VC_X.

Table 11.2. Creep rupture strength of several alloys (after [39, 125, 129]). The creep rupture strength $R_{m/100\,000/T}$ i. e., the stress needed to cause fracture in a specimen at temperature T after 10^5 hours (creep rupture time), is stated. The creep resistance of the ferritic steels with large amounts of vanadium and chromium is significantly larger than that of simpler steels because vanadium and chrome carbides have a better temperature stability. Due to their close-packed face-centred cubic structure, the creep resistance of austenitic steels is larger. The creep strength of the nickel-base superalloys IN 738 (polycrystalline) and SC 16 (single crystalline) were estimated from Larson-Miller data

temperature in °C	$R_{m/100\,000/T}$/MPa							
	420	450	500	550	600	700	800	900
ferritic steels								
C 35	108	69	34					
19 Mn 5	136	85	41					
24 CrMo 5	308	226	118	36				
10 CrMo 9-10		221	135	68				
13 CrMo 4-4		285	137	49				
21 CrMoV 5-11	410	349	212	92				
austenitic steels								
X 5 CrNi 18-10				127	74	30		
X 10 CrNiNb 18-9			300	205	131	55	18	
X 5 CrNiMo 17-12-2					145	52	23	
nickel-base superalloys								
IN 738						360	155	
SC 16							240	110˙

increased further by using austenitic steels like X 5 CrNi 18-10. This is not only due to the higher content of alloying elements, but also to the lower diffusivity in the face-centred cubic lattice.

Grain boundary sliding, discussed in section 11.2.4, can be impeded by adding discrete particles (e. g., carbides) at the grain boundaries. In this case, sliding of the grain boundaries requires material transport from the side of the particle under compression to the side loaded in tension, slowing down the process.

In cast magnesium alloys, which are increasingly used in automotive industry, the influence of grain boundary sliding is rather strong, partially due to the fine-grained structure of these alloys. To improve the creep resistance, silicon can be added because it forms a Mg_2Si phase on the grain boundaries and thus impedes sliding. A similar effect can be achieved by adding rare-earth metals which also cause precipitation hardening [111].

12

Exercises

In this chapter, we present exercises to elaborate upon the topics discussed in this book. Simple memorising exercises that can be solved by looking up the topics are not given. Complete solutions to the exercises are provided in chapter 13.

Exercise 1: Packing density of crystals

Frequently, crystal structures are visualised by drawing the atoms as touching spheres. Using this assumption, calculate the packing density of

a) face-centred cubic,
b) body-centred cubic, and
c) hexagonal close-packed crystals!

Exercise 2: Macromolecules

Consider a polyethylene molecule with a degree of polymerisation of 10^4.

a) Calculate the molar mass of the molecule! The molar mass of carbon is $12.01\,\mathrm{g/mol}$, that of hydrogen is $1.01\,\mathrm{g/mol}$.
b) Calculate the length of the chain, assuming that it is in a straight conformation! The bond length between two carbon atoms is $0.154\,\mathrm{nm}$, the bond angle is $109°$.

Exercise 3: Interaction between two atoms

Assume the following functions for the interaction energy between two atoms of a molecule of common salt (NaCl):

$$U_A = -\frac{1.436}{r}\,\mathrm{eV\,nm}\,, \quad U_R = \frac{5.86\times 10^{-6}}{r^9}\,\mathrm{eV\,nm^9}\,.$$

a) Calculate the bond length of the diatomic molecule!
b) How large is the binding energy?
c) Compare the result of the calculation with the interatomic distance calculated for a NaCl crystal with a density of $\varrho = 2.165\,\mathrm{g/cm^3}$! The molar mass of Na is $23\,\mathrm{g/mol}$, that of chlorine is $35.4\,\mathrm{g/mol}$. Avogadro's constant (the number of molecules in a mol) is $N_A = 6.022\times 10^{23}\,\mathrm{mol^{-1}}$.
d) Estimate Young's modulus of NaCl in the $\langle 100\rangle$ direction! Neglect the bonds between next-nearest neighbours and those even further away! *Note:* $1\,\mathrm{eV} = 1.602\times 10^{-19}\,\mathrm{J}$.
e) The elastic constants of NaCl are $C_{11} = 48.7\,\mathrm{GPa}$, $C_{12} = 12.6\,\mathrm{GPa}$ and $C_{44} = 12.75\,\mathrm{GPa}$. Use these values to calculate $E_{\langle 100\rangle}$! Compare this value to your estimate!

Exercise 4: Bulk modulus

The *bulk modulus* K is a measure of the pressure Δp needed to change a material's volume V_0 by ΔV:

$$\Delta p = -K\cdot\frac{\Delta V}{V_0}\,. \tag{12.1}$$

The minus sign accounts for the fact that positive pressure usually causes a reduction in the volume.

a) Derive the relation between K and the elastic constants E, G, and ν in an isotropic material at small deformations!
b) Calculate the bulk modulus of a material with a Poisson's ratio of $\nu_1 = 0$, $\nu_2 = 1/3$, and $\nu_3 = 0.5$! How does the volume of a tensile specimen change with the uniaxial stress σ in the three cases?
c) Some rare materials possess a negative Poisson's ratio. What is the transversal strain for a positive normal strain in this case?

Exercise 5: Relation between the elastic constants

In section 2.4.3, we introduced equation (2.23), $C_{44} = (C_{11}-C_{12})/2$, specifying the relation between the components C_{11}, C_{12}, and C_{44} of the elasticity matrix $(C_{\alpha\beta})$ of an isotropic material. Check this equation by prescribing a strain tensor

$$(\varepsilon_{ij}) = \begin{pmatrix} -\varepsilon & 0 & 0 \\ 0 & \varepsilon & 0 \\ 0 & 0 & 0 \end{pmatrix}$$

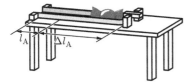 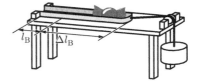

(a) Booth A. The rubber band is elon- (b) Booth B. The rubber band is loaded
gated by the same distance Δl_A always with the same force F_B always

Fig. 12.1. Candy catapults, shown in the loaded state

and calculating the required stress state in the un-rotated x_i coordinate system
and in a coordinate system with axes $x_{i'}$, rotated by 45°. To do so, use Hooke's
law twice for the strain tensor, once in the x_i and once in the $x_{i'}$ coordinate
system!

Exercise 6: Candy catapult

At a child's fair, two booths present almost identical candy catapults. At both
booths, candies are accelerated on a horizontal plane using rubber bands. At
booth A, the rubber band is stressed for each shot by lengthening it from
the initial length $l_A = l_B$ by $\Delta l_A = $ const (figure 12.1(a)), whereas booth B
stresses the rubber using a rope, a pulley, and a weight loading the band with
a force $F_B = $ const.

The cross section of both rubber bands is identical (A). Assume that
both rubber bands are linear-elastic. Young's modulus of the rubber band
of booth B is twice as large as that at booth A: $E_B = 2E_A$.

At both booths, the take-off velocity of the candies is disappointingly small.
We want to find a way to increase the velocity without using additional material or changing the construction of the catapults.

a) Start by deriving equations for the stored elastic energy $W^{(el)}$ when the
   bands are stretched!
b) How does the stored energy change when Young's modulus is increased or
   decreased?
c) Derive an equation for the take-off velocity of the candies (mass m) at
   both booths! Neglect the mass of the rubber bands and any occurring
   friction!
d) Suggest a *simple* method to increase the take-off velocity! By what factor
   does the velocity increase?

Exercise 7:   True strain

To compare *nominal strain* and *true strain*, we investigate two different deformation processes of two rods: The first rod is strained in two steps Δl_1 and Δl_2, the second in a single step $\Delta l_{1+2} = \Delta l_1 + \Delta l_2$.

a) Prove the inequality $\varepsilon_1 + \varepsilon_2 \neq \varepsilon_{1+2}$ for the strains!
b) Estimate how the strain measures differ when the length change Δl_i is small and large, respectively! To do so, sketch a diagram of the difference $\Delta\varepsilon$ versus $\varepsilon_1/\varepsilon_{1+2}$!
c) Show that, for the true strains, the equation $\varphi_1 + \varphi_2 = \varphi_{1+2}$ holds!

Exercise 8:   Interest calculation

This exercise illustrates that the question of what initial value is used to calculate a quantity is not only important in calculating strains, but also in calculating interests.

The customer of a bank invests an initial amount of $G_0 = 10\,000\,€$ at his bank. He wants to double his money within ten years (G_{10}).

a) Calculate the interest rate z_0 required if the interest is always calculated relative to the initial deposit G_0!
b) How large is the required interest rate z if the interest is always paid on the current deposit G_i?

Exercise 9:   Large deformations

As in the solution to exercise 4 a), we again consider the deformation of a brick with edge lengths l_i to the new lengths $l_i + \Delta l_i$. This time, however, we want to account for large deformations.

a) Calculate Green's strain tensor $\underline{\underline{G}}$ for this deformation!
b) Compare Green's strain tensor $\underline{\underline{G}}$ with the strain tensor $\underline{\underline{\varepsilon}}$ for large and small deformations!

Exercise 10:   Yield criteria

A component made of a polycrystalline aluminium alloy with yield strength $R_{p0.2} = 200\,\text{MPa}$ is loaded in a plane-stress state. The stress components are $\sigma_{11} = 155\,\text{MPa}$, $\sigma_{22} = 155\,\text{MPa}$, and $\tau_{12} = 55\,\text{MPa}$.

a) Calculate the principal stresses!
b) Use the Tresca yield criterion to decide whether the material yields!

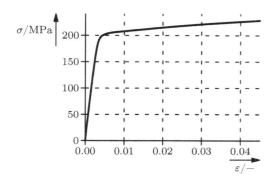

Fig. 12.2. Stress-strain curve of an aluminium alloy

c) Use the von Mises yield criterion to decide whether the material yields!
d) Can you decide which of the two results is correct? Justify your answer!
e) In experiments on single crystals, the yield strength of the slip systems was determined as $\tau_{\text{crit}} = 60\,\text{MPa}$. Use the von Mises yield criterion to check whether a significant amount of slip systems in the polycrystal is activated at the stress value given! The Taylor factor is $M = 3.1$.
f) Calculate the stress deviator $\underline{\underline{\sigma}}'$ for the given stress state!

Exercise 11:   Yield criteria of polymers

In a thermoplastic, a yield strength of $R_{\text{p}} = 40\,\text{MPa}$ was measured in tension, whereas the stress in a compressive test is $R_{\text{c}} = 50\,\text{MPa}$. The conically and parabolically modified yield criteria are to be compared for several load cases.

a) At what purely hydrostatic stress does yielding occur according to the two criteria?
b) A component made of this polymer is loaded with a stress state $\sigma_{11} = \sigma_{22} = -\sigma_{33} = 0.56\,R_{\text{p}}$, $\sigma_{23} = \sigma_{13} = \sigma_{12} = 0$. Use both criteria to check whether the material yields!
c) Does the material yield at a state $\sigma_{11} = -\sigma_{22} = -\sigma_{33} = 0.56\,R_{\text{p}}$, $\sigma_{23} = \sigma_{13} = \sigma_{12} = 0$?

Exercise 12:   Design of a notched shaft

A shaft with a circumferential round notch (see figure 4.3) with dimensions $D = 20\,\text{mm}$, $d = 16\,\text{mm}$, and $\varrho = 4\,\text{mm}$ is loaded in tension. It has to be checked whether it can be used at a service load of $F = 40\,\text{kN}$. The shaft is made of an aluminium alloy with Young's modulus of $E = 68\,000\,\text{MPa}$ and the stress-strain curve shown in figure 12.2 ($R_{\text{p}} = 202\,\text{MPa}$, $R_{\text{m}} = 280\,\text{MPa}$).

a) Determine the stress concentration factor K_{t} using diagram 4.3!

b) How large would the maximum stress at the notch root, σ_{max}, be if the material were linear-elastic? Could the shaft be used in this case?
c) Determine the Neuber's hyperbola for the shaft under the given load! Add it to the stress-strain diagram!
d) Determine the maximum stress and strain at the notch root from the diagram!
e) Can the shaft be used? Justify your answer!

Exercise 13:   Estimating the fracture toughness K_{Ic}

The stress in front of the crack tip shows a singularity. In this exercise, we investigate why the material does not fail immediately although the stress is numerically infinite.

To do so, we consider a sodium chloride crystal as in exercise 3. The crystal contains a crack of length a. Assume that, even on the atomic scale, the stress field at the crack tip can be described by equation (5.1). The crack is assumed to propagate in the [100] direction.

a) Calculate the force on a single atomic bond situated directly in front of the crack tip as a function of K_{Ic}! The lattice constant of NaCl is $a_{NaCl} = 0.282 \times 10^{-9}\,\text{m}$.
b) According to exercise 3, the spring stiffness between the atoms in the NaCl-crystal is approximately $k = 85\,\text{N/m}$. We make the arbitrary assumption that the bond breaks when it is strained by one-tenth of the lattice constant. Calculate K_{Ic} with this assumption!
c) Check the assumptions made in this calculation! Do you believe them to be correct? Why?

Exercise 14:   Determination of the fracture toughness K_{Ic}

The fracture toughness of AlCuMg 2 is to be determined using a CT specimen. The dimensions of the specimen according to standard ASTM E 399 (see figure 5.14(b)) are $G = 62.5\,\text{mm}$, $W = 50\,\text{mm}$, $H = 60\,\text{mm}$, $B = 25\,\text{mm}$. AlCuMg 2 has the following material parameters: $R_{p0.2} = 510\,\text{MPa}$, $R_m = 590\,\text{MPa}$. The initial crack introduced by cyclic loading was measured after the test. Its length was $a = 25\,\text{mm}$. The stress intensity factor is calculated using equation (5.30), with a geometry factor determined by equation (5.31). Figure 12.3 shows a plot of the force versus the displacement of the specimen. Use the method from section 5.2.7 and figure 12.3 to determine K_{Ic}!

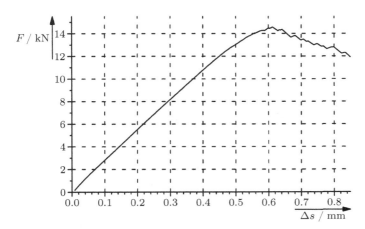

Fig. 12.3. Force F versus displacement Δs to determine the fracture toughness

Exercise 15:   Static design of a tube

In a power plant, a tube made of a newly developed austenitic steel is to be used. Young's modulus of the material is $E = 200\,000$ MPa, its yield strength is $R_{\mathrm{p}0.2} = 1420$ MPa, the cleavage strength is $\sigma_{\mathrm{C}} = 2200$ MPa and the fracture toughness is $K_{\mathrm{Ic}} = 90$ MPa$\sqrt{\mathrm{m}}$. At the moment, a tube diameter of $D = 1000$ mm with a wall thickness of $t = 5$ mm is planned. The pressure within the tube is $p = 12$ MPa. Ultrasonic measurements can limit the largest crack in the material to a size smaller than $2a = 3$ mm.

Hints: The stress state in a pressurised thin-walled tube is as follows: longitudinal stress $\sigma_{\mathrm{l}} = 0$, radial stress $\sigma_{\mathrm{u}} = pD/(2t)$, circumferential stress $\sigma_{\mathrm{r}} = 0$ [18]. The geometry factor can be assumed as $Y = 1$.

a) Can the tube be used with the intended material and dimensions? Design against yielding, cleavage fracture, and crack propagation!
b) Does your answer depend on whether the yield criterion of von Mises or Tresca is used? Justify your answer!
c) If the pressure is increased from zero until a failure criterion is met, which criterion is this? What is the corresponding pressure?
d) At what crack length are yield strength and fracture toughness reached simultaneously?
e) Plot the failure criteria 'yielding' and 'crack propagation' in a failure-assessment diagram! Sketch a more realistic curve together with the idealised one!
f) Calculate the crack opening of a crack of length $2a = 3$ mm loaded in mode I with the load specified above!

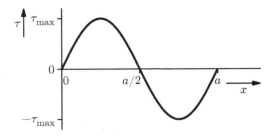

Fig. 12.4. Dependence of the shear stress on the relative displacement x of one atomic layer to the other

Exercise 16:   Theoretical strength

We want to estimate the shear stress required to shift a whole layer of atoms in a metal versus a neighbouring layer as sketched in figure 6.1. We assume a simple cubic crystal with lattice constant a. We make the simplifying assumption that the shear stress required to shift the layer can be represented by a sine function as shown in figure 12.4.

a) Estimate the stress τ_F required to shift an atomic layer, assuming that Hooke's law $\tau = G\gamma$ can be used for small shear values!
b) Explain why the atomic configuration is unstable at a displacement of $x = a/2$ although the stress is $\tau = 0$!

Exercise 17:   Estimating the dislocation density

A metal cube of edge length $a = 10\,\text{mm}$ is to be sheared plastically, resulting in a total relative displacement of $s = 0.1\,\text{mm}$ of its top versus its bottom side. The Burgers vector is $0.286\,\text{nm}$.

a) Estimate the minimum number of dislocations in the cube needed to allow this deformation! Assume the cube to be a simple cubic single crystal with crystal orientation parallel to the cube edges!
b) Calculate the resulting dislocation density!
c) How do the results change if we assume a polycrystalline material with grain size $d = 100\,\mu\text{m}$?
d) If we were to string together all dislocations in the cube, what distance would they cover?

Exercise 18:   Thermally activated dislocation generation

Calculate the probability of a dislocation loop in aluminium being generated by thermal activation! Assume that the minimum possible loop size is six Burgers vectors so that at least one atom is enclosed by the loop! Calculate the probability at temperatures of $300\,\text{K}$ and $900\,\text{K}$! The shear modulus of aluminium is $G = 26\,\text{GPa}$, its Burgers vector is $b = 2.86 \times 10^{-10}\,\text{m}$.

Table 12.1. Measured failure stresses in a ceramic

Experiment	1	2	3	4
Failure stress σ/MPa	244.69	54.60	665.15	90.02

Exercise 19:   Work hardening

A sheet of aluminium is rolled from an initial thickness of 10 mm to a final thickness of 5 mm. Using a transmission electron microscope, it was found that the dislocation density increases from $\varrho_0 = 10^{12}\,\mathrm{m}^{-2}$ to $\varrho_1 = 10^{16}\,\mathrm{m}^{-2}$. The prefactor in equation (6.20), needed to calculate the hardening contribution, is $k_V = 0.1$. The length of the Burgers vector in a face-centred cubic lattice is $b = \sqrt{2}/2 \cdot a$. The lattice constant of pure aluminium is $a = 4.049 \times 10^{-7}\,\mathrm{mm}$, the shear modulus is $G = 26\,200\,\mathrm{MPa}$, and the Taylor factor is $M = 3.1$. Calculate the increase in strength due to this deformation!

Exercise 20:   Grain boundary strengthening

Your task is to increase the yield strength of pure aluminium by grain boundary strengthening. Your starting material is pure aluminium with a grain size of $d_{\mathrm{coarse}} = 100\,\mu\mathrm{m}$ (Hall-Petch constant $k = 3.5\,\mathrm{N/mm}^{3/2}$). In a tensile test, the yield strength is found to be $R_{\mathrm{p0.2}} = 20\,\mathrm{MPa}$. What grain size is required to raise the yield strength to a value of $100\,\mathrm{MPa}$?

Exercise 21:   Precipitation hardening

The yield strength of an aluminium-copper alloy is to be increased by precipitation hardening by $\Delta R_{\mathrm{p0.2}} = 600\,\mathrm{MPa}$.

a) Calculate the required particle spacing of incoherent particles!
b) Calculate the particle radius, assuming a copper content of 4 vol-%! To simplify the calculation, neglect the solubility of copper in the aluminium matrix!

Exercise 22:   Weibull statistics

The Weibull modulus m and the stress σ_0 of a ceramic material are to be determined. To do so, the failure strength of four identical specimens has been measured (table 12.1). Determine the two parameters for this material graphically, using the instructions from section 7.3.3!

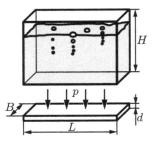

Fig. 12.5. Tank used in chemical engineering

Table 12.2. Material parameters

Young's modulus	70 GPa
failure stress in bending σ_B	100 MPa
Weibull modulus m	15

Exercise 23:   Design of a fluid tank

You are the design engineer of a company working in chemical engineering. Your task is to design the bearing plate of a tank used to store acids. The tank has a capacity of 200 L with dimensions of $L \times B \times H = 1000$ mm × 400 mm × 510 mm. The bearing plate must bear a total weight of 250 kg.

The bearing plate is to be made of a ceramic and is fixed at its edges using a metal frame (see figure 12.5). According to the supplier, the ceramic material has the mechanical properties given in table 12.2. To avoid claims for damages, the failure probability at full service load is to be limited to $P_f = 10^{-4}$.

The maximum bending stress of a rectangular plate supported at all edges with homogeneous pressure p is approximately given by $\sigma = 2pL^2/d^2$, with L being the length of the longer edge and d being the plate's thickness.

a) Calculate the pressure load p, using a value of $g = 9.81 \, \text{m/s}^2$!
b) Use the formula for the volume-independent Weibull statistics, equation (7.6) with $V = V_0$, to calculate the thickness of the ceramic plate required to bear the calculated pressure with the given failure probability!
c) The supplier tells you that the failure stress of the material was measured in bending using specimens with a volume of $V_{\text{spec}} = 5 \times 10^5 \, \text{mm}^3$. Correct your design for this!
d) After the series production of the tank has already started, you get a telephone call from the supplier who tells you that the failure stress was not measured in bending, but in tension. Do you have to stop the production? Justify your answer!
e) If the bearing plate were made of a metal with a yield strength of $R_p = 100$ MPa, how thick would the plate have to be?

Exercise 24:   Subcritical crack growth of a ceramic component

In a plant producing common salt, a connecting pin is to be used that has to service in concentrated saline solution at a temperature of 70°C. The pin is loaded at a stress of $\sigma_{\text{service}} = 100\,\text{MPa}$. The component has to serve for 25 000 hours with a failure probability of 0.5% at most. Due to the aggressive environment, a metal cannot be used for the pin. Instead, hot isostatic pressed aluminium oxide (Al_2O_3) is to be used, characterised by the following parameters:

$$K_{\text{Ic}} = 3.2\,\text{MPa}\sqrt{\text{m}}\,,$$
$$m = 22\,,$$
$$\sigma_0 = 375\,\text{MPa}\,.$$

The geometry factor of the connecting pin is $Y = 1.3$. The maximum crack growth resistance has been measured to

$$K_{\text{IR}} = 3.5\,\text{MPa}\sqrt{\text{m}}\,.$$

a) In experiments under the same environmental conditions, failure occurred at a load of 140 MPa after 375.2 h and at a load of 150 MPa after 94.4 h. The dependence between the crack propagation rate and the stress in the pin is to be approximated by equation (7.2). Determine the parameters in this equation using the experimental results! The inert strength has been measured to be 355 MPa in an identical specimen. To solve this sub-exercise, assume that there is no scatter in the material parameters so that using Weibull statistics is unnecessary!

b) Now we consider the scatter of the parameters, using a Weibull statistics for the failure probability. Can the ceramic be used in its present form to guarantee the required service time at the stated failure probability? Justify your answer!

c) What possible method do you suggest to enable using the component nevertheless? Changing the design or switching to another material is not possible due to severe time and cost restrictions.

d) Calculate the minimum proof stress in a proof test needed to exactly meet the required failure probability! Derive a relation analogous to equation (7.16), replacing the failure probability $P_{\text{f}}(\sigma)$ by the probability $P_{\text{f}}(t_{\text{f}})$ from equation (7.10)! Solve the equation for the proof stress!

e) How large is the fraction of components that fail during the proof test?

Exercise 25:   Mechanical models of viscoelastic polymers

As explained in section 8.2, the time-dependent behaviour of polymers can be described using spring and dashpot elements. The behaviour of a spring element can be described with the equation $\sigma = E\varepsilon$, with σ being the stress, ε the strain, and E Young's modulus. A dashpot element behaves according to $\dot{\varepsilon} = \sigma/\eta$, with strain rate $\dot{\varepsilon}$ and viscosity η.

a) Start considering the Kelvin model from figure 8.7(a). Calculate the strain as a function of time in a retardation experiment with prescribed stress σ!
b) What is the result if you perform a relaxation experiment instead?
c) In real-world polymers, part of the elastic deformation is time-independent. Use a three-parameter model according to figure 8.7(b), assuming am infinite viscosity of the dashpot element in series, to describe the behaviour in a relaxation and retardation experiment!
d) Calculate the creep modulus and the relaxation modulus as function of time, and calculate the relaxation and retardation time!

Exercise 26: Elastic damping

The time-dependence of elastic deformation causes *elastic damping,* a phenomenon investigated in this exercise.

Assume that a component, for example a tensile specimen, made of a viscoelastic material is loaded cyclically with angular frequency ω. After some initial transient effects, the strain will also oscillate with the same angular frequency ω. Due to the time-dependence of elastic deformation, stress and strain are out of phase because the strain follows the current stress only with some delay. The following time-dependence is assumed for stress and strain:

$$\sigma(t) = \sigma_0 \sin(\omega t + \delta)\,,$$
$$\varepsilon(t) = \varepsilon_0 \sin \omega t\,. \tag{12.2}$$

a) Sketch the time-dependence of stress and strain and explain the meaning of the parameter δ!
b) Write the stress as a function of the strain! Use an addition theorem to split the stress into two components, one in-phase with the strain, the other phase-shifted by 90°!
c) Draw a stress-strain diagram for a complete cycle!
d) Since stress and strain are out of phase, elastic energy is dissipated during each cycle. Calculate the energy dissipated per volume in each cycle!
   Hint: Use the relation

$$\int \cos(\arcsin x)\,\mathrm{d}x = \frac{1}{2}\left(x\sqrt{1-x^2} + \arcsin x\right).$$

Exercise 27: Eyring plot

Estimate the activation energy of the relaxation process responsible for the deformation of polycarbonate from figure 8.9!

Exercise 28:   Elasticity of fibre composites

We want to calculate Young's modulus of a fibre composite loaded in parallel and perpendicular to the fibre direction (see sections 9.2.1 and 9.2.2). We start by considering a polymer matrix composite with perfectly aligned, 'infinitely' long, uniaxial fibres.

a) What are the relations between stress and strain in the cases
   • mechanical load parallel to the fibres,
   • mechanical load perpendicular to the fibres?
   Make the simplifying assumption in the case of loading perpendicular to the fibres that the fibres are plates extending throughout the volume (see figures 9.1(a) and 9.2(a))! Neglect the transversal contraction!
b) Use the relations derived in subtask a) to find an equation for the resulting Young's modulus for each case!

A polymer matrix composite comprises a polyester matrix ($E_m = 1500\,\text{MPa}$) and carbon fibres ($E_f = 390\,000\,\text{MPa}$).

c) Sketch the dependence of Young's modulus on the fibre volume fraction (between $0\,\%$ and $100\,\%$) in both cases!

Exercise 29:   Properties of a polymer matrix composite

A polymer matrix composite is made from a duromer matrix containing continuous carbon fibres aligned in the loading direction. Young's modulus is $3\,\text{GPa}$ in the matrix and $350\,\text{GPa}$ in the carbon fibres, the tensile strengths are $60\,\text{MPa}$ and $4900\,\text{MPa}$. The volume fraction of the carbon fibres is $55\,\%$.

a) Estimate Young's modulus of the composite in fibre direction!
b) Estimate the tensile strength of the composite in fibre direction! Start by checking which component will fail first when the load is increased!
c) Assuming that the compressive strength of the matrix is the same as the tensile strength, estimate the compressive strength of the composite!
d) How does your calculation of the tensile strength change if the strengthening fibres are as short as 5 mm, much smaller than the dimension of the component? Assume that the short fibres are still perfectly aligned! The interfacial strength between fibre and matrix is 30 MPa, the fibre diameter is $8\,\mu\text{m}$.

Exercise 30:   Estimating the number of cycles to failure

During a regular inspection, as done all 5000 cycles, a crack of length $a_0 = 1\,\text{mm}$ was detected in the lever gear of a heavy-duty postage metre machine. The following service parameters are known: At the service stress

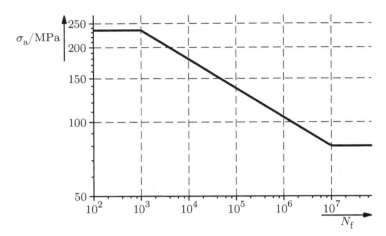

Fig. 12.6. *S-N* diagram for exercise 31

range of $\Delta\sigma = 100\,\text{MPa}$ and the stress ratio $R = -1$, the parameters of the Paris law for the aluminium alloy used are $C = 2 \times 10^{-12}\,\text{MPa}^{-2}\,\text{cycle}^{-1}$, and $n = 2$. The geometry factor depends on the crack length according to $Y(a) = 1 + 0.1\,\text{mm}^{-1} \cdot a$. The critical crack length is $a_f = 10\,\text{mm}$.

a) Calculate the fracture toughness K_{Ic} from the critical crack length!
b) Check whether unstable crack propagation is to be expected!
c) Check whether crack propagation has to be expected, using equation (10.6) with the prefactor 2.75×10^{-5} and a Young's modulus of $E = 70\,000\,\text{MPa}$!
d) Estimate whether the component can stay in service if the crack propagation per cycle $\mathrm{d}a/\mathrm{d}N$ is assumed to be constant!
e) Use equation (10.10) to check whether the machine can stay in service until the next maintenance interval! *Hint:* Use

$$\int \frac{\mathrm{d}x}{(A + Bx)^2 x} = -\frac{1}{A^2}\left(\ln\frac{A + Bx}{x} + \frac{Bx}{A + Bx}\right).$$

Exercise 31:   Miner's rule

The life time of the fixing screw of an industrial robot is to be calculated for the expected load history. In each week, the screw is loaded 10 000 times with $\sigma_a = 60\,\text{MPa}$, 5000 times with 100 MPa, 2000 times with 150 MPa, and 200 times with 200 MPa, always at an R ratio of 0.3. Miner's rule is to be used for the design. The *S-N* curve of the material is shown in figure 12.6.

Calculate the life time of the screw in weeks! For how many cycles will the screw survive? Assume that the loads are distributed evenly over the week!

Exercise 32:   Larson-Miller parameter

The creep properties of a newly developed high-temperature material are to be investigated experimentally. Several experiments are performed.

a) At a stress σ_1, the material fails at a temperature $T_1 = 940°C$ after $t_1 = 23$ hours and at a temperature of $T_2 = 850°C$ after $t_2 = 1017$ hours. Calculate the Larson-Miller parameter C from equation (11.5)!
b) A component made of this material has to serve for 100 000 hours at a stress σ_2. A creep experiment at this stress and a temperature $T_3 = 940°C$ results in a time to failure of $t_3 = 173$ hours. What is the maximum service temperature to meet the life time demands?

Exercise 33:   Creep deformation

In section 8.2.1, the time-dependent behaviour of polymers was described using spring-and-dashpot models.

a) Sketch a spring-and-dashpot model suitable to describe creep deformation!
b) Consider a material with Young's modulus E and the creep law $\dot{\varepsilon} = A\sigma^n$. Calculate the time-dependence of the strain in a retardation experiment.
c) Due to its low melting temperature, lead creeps already at ambient temperatures. A thin-walled lead tube fixed at its ends bends under its own weight in the course of time. Estimate by how much the centre of the tube is displaced within one year!

The maximum stress in the lead tube is

$$\sigma = \frac{\varrho g l^2}{8d} \, ,$$

where $\varrho = 11.4\,\mathrm{g/cm^3}$ is the density of lead, $l = 0.8\,\mathrm{m}$ is the length of the tube, and $d = 0.03\,\mathrm{m}$ is the diameter of the tube. The relation between the displacement h and the strain ε in the middle of the tube is

$$h = \frac{\varepsilon l^2}{4d} \, .$$

The creep law is assumed to be $\dot{\varepsilon} = A\sigma$ since diffusion creep is the dominant mechanism. The creep constant A is $4.11 \times 10^{-18}\,\mathrm{Pa^{-1}s^{-1}}$.

Exercise 34:   Relaxation of thermal stresses by creep

A rod made of a nickel-base superalloy is clamped at low stress between two plates and then heated rapidly from ambient temperature, $T_1 = 23°C$, to a final temperature $T_1 = 1000°C$ and kept at this temperature for $t = 100\,\mathrm{s}$.

Young's modulus is $130\,000\,\text{MPa}$, the coefficient of thermal expansion is $\alpha = 17.5 \times 10^{-6}\,\text{K}^{-1}$ (for simplicity, both parameters are assumed to be temperature-independent). The creep law of the material is $\dot{\varepsilon} = A\sigma^n$, with $n = 3$ and $A = 3 \times 10^{-12}\,\text{MPa}^{-3}\text{s}^{-1}$.

a) Calculate the stress in the rod at the end of the holding time!
b) The rod is rapidly cooled down again. At what temperature does it fall from the clamping plates?

13

Solutions

Solution 1:

a) A unit cell of a face-centred cubic crystal (figure 1.5(a)) has an edge length of a and comprises 4 atoms (8 atoms on the corners, counting as $1/8$ each, and 6 atoms on the side faces, each counted as half an atom). 4 atomic radii are aligned on the face diagonal of length $\sqrt{2}\,a$. Thus, the atomic radius r is given by $4r = \sqrt{2}\,a$. The total volume of the four atoms is

$$V_{\text{atom}} = 4 \cdot \frac{4}{3}\pi r^3 = \frac{\sqrt{2}}{6}\pi a^3 \,.$$

Using the volume of the unit cell, $V_0 = a^3$, we find the relative density

$$f_V = \frac{V_{\text{atom}}}{V_0} = 0.7405 \approx 74\% \,.$$

b) The body-centred cubic unit cell (figure 1.5(b)) contains 2 atoms (8 on the corners, counting $1/8$ each, and one in the centre). The space diagonal of length $\sqrt{3}\,a$ contains 4 atomic radii $4r = \sqrt{3}\,a$. Calculating as above, we find $f_V = 0.6802 \approx 68\%$.

c) The hexagonal close-packed unit cell (figure 1.6) contains 6 atoms (12 atoms, counted $1/6$, 2 atoms on the basal planes, counted as half an atom each, and 3 atoms within the cell). The atomic radius is $r = a/2$. The volume of the atoms is thus $V_{\text{atom}} = \pi a^3$.

Calculating the volume of the unit cell is a bit involved. We start by considering the basal planes: If we draw the length b as in figure 13.1(a), we find $b = a \sin 60°$ or $b = \sqrt{3}\,a/2$. The basal plane comprises 6 equilateral triangles with an area $ab/2 = \sqrt{3}\,a^2/4$. The middle atomic layer is shifted relative to the lower layer by $2b/3$ as shown in figure 13.1(b). The height can be calculated, according to figure 13.1(c), as $a^2 = (c/2)^2 + (2b/3)^2$ or $c = \sqrt{8/3}\,a$. The volume of the unit cell is thus $V_0 = 6 \cdot (\sqrt{3}\,a/2) \cdot (\sqrt{8/3}\,a) = 3\sqrt{8}\,a^3/2$. The relative density follows to

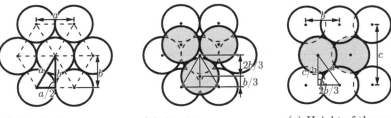

(a) Basal plane   (b) Stacking sequence   (c) Height of the unit cell

Fig. 13.1. Dimensions in a hexagonal close-packed crystal

$$f_V = \frac{V_{atom}}{V_0} = \frac{\pi a^3}{3\sqrt{8}\,a^3/2} = \frac{2\pi}{3\sqrt{8}} = 0.7405 \approx 74\%\,.$$

It would have been much simpler to solve the exercise using the following argument: The face-centred cubic and the hexagonal close-packed crystal are close packed with each atom having 12 nearest neighbours. Therefore, the relative density must be the same in both cases and the result for the face-centred cubic lattice can be used.

Solution 2:

a) Each monomer has the chemical composition C_2H_4, resulting in a molar mass of

$$m_{mono} = 2 \times 12.01\,\text{g/mol} + 4 \times 1.01\,\text{g/mol} = 28.06\,\text{g/mol}\,.$$

The molar mass of the whole molecule is thus $2.8 \times 10^5\,\text{g/mol}$.

b) The horizontal distance between two carbon atoms on a chain as in figure 1.22 is $d = 0.154\,\text{nm} \cdot \cos(90° - 109°/2) = 0.125\,\text{nm}$. The total length of the molecule is thus $L = 2 \times 10^4 \times 0.125\,\text{nm} = 2.507\,\mu\text{m}$.

Solution 3:

a) The interaction force (2.10) must be zero at equilibrium:

$$-\left.\frac{dU_A(r)}{dr}\right|_{r_0} - \left.\frac{dU_R(r)}{dr}\right|_{r_0} = -\frac{1.436}{r_0^2}\,\text{eV nm} + 9 \cdot \frac{5.86 \times 10^{-6}}{r_0^{10}}\,\text{eV nm}^9 = 0\,.$$

$$r_0 = \sqrt[8]{\frac{5.274 \times 10^{-5}}{1.436}}\,\text{nm}^8 = 0.279\,\text{nm}\,.$$

b) The binding energy is the energy at the minimum, $U(r_0)$. This yields

$$U(r_0) = -\frac{1.436}{0.279\,\text{nm}}\,\text{eV nm} + \frac{5.86 \times 10^{-6}}{(0.279\,\text{nm})^9}\,\text{eV nm}^9 = -4.57\,\text{eV}\,.$$

c) The molar mass of a NaCl molecule is $m_{mol} = 58.4\,\mathrm{g/mol}$. The molecular density (number of molecules per cubic metre) is thus

$$n_{NaCl} = \frac{\varrho}{m_{mol}} = \frac{2165\,\mathrm{kg/m^3}}{0.0584\,\mathrm{kg/mol}} = 37\,071.9\,\mathrm{mol/m^3}\,.$$

The number of atoms is twice as large; the lattice constant is therefore (since the lattice is simple cubic)

$$a = \frac{1}{\sqrt[3]{2n_{NaCl}}} = \frac{1}{\sqrt[3]{74\,143.83\,\mathrm{mol/m^3} \cdot N_A}} = 0.282\,\mathrm{nm}\,.$$

This result agrees well with the result from subtask a).

d) We start by calculating the spring constant k of the bond between two atoms, according to section 2.3:

$$k = \frac{d^2 U}{dr^2}\Big|_{r_0} = -2 \times \frac{1.436}{r_0^3}\,\mathrm{eV\,nm} + 90 \times \frac{5.86 \times 10^{-6}}{r_0^{11}}\,\mathrm{eV\,nm^9}$$
$$= 529.20\,\mathrm{eV/nm^2} = 84.778\,\mathrm{J/m^2}\,.$$

If a force F is applied in the $\langle 100 \rangle$ direction to an area A, each bond is loaded by a force $F_b = Fa^2/A$. The resulting displacement Δl per bond is $\Delta l = F_b/k$. The strain is thus

$$\varepsilon = \frac{\Delta l}{a} = \frac{Fa^2}{akA}\,.$$

Using the definition of stress, $\sigma = F/A$, we find Young's modulus E as

$$E = \frac{\sigma}{\varepsilon} = \frac{F/A}{Fa^2/akA} = \frac{k}{a} = \frac{84.778\,\mathrm{J/m^2}}{2.79 \times 10^{-10}\,\mathrm{m}} = 304\,\mathrm{GPa}\,.$$

The reference area A and the value of the applied force cancel from the equation as expected.

e) According to equations (2.35) and (2.39), Young's modulus is

$$E_{\langle 100 \rangle} = \frac{1}{S_{11}} = \frac{(C_{11} - C_{12})(C_{11} + 2C_{12})}{C_{11} + C_{12}} = 43.5\,\mathrm{GPa}\,,$$

The simple estimate has roughly the right order of magnitude, but it is still way too large. This is mainly due to the fact that the repulsion of diagonally neighbouring atoms eases the deformation and has been neglected.

Solution 4:

a) Consider a brick-shaped volume with dimensions $l_1 \times l_2 \times l_3$, enlarged to a size $(l_1 + \Delta l_1) \times (l_2 + \Delta l_2) \times (l_3 + \Delta l_3)$. The ratio of the two volumes is

$$\frac{V_1}{V_0} = 1 + \frac{\Delta V}{V_0} = \frac{(l_1 + \Delta l_1)(l_2 + \Delta l_2)(l_3 + \Delta l_3)}{l_1 l_2 l_3} = (1 + \varepsilon_{11})(1 + \varepsilon_{22})(1 + \varepsilon_{33})$$

$$= 1 + \varepsilon_{11} + \varepsilon_{22} + \varepsilon_{33} + \underbrace{\varepsilon_{11}\varepsilon_{22}}_{\ll 1} + \underbrace{\varepsilon_{11}\varepsilon_{33}}_{\ll 1} + \underbrace{\varepsilon_{22}\varepsilon_{33}}_{\ll 1} + \underbrace{\varepsilon_{11}\varepsilon_{22}\varepsilon_{33}}_{\ll 1},$$

$$\frac{\Delta V}{V_0} = \varepsilon_{11} + \varepsilon_{22} + \varepsilon_{33}. \tag{13.1}$$

The strains can be calculated from (2.33a). Their sum is

$$\varepsilon_{11} + \varepsilon_{22} + \varepsilon_{33} = \frac{1 - 2\nu}{E} \left(\sigma_{11} + \sigma_{22} + \sigma_{33} \right). \tag{13.2}$$

The sum of the normal stresses equals, according to equation (3.25), three times the hydrostatic stress or three times the negative hydrostatic pressure: $-3\Delta p = 3\sigma_{\text{hyd}} = \sigma_{11} + \sigma_{22} + \sigma_{33}$. Inserting this and equation (13.1) into (13.2) yields

$$\Delta p = -\frac{E}{3(1 - 2\nu)} \cdot \frac{\Delta V}{V_0}.$$

Comparing terms with equation (12.1) yields the *bulk modulus*

$$K = \frac{E}{3(1 - 2\nu)}. \tag{13.3}$$

b) In uniaxial tension, the pressure is $\Delta p = -\sigma/3$. From this we find
   $\nu_1 = 0$: $K_1 = E/3$, $\Delta V_1/V_0 = \sigma/E = \varepsilon$.
   $\nu_2 = 1/3$: $K_2 = E$, $\Delta V_2/V_0 = \sigma/(3E) = \varepsilon/3$.
   $\nu_3 = 0.5$: $K_3 = \infty$, $\Delta V_3/V_0 = 0$.
c) A positive normal strain results in a positive transversal strain and thus an increase in the cross section.

Solution 5:

In the x_i coordinate system, the state is one of plane strain as sketched in figure 2.8(a). The strain tensor is

$$(\varepsilon_{ij}) = \begin{pmatrix} -\varepsilon & 0 & 0 \\ 0 & \varepsilon & 0 \\ 0 & 0 & 0 \end{pmatrix}. \tag{13.4}$$

The strain tensor in the $x_{i'}$ coordinate system, rotated by 45°, can be found using the transformation matrix

$$(g_{i'i}) = \frac{\sqrt{2}}{2} \begin{pmatrix} 1 & 1 & 0 \\ -1 & 1 & 0 \\ 0 & 0 & \sqrt{2} \end{pmatrix} \tag{13.5}$$

and the rule of transformation $\varepsilon_{i'j'} = g_{i'i}\,\varepsilon_{ij}\,g_{jj'}$ or $(\varepsilon_{i'j'}) = (g_{i'i})\,(\varepsilon_{ij})\,(g_{j'j})^{\mathrm{T}}$. It is

$$(\varepsilon_{i'j'}) = \frac{\sqrt{2}}{2}\begin{pmatrix} 1 & 1 & 0 \\ -1 & 1 & 0 \\ 0 & 0 & \sqrt{2} \end{pmatrix} \cdot \begin{pmatrix} -\varepsilon & 0 & 0 \\ 0 & \varepsilon & 0 \\ 0 & 0 & 0 \end{pmatrix} \cdot \frac{\sqrt{2}}{2}\begin{pmatrix} 1 & -1 & 0 \\ 1 & 1 & 0 \\ 0 & 0 & \sqrt{2} \end{pmatrix}$$

$$= \begin{pmatrix} 0 & \varepsilon & 0 \\ \varepsilon & 0 & 0 \\ 0 & 0 & 0 \end{pmatrix}. \tag{13.6}$$

Using Hooke's law in the Voigt matrix notation, we can calculate the stress state in each coordinate system. We initially assume that the material parameters might differ in the different systems. In the un-primed system, we find $\sigma_\alpha = C_{\alpha\beta}\,\varepsilon_\beta$, in the primed system, the stress is $\sigma_{\alpha'} = C_{\alpha'\beta'}\,\varepsilon_{\beta'}$. This yields

$$(\sigma_\alpha) = \begin{pmatrix} \sigma_{11} \\ \sigma_{22} \\ \sigma_{33} \\ \sigma_{23} \\ \sigma_{13} \\ \sigma_{12} \end{pmatrix} = \begin{pmatrix} C_{11} & C_{12} & C_{12} & & & \\ C_{12} & C_{11} & C_{12} & & & \\ C_{12} & C_{12} & C_{11} & & & \\ & & & C_{44} & & \\ & & & & C_{44} & \\ & & & & & C_{44} \end{pmatrix}\begin{pmatrix} -\varepsilon \\ \varepsilon \\ 0 \\ 0 \\ 0 \\ 0 \end{pmatrix} = \varepsilon\begin{pmatrix} -(C_{11}-C_{12}) \\ C_{11}-C_{12} \\ 0 \\ 0 \\ 0 \\ 0 \end{pmatrix},$$

$$(\sigma_{ij}) = \varepsilon\begin{pmatrix} -(C_{11}-C_{12}) & 0 & 0 \\ 0 & C_{11}-C_{12} & 0 \\ 0 & 0 & 0 \end{pmatrix}, \tag{13.7}$$

$$(\sigma_{\alpha'}) = \begin{pmatrix} \sigma_{1'1'} \\ \sigma_{2'2'} \\ \sigma_{3'3'} \\ \sigma_{2'3'} \\ \sigma_{1'3'} \\ \sigma_{1'2'} \end{pmatrix} = \begin{pmatrix} C_{1'1'} & C_{1'2'} & C_{1'2'} & & & \\ C_{1'2'} & C_{1'1'} & C_{1'2'} & & & \\ C_{1'2'} & C_{1'2'} & C_{1'1'} & & & \\ & & & C_{4'4'} & & \\ & & & & C_{4'4'} & \\ & & & & & C_{4'4'} \end{pmatrix}\begin{pmatrix} 0 \\ 0 \\ 0 \\ \varepsilon \\ 0 \\ 0 \end{pmatrix} = \varepsilon\begin{pmatrix} 0 \\ 0 \\ 0 \\ 0 \\ 0 \\ C_{4'4'} \end{pmatrix},$$

$$(\sigma_{i'j'}) = 2\varepsilon\begin{pmatrix} 0 & C_{4'4'} & 0 \\ C_{4'4'} & 0 & 0 \\ 0 & 0 & 0 \end{pmatrix}. \tag{13.8}$$

The matrices of coefficients, (σ_{ij}) and $(\sigma_{i'j'})$, must represent the same stress state and thus the same stress tensor $\underline{\underline{\sigma}}$ because we consider the same physical system. Therefore, a coordinate transformation must get us from one state to the other. Using the transformation rules $\sigma_{i'j'} = g_{i'i}\,\sigma_{ij}\,g_{jj'}$, or $(\sigma_{i'j'}) = (g_{i'i})\,(\sigma_{ij})\,(g_{j'j})^{\mathrm{T}}$, we find $(\sigma_{i'j'})$ from (σ_{ij}) according to equation (13.7):

$$(\sigma_{i'j'}) = \varepsilon\begin{pmatrix} 0 & (C_{11}-C_{12}) & 0 \\ (C_{11}-C_{12}) & 0 & 0 \\ 0 & 0 & 0 \end{pmatrix}.$$

Comparing terms with equation (13.8) yields the condition

$$2C_{4'4'} = C_{11} - C_{12}. \tag{13.9}$$

If the material is isotropic, its properties must not change when the frame of reference is changed. Therefore, we know $C_{1'1'} = C_{11}$, $C_{1'2'} = C_{12}$ and $C_{4'4'} = C_{44}$. This yields the proposed equation

$$C_{44} = \frac{C_{11} - C_{12}}{2} .$$ (13.10)

Solution 6:

a) The stored elastic strain energy $W^{(el)}$ equals the external work done during the deformation, $\int F(\Delta l)\, d(\Delta l)$. In both geometries, the cross section is A.
   At **booth A**, we find, using $\sigma = E_A \cdot \varepsilon$:

$$F(\Delta l) = AE_A \frac{\Delta l}{l_A} ,$$

$$W_A^{(el)} = \int_0^{\Delta l_A} F(\Delta l)\, d(\Delta l) = \frac{AE_A}{l_A} \int_0^{\Delta l_A} \Delta l\, d(\Delta l) = \frac{AE_A (\Delta l_A)^2}{2l_A} .$$

Booth B: Because the material is assumed to be linear elastic, we can integrate over dF instead of dl: $\int \Delta l(F)\, dF$. Using $\varepsilon = \sigma/E_B$ yields

$$\Delta l(F) = l_B \frac{F}{AE_B} ,$$

$$W_B^{(el)} = \int_0^{F_B} \Delta l(F)\, dF = \frac{l_B}{AE_B} \int_0^{F_B} F\, dF = \frac{l_B (F_B)^2}{2AE_B} .$$

b) **Booth A:** $W_A^{(el)} \propto E_A$. Increasing E_A increases $W_A^{(el)}$ and vice versa.
   Booth B: $W_B^{(el)} \propto 1/E_B$. Increasing E_B decreases $W_B^{(el)}$ and vice versa.
   Figure 13.2 shows the stored elastic energy as area in the stress-strain diagram. In strain-controlled loading, the rubber band with the larger stiffness stores more energy, in stress-controlled loading, it stores less energy.

c) The take-off velocity v is reached when all of the elastic energy is converted to kinetic energy: $W^{(kin)} = W^{(el)}$. The kinetic energy is $W^{(kin)} = mv^2/2$. The take-off speed is thus $v = \sqrt{2W^{(kin)}/m}$.
   Booth A:

$$v_A = \sqrt{\frac{2W_A^{(kin)}}{m}} = \sqrt{\frac{2W_A^{(el)}}{m}} = \sqrt{\frac{AE_A}{ml_A}} \Delta l_A .$$ (13.11)

   Booth B:

$$v_B = \sqrt{\frac{2W_B^{(kin)}}{m}} = \sqrt{\frac{2W_B^{(el)}}{m}} = \sqrt{\frac{l_B}{mAE_B}} F_B .$$ (13.12)

d) The following quantities cannot be changed: A, m, l_A, Δl_A, l_B, F_B. Young's moduli E_A and E_B in equations (13.11) and (13.12) can be changed by exchanging the rubber bands.
   Booth A:

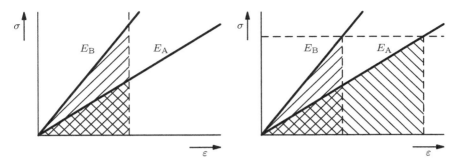

(a) Strain-controlled. The stored
energy is larger in the stiffer material

(b) Stress-controlled. The stored
energy is smaller in the stiffer material

Fig. 13.2. stored elastic strain energy in the rubber bands in strain- and stress-controlled loading

$$v_A^{(new)} = \sqrt{\frac{AE_B}{ml_A}}\,\Delta l_A = \sqrt{\frac{A \cdot 2E_A}{ml_A}}\,\Delta l_A = \sqrt{2} \cdot v_A .$$

Booth B:

$$v_B^{(new)} = \sqrt{\frac{l_B}{mAE_A}}\,F_B = \sqrt{\frac{l_B}{mAE_B/2}}\,F_B = \sqrt{2} \cdot v_B .$$

Conclusion: In strain-controlled loading (booth A), increasing the stiffness increases the elastically stored energy, whereas decreasing the stiffness increases the elastically stored energy in stress-controlled loading (booth B), see figure 13.2.

Solution 7:

a) The initial length for the first elongation Δl_1 is l. The strain is

$$\varepsilon_1 = \frac{\Delta l_1}{l} .$$

When adding the second elongation Δl_2, we have to relate the strain to the current length $l + \Delta l_1$:

$$\varepsilon_2 = \frac{\Delta l_2}{l + \Delta l_1} .$$

The total nominal strain is the sum of the two contributions:

$$\varepsilon_1 + \varepsilon_2 = \frac{\Delta l_1}{l} + \frac{\Delta l_2}{l + \Delta l_1} = \frac{\Delta l_1(l + \Delta l_1) + \Delta l_2 l}{l(l + \Delta l_1)} = \frac{\Delta l_1 l + \Delta l_1^2 + \Delta l_2 l}{l(l + \Delta l_1)} .$$

$$(13.13)$$

If we elongate the rod in one step by $\Delta l_1 + \Delta l_2$, the reference length is l:

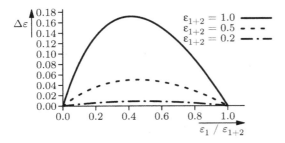

Fig. 13.3. Deviation between nominal strains in one and two steps as function of the step size fractions

$$\varepsilon_{1+2} = \frac{\Delta l_1 + \Delta l_2}{l} .$$

The difference between the two total strains is $\Delta\varepsilon$:

$$\Delta\varepsilon = \varepsilon_{1+2} - (\varepsilon_1 + \varepsilon_2) = \frac{(\Delta l_1 + \Delta l_2)(l + \Delta l_1)}{l(l + \Delta l_1)} - \frac{\Delta l_1 l + \Delta l_1^2 + \Delta l_2 l}{l(l + \Delta l_1)}$$

$$= \frac{\Delta l_1 \Delta l_2}{l(l + \Delta l_1)} = \varepsilon_1\,\varepsilon_2 .$$

b) Figure 13.3 shows the dependence of the strain difference for a rod of length $l = 100\,\mathrm{mm}$ on the division of the total length change Δl into the two partial steps Δl_1 and Δl_2. The quantity on the abscissa, $\varepsilon_1/\varepsilon_{1+2}$, is equal to $\Delta l_1/(\Delta l_1 + \Delta l_2)$. At $\varepsilon_1/\varepsilon_{1+2} = 0$ and $\varepsilon_1/\varepsilon_{1+2} = 1$, the deviation vanishes because the total deformation is done in one step. The maximum difference occurs if about 40% of the deformation are done in the first step. It can also be seen that the deviation increases quadratically with the total strain ε_{1+2}.

c) Using the definition of the true strain, equation (3.3), we find for the first step

$$\varphi_1 = \ln \frac{l + \Delta l_1}{l} ,$$

and for the second step

$$\varphi_2 = \ln \frac{l + \Delta l_1 + \Delta l_2}{l + \Delta l_1} .$$

The sum of the two steps is

$$\varphi_1 + \varphi_2 = \ln \frac{l + \Delta l_1}{l} + \ln \frac{l + \Delta l_1 + \Delta l_2}{l + \Delta l_1} = \ln \frac{l + \Delta l_1 + \Delta l_2}{l} . \tag{13.14}$$

If we perform the deformation in a single step, the strain is

$$\varphi_{1+2} = \ln \frac{l + \Delta l_1 + \Delta l_2}{l} . \tag{13.15}$$

Comparing equations (13.14) and (13.15), we see that $\varphi_{1+2} = \varphi_1 + \varphi_2$ holds.

Solution 8:

a) In this case, we find $G_{10} = G_0 \cdot (1 + 10\,a \cdot z_0)$. The rate of interest is thus $z_0 = 0.1\,a^{-1} = 10\,\%/a$.

b) Now, we have $G_1 = (1+z)G_0$, $G_2 = (1+z)G_1 = (1+z)^2 G_0, \ldots, G_n = (1+z)^n G_0$. This yields $z = \sqrt[n]{G_n/G_0} - 1$ and $z = \sqrt[10]{G_{10}/G_0} - 1 = 7.2\,\%/a$.

   Conclusion: If the current deposit is used to calculate the interest, the rate of interest needed to get the same final amount is smaller than in the other case. Accordingly, the true strain φ is smaller than the nominal strain ε at the same elongation.

Solution 9:

a) It is easy to find the relation between $\underline{\xi}$ and $\underline{x}$:

$$x_i = \xi_i \cdot \frac{l_i + \Delta l_i}{l_i} = \xi_i \cdot \left(1 + \frac{\Delta l_i}{l_i} \right).$$

Using equation (3.4) yields the deformation gradient

$$F_{ij} = \frac{\partial x_i(\underline{\xi})}{\partial \xi_j} = \begin{cases} \left(1 + \frac{\Delta l_i}{l_i} \right) & \text{for } i = j, \\ 0 & \text{for } i \neq j. \end{cases}$$

F_{ij} thus contains only diagonal entries. Green's strain tensor $\underline{\underline{G}}$ can be calculated using equation (3.7):

$$G_{ij} = \frac{1}{2} \cdot \begin{cases} \left(1 + \frac{\Delta l_i}{l_i} \right)^2 - 1 & \text{for } i = j, \\ 0 & \text{for } i \neq j \end{cases}$$

$$= \begin{cases} \frac{\Delta l_i}{l_i} + \frac{1}{2} \left(\frac{\Delta l_i}{l_i} \right)^2 & \text{for } i = j, \\ 0 & \text{for } i \neq j. \end{cases} \tag{13.16}$$

b) If we insert the normal strains $\varepsilon_{ii} = \Delta l_i / l_i$ into equation (13.16), we find

$$\underline{\underline{G}} = \begin{pmatrix} \varepsilon_{11} + \varepsilon_{11}^2/2 & 0 & 0 \\ 0 & \varepsilon_{22} + \varepsilon_{22}^2/2 & 0 \\ 0 & 0 & \varepsilon_{33} + \varepsilon_{33}^2/2 \end{pmatrix}$$

For small deformations, the terms ε_{ii} are small and ε_{ii}^2 can be neglected. In this case, Green's strain tensor $\underline{\underline{G}}$ approaches the strain tensor $\underline{\underline{\varepsilon}}$. The deviation increases with increasing deformation.

Solution 10:

a) The principal stresses are the eigenvalues of the stress tensor (calculated without writing units):

$$\det \begin{pmatrix} 155 - \lambda & 55 & 0 \\ 55 & 155 - \lambda & 0 \\ 0 & 0 & -\lambda \end{pmatrix} = -\lambda \left[(155 - \lambda)^2 - 55^2 \right] = 0 ,$$

$$\Rightarrow \lambda_1 = 0, \ \lambda_2 = 210, \ \lambda_3 = 100 .$$

The result is: $\sigma_{\mathrm{I}} = 210\,\mathrm{MPa}$, $\sigma_{\mathrm{II}} = 100\,\mathrm{MPa}$, $\sigma_{\mathrm{III}} = 0\,\mathrm{MPa}$.

b) $\sigma_{\mathrm{eq,T}} = \sigma_{\mathrm{I}} - \sigma_{\mathrm{III}} = 210\,\mathrm{MPa} > R_{\mathrm{p0.2}}$. The material yields.

c) $\sigma_{\mathrm{eq,M}} = \sqrt{\frac{1}{2} \left[(\sigma_{\mathrm{I}} - \sigma_{\mathrm{II}})^2 + (\sigma_{\mathrm{II}} - \sigma_{\mathrm{III}})^2 + (\sigma_{\mathrm{III}} - \sigma_{\mathrm{I}})^2 \right]} = 181.93\,\mathrm{MPa} < R_{\mathrm{p0.2}}$.
The material does not yield

d) It is not possible to decide because both yield criteria are only approximately true.

e) $\tau = \sigma_{\mathrm{eq,M}}/M = 58.7\,\mathrm{MPa} < \tau_{\mathrm{F}}$. No significant activation of dislocation movement.

f) The deviator can be calculated from $\underline{\underline{\sigma}}' = \underline{\underline{\sigma}} - \underline{\underline{1}}\,\sigma_{\mathrm{hyd}}$, using $\sigma_{\mathrm{hyd}} = \mathrm{tr}\,\underline{\underline{\sigma}}/3 = (155 + 155 + 0)/3\,\mathrm{MPa} = 103.\bar{3}\,\mathrm{MPa}$. This results in

$$\underline{\underline{\sigma}}' = \begin{pmatrix} 51.\bar{6} & 55 & 0 \\ 55 & 51.\bar{6} & 0 \\ 0 & 0 & -103.\bar{3} \end{pmatrix} \mathrm{MPa} .$$

Solution 11:

The parameter m from equation (3.36) is $m = 50\,\mathrm{MPa}/40\,\mathrm{MPa} = 1.25$.

a) The hydrostatic stress state is characterised by $\sigma_{11} = \sigma_{22} = \sigma_{33} = \sigma_{\mathrm{hyd}}$ and $\sigma_{23} = \sigma_{13} = \sigma_{12} = 0$.
Parabolic: According to equation (3.37), we find at yielding

$$R_{\mathrm{p}} = \frac{m-1}{2m} \cdot 3\,\sigma_{\mathrm{eq,pM,F}} + \sqrt{\left[\frac{m-1}{2m}\,3\,\sigma_{\mathrm{eq,pM,F}} \right]^2 + 0}$$

$$= \frac{m-1}{m} \cdot 3\,\sigma_{\mathrm{eq,pM,F}} = \frac{3}{5}\,\sigma_{\mathrm{eq,pM,F}} .$$

This results in a 'hydrostatic yield strength' of $\sigma_{\mathrm{eq,pM,F}} = 5/3 \cdot R_{\mathrm{p}} = 66.7\,\mathrm{MPa}$.
Conical: According to equation (3.39), we find at yielding

$$R_{\mathrm{p}} = \frac{1}{2m} \cdot \left[(m-1) \cdot 3\,\sigma_{\mathrm{eq,cM,F}} + 0 \right] = \frac{3}{10}\,\sigma_{\mathrm{eq,cM,F}} .$$

This results in a 'hydrostatic yield strength' of $\sigma_{\mathrm{eq,cM,F}} = 10/3 \cdot R_{\mathrm{p}} = 133.3\,\mathrm{MPa}$.

b) **Parabolic:**

$$\sigma_{\mathrm{eq,pM}} = \frac{0.25}{2.5}\,\sigma_{11} + \sqrt{\left[\frac{0.25}{2.5}\,\sigma_{11} \right]^2 + \frac{8}{2.5}\,\sigma_{11}^2}$$

$$= \frac{1}{10}\,\sigma_{11} + \sqrt{\frac{321}{100}\,\sigma_{11}^2} = 1.89\,\sigma_{11} = 1.058\,R_{\mathrm{p}} .$$

The material yields.

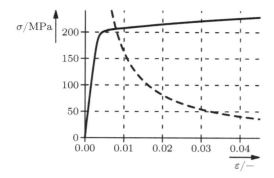

Fig. 13.4. Neuber's hyperbola

Conical:

$$\sigma_{\text{eq,cM}} = \frac{1}{2.5}\left[0.25\,\sigma_{11} + 2.25\,\sqrt{4\sigma_{11}^2}\right] = 1.9\,\sigma_{11} = 1.064\,R_{\text{p}}\,.$$

The material yields.

c) **Parabolic:**

$$\sigma_{\text{eq,pM}} = \frac{0.25}{2.5}\,(-\sigma_{11}) + \sqrt{\left[\frac{0.25}{2.5}\,\sigma_{11}\right]^2 + \frac{8}{2.5}\,\sigma_{11}^2} = 1.69\,\sigma_{11} = 0.947\,R_{\text{p}}\,.$$

The material does not yield.

Conical:

$$\sigma_{\text{eq,cM}} = \frac{1}{2.5}\left[0.25\,(-\sigma_{11}) + 2.25\cdot\sqrt{4\sigma_{11}^2}\right] = 1.7\,\sigma_{11} = 0.952\,R_{\text{p}}\,.$$

The material does not yield.

Solution 12:

a) Reading off from diagram 4.3, we find a stress concentration factor of $K_{\text{t}} = 1.67$.

b) The nominal stress at the notch root is

$$\sigma_{\text{nss}} = \frac{F}{\pi(d/2)^2} = 198.94\,\text{MPa}\,.$$

According to equation (4.1), we find $\sigma_{\max} = K_{\text{t}}\,\sigma_{\text{nss}} = 332\,\text{MPa}$. $\sigma_{\max}$ is above R_{p} and R_{m}. Thus, the component could not be used.

c) Using equation (4.5) yields

$$\sigma_{\max}\,\varepsilon_{\max} = \frac{\sigma_{\text{nss}}^2}{E}\,K_{\text{t}}^2 = 1.623\,\text{MPa}\,.$$

The corresponding Neuber's hyperbola is shown in figure 13.4.

d) The values can be read off the diagram: $\sigma_{\max} = 210\,\text{MPa}$, $\varepsilon_{\max} = 0.008 = 0.8\%$.

e) The component can be used because the maximum strain is significantly smaller than the strain at necking.

Solution 13:

a) Since the stress is defined as force per area, we have to look at the stress over the width of a lattice constant and have to integrate the stress field in x_1 direction over a distance of one lattice constant. The force is thus

$$F = a_{NaCl} \int_0^{a_{NaCl}} \frac{K_{Ic}}{\sqrt{2\pi}} \frac{1}{\sqrt{r}} = a_{NaCl} \frac{K_{Ic}}{\sqrt{2\pi}} \left[2\sqrt{r}\right]_0^{a_{NaCl}} = \frac{K_{Ic}}{\sqrt{2\pi}} \cdot 2a_{NaCl}^{3/2} .$$

b) The force is $F = kx$ if the bond is strained by a distance x. The force at a strain of $a_{NaCl}/10$ must, according to the assumption, equal the force from subtask a). The fracture toughness can thus be calculated as follows:

$$\frac{k\, a_{NaCl}}{10} = \frac{K_{Ic}}{\sqrt{2\pi}} \cdot 2a_{NaCl}^{3/2} ,$$

$$\frac{k}{10} = \frac{K_{Ic}}{\sqrt{2\pi}} \cdot 2a_{NaCl}^{1/2} ,$$

$$K_{Ic} = \frac{k}{10} \frac{\sqrt{2\pi}}{2} \frac{1}{a_{NaCl}^{1/2}} = 0.634\,\text{MPa}\sqrt{\text{m}} .$$

This value is of the correct order of magnitude for a ceramic crystal.

c) Because we simply used the stress field calculated from continuum mechanics to find the force at one atom, the calculation is incorrect since there can be no stresses in between the atomic positions. The calculation could be improved by using the elastic stress field at some distance from the crack tip and by calculating the displacements of all atoms inside this region using the force law. Furthermore, it would be necessary to quantify the fracture strain of a bond more precisely. Assuming a simple spring force is also a severe approximation because the potential curve is not parabolic if the displacements are large (see, for example, figure 2.6). Calculations accounting for all this can yield realistic values for the fracture toughness of a material. The calculation as presented here can be accepted as a very coarse approximation that mainly serves to show why a stress singularity does not imply a force singularity at the position of the crack tip.

Solution 14:

We start by adding the elastic line and the 95% line to the diagram (figure 13.5). Reading off the forces yields $F_5 = 13.5\,\text{kN}$, $F_{max} = 14.5\,\text{kN}$. F_5 is to the left of F_{max}, corresponding to the case from figure 5.16(b), yielding $F_Q = F_5$. The condition (5.32) has to be met: $F_{max}/F_Q = 1.07 \le 1.1$ (true).

The geometry factor for the initial crack length is $f = 9.66$. Using equation (5.30) yields $K_Q = 23.3\,\text{MPa}\sqrt{\text{m}}$. We finally have to check the inequality (5.33). The right-hand side is $2.5(K_Q/R_p)^2 = 5.2\,\text{mm}$. All required dimensions (B, a, $W - a$) fulfil this condition.

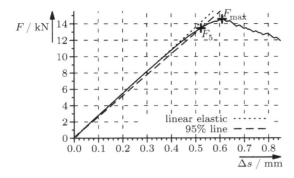

Fig. 13.5. Determination of the forces F_5 and $F_{\max}$ in the load-displacement diagram

Thus, the fracture toughness is $K_{\mathrm{Ic}} = 23.3\,\mathrm{MPa}\sqrt{\mathrm{m}}$.

Solution 15:

The solution is based on section 5.2.3.

a) Design against yielding: The stress state is uniaxial with $\sigma = pD/(2t)$. Thus, the condition $\sigma = pD/(2t) < R_{\mathrm{p0.2}}$ must be met: Since $\sigma = 1200\,\mathrm{MPa} < 1420\,\mathrm{MPa}$, there is no yielding.

   Design against cleavage fracture: $\sigma_{\mathrm{I}} < \sigma_{\mathrm{C}}$: $1200\,\mathrm{MPa} < 2200\,\mathrm{MPa}$. Cleavage fracture is not to be expected.

   Design against crack propagation: $\sigma_{\mathrm{I}} < K_{\mathrm{Ic}}/\sqrt{\pi a}$, where $a = 1.5\,\mathrm{mm}$ is the maximum half crack length to be expected: $1200\,\mathrm{MPa} < 1311\,\mathrm{MPa}$. There will be no crack propagation.

   The tube can be used.

b) Since R_{p} was stated for uniaxial loading, the result is independent of the yield criterion because the service load is also uniaxial.[1]

c) Yielding occurs at a pressure $p = 2tR_{\mathrm{p0.2}}/D = 14.2\,\mathrm{MPa}$.

   Cleavage fracture will be observed at a pressure $p = 2t\sigma_{\mathrm{C}}/D = 22\,\mathrm{MPa}$.

   From the fracture toughness, the stress can be calculated using $\sigma = K_{\mathrm{Ic}}/\sqrt{\pi a}$. The resulting failure pressure is $p = 2tK_{\mathrm{Ic}}/(\sqrt{\pi a}\,D) = 13.1\,\mathrm{MPa}$.

   Thus, the tube will fail by crack propagation at a pressure $p = 13.1\,\mathrm{MPa}$ if a crack of length $a = 1.5\,\mathrm{mm}$ is present.

d) The yield strength and the fracture toughness are reached simultaneously at a value a_{c} from equation (5.28): $a_{\mathrm{c}} = (90\,\mathrm{MPa}\sqrt{\mathrm{m}}/1420\,\mathrm{MPa})^2/\pi = 1.28\,\mathrm{mm}$.

e) See figure 13.6.

f) According to equation (5.3), the crack opening is

$$v_0 = 2\frac{2\sigma a}{E} = 0.036\,\mathrm{mm} = 36\,\mu\mathrm{m}\,,$$

---

[1] If the yield criteria had been assumed equal for pure shear, there would be a difference in uniaxial tension.

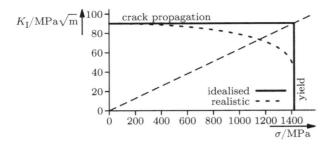

Fig. 13.6. Failure-Assessment diagram for exercise 15

where the additional factor 2 is necessary because equation (5.3) uses the displacement of one crack surface which is half of the crack opening.

Solution 16:

a) At small displacements $a \ll x$, Hooke's law from equation (2.5) can be used with shear strain $\gamma = x/a$:

$$\tau(x) = G\frac{x}{a} \,. \tag{13.17}$$

The shear stress is given by (see figure 12.4)

$$\tau(x) = \tau_{\text{max}} \sin\left(2\pi\frac{x}{a}\right)$$

for all x between $0 \le x \le a$. For small arguments $\alpha \ll 1$, the sine can be approximated as $\sin(\alpha) \approx \alpha$. The result is

$$\tau(x) \approx \tau_{\text{max}} \cdot 2\pi\frac{x}{a} \,.$$

Equalling this with equation (13.17) yields

$$\tau_{\text{max}} = \frac{G}{2\pi} \,, \tag{13.18}$$

where τ_{max} is equal to the theoretical shear stress $\tilde{\tau}_{\text{F}}$. In aluminium, with a shear modulus of $G = 26\,500\,\text{MPa}$, this results in the estimate $\tilde{\tau}_{\text{F}} = 4218\,\text{MPa}$. In reality, pure aluminium has a yield strength of only $R_{\text{p0.2}} \approx 50\,\text{MPa}$, corresponding to a maximum shear stress of $\tau_{\text{F}} = 25\,\text{MPa}$. The simple estimate is thus too large by two orders of magnitude. From this, we can conclude that slip does not occur by shifting layers of atoms simultaneously.

b) The equilibrium position at $x = a/2$ is unstable because the atoms of one layer are situated between those of the other layer, resulting in a maximum strain of the bonds. An infinitesimal displacement from this position would result in the atoms moving to either $x = 0$ or $x = a$. This is due to the fact that the stiffness $C = \mathrm{d}\tau/\mathrm{d}x$ is negative at this point:

$$C = \left.\frac{d\tau}{dx}\right|_{x=a/2} = 2\pi\tau_{\text{max}}\cos\left(2\pi\frac{a/2}{a}\right) = 2\pi\tau_{\text{max}}\cos\pi = -2\pi\tau_{\text{max}}.$$

Solution 17:

a) A dislocation moving from one side of the crystal to the other causes a slip of one Burgers vector b. To shear the crystal by a length s, $N = s/b = 3.5 \times 10^5$ dislocations have to move through the crystal. Because the dislocations are not all at one side of the crystal initially, they can, on average, cover only half the length of the crystal, resulting in twice this value, $N = 7 \times 10^5$.

b) The dislocation density is the dislocation length per volume. To completely shear the crystal over its width, the dislocation has to extend throughout the crystal and thus have a length of $10\,\text{mm}$. The resulting dislocation density is thus

$$\varrho = \frac{aN}{a^3} = \frac{N}{a^2} = 7 \times 10^9\,\text{m}^{-2}.$$

Not all of the dislocations can contribute because of their orientation. Assume that the shear is in the x direction. If we consider screw dislocations, all dislocations with a line vector in the y direction can contribute (one third of all screw dislocations). Of the edge dislocations, only those with line vector in the y direction can contribute that have the additional half-plane in the z direction (one sixth of all edge dislocations). As we are interested in an estimate only, we can assume that about one fifth of all dislocations contribute to the deformation. This results in a final estimate for the dislocation density of about $3.5 \times 10^{10}\,\text{m}^{-2}$.

c) If the grain size is d, the dislocations do not move throughout the crystal, but are limited to one grain because, due to the small amount of deformation, the stresses can be expected to be too small to allow dislocations to pass grain boundaries. The number of dislocations thus increases by a factor $s/d = 100$, resulting in a dislocation density of $\varrho = 3.5 \times 10^{12}\,\text{m}^{-2}$.

d) The total length of all dislocations is $L = \varrho a^3 = 3.5 \times 10^6\,\text{m} = 3500\,\text{km}$.

Solution 18:

The energy per length of a dislocation line is $T \approx Gb^2/2$ according to equation (6.3). Strictly speaking, this is only valid for a straight segment, but we will see that the required energy is so large that this is irrelevant. The energy E of a dislocation loop of length $6b$ is $E = 6Gb^3/2 = 1.8 \times 10^{-18}\,\text{J}$. The probability to form such a dislocation loop is $P = \exp(-E/(kT))$, resulting in $P_{300} \approx 1.5 \times 10^{-189}$ at $300\,\text{K}$ and $P_{900} \approx 1.1 \times 10^{-63}$ at $900\,\text{K}$. The thermally activated generation of dislocation loops is thus practically impossible.

Solution 19:

To calculate the increase in strength, we can use equation (6.20), with the increase being the difference of the strengthening contribution in both states:

Table 13.1. Determination of the Weibull modulus

i	$\tilde{P}_{\mathrm{f},i}$	$\ln\ln\big(1/(1-\tilde{P}_{\mathrm{f},i})\big)$	σ_i/MPa	$\ln(\sigma_i/\mathrm{MPa})$
1	0.125	-2.013	54.60	4.0
2	0.375	-0.755	90.02	4.5
3	0.625	$-0.019\,4$	244.69	5.5
4	0.875	0.732	665.15	6.5

$$\Delta\sigma = k_{\mathrm{d}} M G b \left(\sqrt{\varrho_1} - \sqrt{\varrho_0}\right) = 230\,\mathrm{MPa}\,.$$

Solution 20:

Using the Hall-Petch equation (6.25), the contribution to strengthening is $\Delta\sigma_{\mathrm{coarse}} = k/\sqrt{d_{\mathrm{coarse}}}$ in the coarse-grained and $\Delta\sigma_{\mathrm{fine}} = k/\sqrt{d_{\mathrm{fine}}}$ in the fine-grained material. Reducing the grain size strengthens by the difference of these two contributions:

$$\Delta\sigma = \frac{k}{\sqrt{d_{\mathrm{fine}}}} - \frac{k}{\sqrt{d_{\mathrm{coarse}}}}\,.$$

Solving for the new grain size, we find for $\Delta\sigma = 80\,\mathrm{MPa}$

$$d_{\mathrm{fine}} = \left(\frac{\Delta\sigma}{k} + \frac{1}{\sqrt{d_{\mathrm{coarse}}}}\right)^{-2} = 1.48\,\mathrm{\mu m}\,.$$

Solution 21:

a) The Orowan stress is $\tau = Gb/2\lambda$ according to equation (6.17). The normal stress σ and the shear stress τ are related by the Taylor factor which takes a value of $M = 3.1$ in face-centred cubic metals. This results in

$$2\lambda = \frac{GbM}{\Delta R_{\mathrm{p}0.2}} = 39\,\mathrm{nm}\,.$$

b) According to equation (6.28), we find

$$r = 2\lambda\sqrt{\frac{f_{\mathrm{V}}}{2}} = 5.5\,\mathrm{nm}\,.$$

Solution 22:

We start by sorting the data with increasing size and assigning approximate failure probabilities according to equation (7.14) (see table 13.1). We also enter the quantities $\ln\ln\big(1/(1-\tilde{P}_{\mathrm{f},i})\big)$ and $\ln(\sigma_i/\mathrm{MPa})$ into the table to enable drawing the diagram in figure 13.7. From this diagram, we can read off the Weibull modulus $m = 1$ which equals the slope. The intersection with the axis is $-m\ln(\sigma_0/\mathrm{MPa}) = -5.6$, yielding $\sigma_0 = 270\,\mathrm{MPa}$.

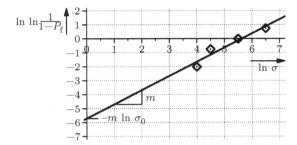

Fig. 13.7. Graphical determination of m and σ_0 in a diagram analogous to figure 7.17

Solution 23:

a) The pressure is $p = F/A = mg/(LB) = 6.131\,25 \times 10^{-3}\,\mathrm{N/mm^2}$.

b) Equation (7.6) yields

$$1 - P_f = \exp\left[-\frac{V}{V_0}\left(\frac{\sigma_{\mathrm{limit}}}{\sigma_0}\right)^m\right],$$

$$\ln(1 - P_f) = -\frac{V}{V_0}\left(\frac{\sigma_{\mathrm{limit}}}{\sigma_0}\right)^m,$$

$$\frac{\sigma_{\mathrm{limit}}}{\sigma_0} = \sqrt[m]{-\frac{V_0}{V}\ln(1 - P_f)}. \tag{13.19}$$

Using $V/V_0 = 1$, we thus find $\sigma_{\mathrm{limit}} = 0.541\,\sigma_0$. Relating this to the pressure, $\sigma_{\mathrm{limit}} = 2pL^2/d_{\mathrm{min}}^2$, we find for the thickness

$$d_{\mathrm{min}} = \sqrt{\frac{2pL^2}{0.541\,\sigma_0}} = 15.1\,\mathrm{mm}. \tag{13.20}$$

c) From equation (13.19), we find, using $V_1/V_0 = LBd_{\mathrm{min}}/V_{\mathrm{spec}} = 12.08$, the new value of the maximum stress $\sigma_{\mathrm{limit},1} = 0.458\,\sigma_0$. With the help of equation (13.20), the thickness is calculated to be $d_{\mathrm{min},1} = 16.35\,\mathrm{mm}$.

This, however, changes the specimen volume V, changing the allowed stress σ_{limit} from equation (13.19). Using the current thickness value $d_{\mathrm{min},1} = 16.35\,\mathrm{mm}$ yields a permitted stress of $\sigma_{\mathrm{limit},2} = 0.456\,\sigma_0$. Using again equation (13.20) results in the new thickness $d_{\mathrm{min},2} = 16.40\,\mathrm{mm}$.

The change from $d_{\mathrm{min},1}$ to $d_{\mathrm{min},2}$ is rather small, making further iterations of the procedure unnecessary.

d) Production needs not to be stopped because all of the material volume is maximally stressed in tension, but only a small part of it in bending. The strength in a tensile test is thus smaller than in a bending test. Thus, the safety of the product is increased by this error.

e) $d = \sqrt{2pL^2/R_p} = 11.07\,\mathrm{mm}$.

Solution 24:

a) We want to determine the parameters B^* and n from equation (7.2). To get a system of linear equations for the parameters, we write this equation as

$$t_f = B^* \sigma^{-n} \Rightarrow \ln(t_f) = \ln B^* - n \ln \sigma .$$

Using $\sigma_1 = 140\,\text{MPa}$, $t_{f1} = 375.2\,\text{h}$, $\sigma_2 = 150\,\text{MPa}$, and $t_{f2} = 94.4\,\text{h}$ allows to write the system of equations:

$$\ln(t_{f1}) = \ln B^* - n \ln \sigma_1 \quad \text{and} \quad \ln(t_{f2}) = \ln B^* - n \ln \sigma_2 ,$$

$$\ln \frac{t_{f1}}{t_{f2}} = n \ln \frac{\sigma_2}{\sigma_1} .$$

The result is $n = 20.0$ and $B^* = 3.1529 \times 10^{45}\,\text{MPa}^{20}\text{h}$. Using the provided value of the inert strength, we find $B = B^*/\sigma_c^{n-2} = 0.3912\,\text{MPa}^2\text{h}$.

Note: Due to the large exponents in this calculation, your results may differ from those stated here by several percent. The values here result if the exact values are used.

b) The failure probability can be calculated from equation (7.10), with $V/V_0 = 1$, $m^* = m/(n-2) = 1.2222$, and $t_0(\sigma) = B^* \sigma^{-n} = 3.1529 \times 10^{45}\,\text{MPa}^{20}\text{h} \cdot (100\,\text{MPa})^{-20} = 314\,016\,\text{h}$:

$$P_f(25\,000\,\text{h}) = 1 - \exp\left[-\left(\frac{t_f}{t_0(\sigma)}\right)^{m^*}\right]$$

$$= 1 - \exp\left[-\left(\frac{25\,000\,\text{h}}{314\,016\,\text{h}}\right)^{1.2222}\right] = 4.4\% .$$

The failure probability is larger than the design value 0.5%. The component cannot be used.

c) The failure probability can be reduced using a proof test.

d) The calculation is analogous to the derivation of equation (7.16):

$$G_f(t_f, \sigma) = \frac{\left\{1 - \exp\left[-\left(\frac{t_f}{t_0(\sigma)}\right)^{m^*}\right]\right\} - \left\{1 - \exp\left[-\left(\frac{\sigma_p}{\sigma_0}\right)^{m}\right]\right\}}{1 - \left\{1 - \exp\left[-\left(\frac{\sigma_p}{\sigma_0}\right)^{m}\right]\right\}}$$

$$= 1 - \exp\left[-\left(\frac{t_f}{t_0(\sigma)}\right)^{m^*} + \left(\frac{\sigma_p}{\sigma_0}\right)^{m}\right] . \tag{13.21}$$

The proof stress can be calculated from equation (13.21):

$$\sigma_p = \sigma_0 \left[\ln(1 - G_f) + \left(\frac{t_f}{t_0(\sigma)}\right)^{m^*}\right]^{1/m}$$

$$= 375\,\text{MPa}\left[\ln(1-0.005) + \left(\frac{25\,000\,\text{h}}{314\,016\,\text{h}}\right)^{1.2222}\right]^{1/22}.$$

The result is $\sigma_{\text{p}} = 324.1\,\text{MPa}$.

e) The fraction of scrapped parts is calculated using equation (7.3):

$$P_{\text{f}}(324.1\,\text{MPa}) = 1 - \exp\left[-\left(\frac{324.1\,\text{MPa}}{375\,\text{MPa}}\right)^{22}\right] = 4.0\%.$$

Solution 25:

a) Because of the parallel connection of the elements, the strain ε is the same in both of them. The stress $\sigma_{\text{S}}(t)$ in the spring and $\sigma_{\text{D}}(t)$ in the dashpot element are $\sigma = \sigma_{\text{S}}(t) + \sigma_{\text{D}}(t)$. The strain rate in the dashpot element is thus

$$\frac{d\varepsilon}{dt} = \frac{\sigma_{\text{D}}}{\eta} = \frac{\sigma - E\varepsilon}{\eta}.$$

This first-order differential equation can be solved by separation of variables:

$$\frac{d\varepsilon}{\sigma - E\varepsilon} = \frac{dt}{\eta},$$

$$\int \frac{d\varepsilon}{\sigma - E\varepsilon} = \int \frac{dt}{\eta},$$

$$-\frac{1}{E}\ln(\sigma - E\varepsilon) = \frac{t}{\eta} + C,$$

where C is a constant of integration. Solving for ε yields

$$\varepsilon = \frac{1}{E}\left[\sigma - \exp\left(-\frac{E}{\eta}t\right)C'\right].$$

The constant of integration C' can be determined by the fact that the strain ε is zero at time $t = 0$ because the dashpot element cannot react immediately to the stress. Thus, we find $C' = \sigma$ and

$$\varepsilon = \frac{\sigma}{E}\left[1 - \exp\left(-\frac{E}{\eta}t\right)\right].$$

The strain increases with time and approaches a value σ/E because the dashpot element will have relaxed completely and all of the stress is transferred by the spring.

b) In a relaxation experiment, the strain is to be increased discontinuously by a finite value. This causes an infinite stress in the dashpot element in this model. A relaxation experiment can therefore not be modelled with this approach.

c) The three elements in the three-parameter model are denoted as follows: Element 1 is the spring element in series, element 2 is the parallel spring element, and element 3 is the dashpot element. This yields the following relations for stresses and strains:

$$\varepsilon = \varepsilon_1 + \varepsilon_2 \,, \qquad \varepsilon_2 = \varepsilon_3 \,, \qquad \sigma = \sigma_1 = \sigma_2 + \sigma_3 \,,$$

$$\varepsilon_1 = \frac{\sigma_1}{E_1} \,, \qquad \varepsilon_2 = \frac{\sigma_2}{E_2} \,, \qquad \dot{\varepsilon}_3 = \frac{\sigma_3}{\eta} \,.$$

In general, the strain rate of the dashpot element is

$$\frac{d\varepsilon_3}{dt} = \frac{\sigma_1 - \sigma_2}{\eta} = \frac{\sigma_1 - E_2\varepsilon_2}{\eta} \,.$$

In a retardation experiment, the stress σ is constant. We thus find

$$\frac{d\varepsilon_3}{dt} = \frac{\sigma - E_2\varepsilon_2}{\eta} \,,$$

identical to subtask a). The total strain is thus

$$\varepsilon = \varepsilon_1 + \varepsilon_2 = \frac{\sigma}{E_1} + \frac{\sigma}{E_2} \left[1 - \exp\left(-\frac{E}{\eta}t\right)\right] \,.$$

For retardation, adding a spring does not change anything.

If the strain ε is constant, we find, using $\sigma_1 = E_1\varepsilon_1 = E_1(\varepsilon - \varepsilon_3)$,

$$\frac{d\varepsilon_3}{dt} = \frac{E_1(\varepsilon - \varepsilon_3) - E_2\varepsilon_3}{\eta} = \frac{E_1\varepsilon - (E_1 + E_2)\varepsilon_3}{\eta} \,.$$

This can again be solved by separating variables:

$$\frac{d\varepsilon_3}{E_1\varepsilon - (E_1 + E_2)\varepsilon_3} = \frac{dt}{\eta} \,,$$

$$E_1\varepsilon - (E_1 + E_2)\varepsilon_3 = \exp\left(-\frac{E_1 + E_2}{\eta}t\right) C' \,.$$

The constant of integration C' can be determined as before from $\varepsilon_3(t = 0) = 0$, yielding $C' = E_1\varepsilon$. The result is

$$\varepsilon_3 = \frac{E_1}{E_1 + E_2}\varepsilon \left[1 - \exp\left(-\frac{E_1 + E_2}{\eta}t\right)\right] \,.$$

The strain in the dashpot element approaches a value of $E_1/(E_1 + E_2) \cdot \varepsilon$ at large times.

d) We can read off the creep modulus E_c and the relaxation modulus E_r from the previous part of the exercise as

$$E_c = \frac{E_1 E_2}{E_2 + E_1 \left[1 - \exp\left(-\frac{E_2}{\eta}t\right)\right]} \,,$$

$$E_r = E_1 - \frac{E_1^2}{E_1 + E_2} \left[1 - \exp\left(-\frac{E_1 + E_2}{\eta}t\right)\right] \,.$$

In both cases, the modulus is E_1 at $t = 0$ and $E_1 E_2/(E_1 + E_2)$ at $t = \infty$. Figure 13.8 shows the time-dependence of both moduli. The creep modulus is always larger than the relaxation modulus.

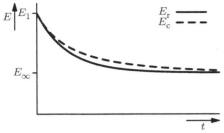

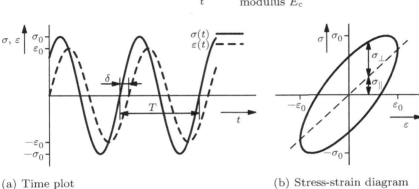

Fig. 13.8. Time-dependence of the relaxation modulus E_r and the creep modulus E_c

(a) Time plot   (b) Stress-strain diagram

Fig. 13.9. Stress and strain in a viscoelastic material under oscillation load

The retardation and relaxation times are the inverse of the prefactor of the variable t in the exponential function. Thus, they are $t_c = \eta/E_2$ and $t_r = \eta/(E_1 + E_2)$, respectively. The relaxation time is always smaller than the retardation time.

Solution 26:

a) The time-dependence is shown in figure 13.9(a).

The parameter δ describes the time-shift between stress and strain. If $\delta = 0$, stress and strain are in phase, and the material is not viscoelastic; if $\delta = 90°$, the strain is at its minimum or maximum when the stress is zero. In real materials, the value of δ depends on the frequency ω. In polymers, δ can take values of a few degrees.

b) Using the addition theorem $\sin(a - b) = \sin a \cos b + \cos a \sin b$ and $\omega t = \arcsin(\varepsilon/\varepsilon_0)$ (where we have to keep in mind for later that values of the arc sine are limited to $[-\pi/2, \pi/2]$) we get

$$\sigma = \sigma_0 \cos \delta \sin \omega t + \sigma_0 \sin \delta \cos \omega t$$

$$= \sigma_0 \cos \delta \cdot \sin \left(\arcsin \frac{\varepsilon}{\varepsilon_0} \right) + \sigma_0 \sin \delta \cdot \cos \left(\arcsin \frac{\varepsilon}{\varepsilon_0} \right)$$

$$= \underbrace{\sigma_0 \cos \delta \cdot \frac{\varepsilon}{\varepsilon_0}}_{\sigma_\parallel} + \underbrace{\sigma_0 \sin \delta \cdot \cos \left(\arcsin \frac{\varepsilon}{\varepsilon_0} \right)}_{\sigma_\perp} . \qquad (13.22)$$

The first term is in phase with the strain, the second is out of phase.

c) The stress-strain diagram is shown in figure 13.9(b). It has to be noted that the relation between stress and strain is not unique because there are two possible stress values for any strain. This is due to the fact that the cosine in equation (13.22) is always positive when arguments in the interval $[-\pi/2 : \pi/2]$ are used. For a full circle ωt, negative values of the cosine may also result. The full stress-strain diagram results when we replace the +-sign in equation (13.22) by a $\pm$.

The same result can be achieved if both parts from equation (12.2) are considered as a parametric function and plotted in a diagram for time values $0 \le t \le 2\pi$.

d) The energy dissipated in each cycle is the area enclosed by the stress-strain curve. To determine this area, equation (13.22) can be exploited. Its first term describes the part of the stress that is in phase with the strain and thus causes no dissipation. Thus, only the second term has to be considered in calculating the energy. The enclosed area – and thus the specific work done – is equal to twice the area above the dashed line in figure 13.9(b). It can be calculated as follows:

$$w = 2 \int_{-\varepsilon_0}^{\varepsilon_0} \sigma_\perp \, d\varepsilon = 2 \int_{-\varepsilon_0}^{\varepsilon_0} \sigma_0 \sin \delta \cdot \cos \left(\arcsin \frac{\varepsilon}{\varepsilon_0} \right) d\varepsilon$$

$$= \sigma_0 \sin \delta \cdot \varepsilon_0 \left[\frac{\varepsilon}{\varepsilon_0} \sqrt{1 - \left(\frac{\varepsilon}{\varepsilon_0} \right)^2} + \arcsin \frac{\varepsilon}{\varepsilon_0} \right]_{-\varepsilon_0}^{\varepsilon_0}$$

$$= \sigma_0 \sin \delta \cdot \varepsilon_0 \cdot \pi .$$

This is the area of an ellipse with major and minor axis ε_0 and $\sigma_0 \sin \delta$, for if we shear the area vertically to move the diagonal to the ε-axis, it is this ellipse that results.

Solution 27:

The activation energy can be determined using equation (8.7). If we compare the strain rates $\dot{\varepsilon}$ at different temperatures T_1 and T_2 and stresses σ_1 and σ_2 at identical values of σ/T, we can divide the strain rates to find

$$\frac{\dot{\varepsilon}_1}{\dot{\varepsilon}_2} = \frac{\exp \left(-\frac{Q}{kT_1} \right)}{\exp \left(-\frac{Q}{kT_2} \right)} .$$

The unknown activation volume cancels. It could be determined in the same way as Q. Solving for Q yields

$$Q = -\frac{k \ln \frac{\dot{\varepsilon}_1}{\dot{\varepsilon}_2}}{\frac{1}{T_1} - \frac{1}{T_2}}.$$

Looking at the diagram and using $\sigma/T = 0.2$, we can read off a strain rate of $2 \times 10^{-5}\,\mathrm{s}^{-1}$ at $T = 21.5°\mathrm{C}$ and $2 \times 10^{-2}\,\mathrm{s}^{-1}$ at $T = 40°\mathrm{C}$. If we insert this into the formula, we find $Q \approx 290\,\mathrm{kJ/mol}$.

Solution 28:

a) **Parallel connection:**

$$\varepsilon_\mathrm{m} = \varepsilon_\mathrm{f}, \quad \sigma_\mathrm{m} \neq \sigma_\mathrm{f}.$$

Serial connection:

$$\varepsilon_\mathrm{m} \neq \varepsilon_\mathrm{f}, \sigma_\mathrm{m} = \sigma_\mathrm{f}.$$

b) **Parallel connection:** The total applied force F can be divided into two parts, one for the fibre and one for the matrix: $F = F_\mathrm{f} + F_\mathrm{m}$. If we call the total cross section A and the cross section of fibre and matrix A_f and A_m, respectively, we find for the stresses:

$$\sigma A = \sigma_\mathrm{f} A_\mathrm{f} + \sigma_\mathrm{m} A_\mathrm{m},$$
$$\sigma = \sigma_\mathrm{f} \frac{A_\mathrm{f}}{A} + \sigma_\mathrm{m} \frac{A_\mathrm{m}}{A}.$$

The stress σ is averaged over fibre and matrix and will not occur at any point in the component. A_f/A and A_m/A are the area fractions of fibre and matrix and thus equal the volume fractions f_f and f_m because the fibres extend throughout the component. If we insert f_f and $f_\mathrm{m} = 1 - f_\mathrm{f}$ into the equation, we find the isostrain rule of mixtures, equation (9.2):

$$\sigma = \sigma_\mathrm{f} f_\mathrm{f} + \sigma_\mathrm{m}(1 - f_\mathrm{f}).$$

Dividing this equation by the strain $\varepsilon = \varepsilon_\mathrm{f} = \varepsilon_\mathrm{m}$, and taking into account equation (9.1), we find Young's modulus (equation (9.3)):

$$E_\| = \frac{\sigma}{\varepsilon} = \frac{\sigma_\mathrm{f}}{\varepsilon_\mathrm{f}} f_\mathrm{f} + \frac{\sigma_\mathrm{m}}{\varepsilon_\mathrm{m}}(1 - f_\mathrm{f}) = E_\mathrm{f} f_\mathrm{f} + E_\mathrm{m}(1 - f_\mathrm{f}) = E_\mathrm{m} \left[1 + f_\mathrm{f}\left(\frac{E_\mathrm{f}}{E_\mathrm{m}} - 1\right)\right].$$
$$(13.23)$$

Serial connection: The total length is the sum of the lengths of fibre and matrix: $l = l_\mathrm{f} + l_\mathrm{m}$. Since this condition holds even after deformation, we can also write $\Delta l = \Delta l_\mathrm{f} + \Delta l_\mathrm{m}$. Using the definition of strain $\varepsilon = \Delta l/l$ or $\Delta l = \varepsilon l$ separately for the three length changes, we find

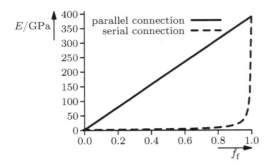

Fig. 13.10. Young's modulus in a fibre composite for different fibre arrangements

$$\varepsilon l = \varepsilon_f l_f + \varepsilon_m l_m \,,$$

$$\varepsilon = \varepsilon_f \frac{l_f}{l} + \varepsilon_m \frac{l_m}{l} \,.$$

Since the fibres are assumed to be plate-shaped, the volume fractions are $f_f = l_f/l$ and $f_m = 1 - f_f = l_m/l$. We thus find the isostress rule of mixtures, equation (9.5):

$$\varepsilon = \varepsilon_f f_f + \varepsilon_m (1 - f_f) \,.$$

Applying Hooke's law to this equation together with the condition (9.4) yields

$$\varepsilon = \frac{\sigma}{E_f} f_f + \frac{\sigma}{E_m} (1 - f_f) = \sigma \cdot \frac{E_m f_f + E_f(1 - f_f)}{E_f E_m} \,.$$

If we solve for $E = \sigma/\varepsilon$, we find Young's modulus (equation (9.6)):

$$E_\perp = \frac{E_m}{1 + f_f \left(\frac{E_m}{E_f} - 1 \right)} \,.$$

c) Both curves are plotted in figure 13.10.

Solution 29:

a) Young's modulus can be found using the isostrain rule of mixtures for fibres parallel to the external load, equation (9.3),

$$E_\| = E_m(1 - f_f) + E_f f_f = 0.45 \times 3 \,\text{GPa} + 0.55 \times 350 \,\text{GPa} = 194 \,\text{GPa} \,.$$

b) Because the fibres are long, the considerations from section 9.3.1 apply. We first have to check whether the fibres or the matrix will fail first. If we assume, in a reasonable approximation, that both materials are linear-elastic until fracture, the fracture strains are

$$\varepsilon_m = \frac{\sigma_m}{E_m} = 0.02 \,, \qquad \varepsilon_f = \frac{\sigma_f}{E_f} = 0.014 \,.$$

Thus, the fracture strain of the fibre is reached first. The failure stress can be estimated using equation (9.7),

$$\sigma = \sigma_f f_f + \sigma_m (1 - f_f),$$

inserting the stresses at a strain of 0.014. The fibre stress is $\sigma_f = 4900\,\text{MPa}$, whereas the matrix stress at strain 0.014 is only $42\,\text{MPa}$. Altogether, we find a tensile strength of

$$4900\,\text{MPa} \times 0.55 + 42\,\text{MPa} \times 0.45 = 2714\,\text{MPa}.$$

c) The compressive strength is, according to equation (9.11),

$$R_{c,\text{in phase}} = \sqrt{\frac{f_f \sigma_{m,F} E_f}{3(1 - f_f)}} = \sqrt{\frac{0.55 \times 60\,\text{MPa} \times 350\,000\,\text{MPa}}{3(1 - 0.55)}} = 2925\,\text{MPa}.$$

$$(13.24)$$

d) We have to check first whether the fibres are larger than the critical length, $l_c = d\sigma_f / 2\tau_i = 0.65\,\text{mm}$. The fibres are much longer than this, ensuring an effective load transfer onto the fibres. However, the strain increases locally near the fibres by a factor of about two relative to the global strain (see section 9.3.2). According to subtask b), the fracture strain in the matrix is 2%. Therefore, the total strain in the structure must not exceed 1%. At this strain, the matrix and fibre stresses are 30 MPa and 3500 MPa, respectively. Using the isostrain rule of mixtures, we find a tensile strength of 1939 MPa.

Solution 30:

a) For $R = -1$, the maximum stress is half of the stress range: $\sigma_{\max} = \Delta\sigma/2$. Using equation (5.2), we find

$$K_{\text{Ic}} = \sigma_{\max} \sqrt{\pi a_f}\, Y(a_f) = 17.72\,\text{MPa}\sqrt{\text{m}}.$$

b) At the current crack length, we thus find $K_{\max} = 3.08\,\text{MPa}\sqrt{\text{m}}$. Because the maximum stress intensity factor $K_{\max}$ is clearly below K_{Ic}, the component will not fail statically.

c) At $R = -1$, equation (10.6) yields

$$\Delta K_{\text{th}} = E \cdot 2.75 \times 10^{-5} \cdot 2^{0.31}\,\sqrt{\text{m}} = 2.39\,\text{MPa}\sqrt{\text{m}}.$$

Since the current range of the stress intensity factor is $\Delta K = 2K_{\max} = 6.16\,\text{MPa}\sqrt{\text{m}}$ according to subtask b), stable crack propagation must be expected.

d) The crack growth per cycle at the current crack length can be calculated from equation (10.8):

$$\begin{aligned}
\mathrm{d}a/\mathrm{d}N = C\Delta K^n &= 2 \times 10^{-12}\,\text{MPa}^{-2}\,\text{cycle}^{-1} \times (6.16\,\text{MPa}\sqrt{\text{m}})^2 \\
&= 7.6 \times 10^{-11}\,\text{m/cycle} = 7.6 \times 10^{-8}\,\text{mm/cycle}.
\end{aligned}$$

If the crack-growth rate would stay constant at this value, the critical crack length would be reached after

$$\tilde{N}_f = \frac{a_f - a_0}{da/dN} = 1.2 \times 10^8$$

cycles. In this case, the component could be cleared for further use.

e) Equation (10.10) is

$$N_f = \frac{1}{C} \left(\frac{1}{\Delta\sigma\sqrt{\pi}} \right)^n \int_{a_0}^{a_f} \frac{1}{(Y\sqrt{a})^n} \, da \, .$$

Inserting $n = 2$, $Y(a) = 1 + ba$, and $b = 0.1\,\mathrm{mm}^{-1}$ yields

$$N_f = \frac{1}{C\Delta\sigma^2\pi} \int_{a_0}^{a_f} \frac{1}{(1 + ba)^2 \, a} da \, .$$

Using the integral formula provided and setting $A = 1$ and $B = b$, we find

$$N_f = -\frac{1}{C\Delta\sigma^2\pi} \left[\ln \frac{1 + ba}{a} + \frac{ba}{1 + ba} \right]_{a_0}^{a_f}$$

$$= -\frac{1}{C\Delta\sigma^2\pi} \left[\ln \frac{a_0(1 + ba_f)}{a_f(1 + ba_0)} + \frac{ba_f}{1 + ba_f} - \frac{ba_0}{1 + ba_0} \right] \, .$$

Putting in numerical values yields the final result

$$N_f = -\frac{1}{2 \times 10^{-5} \cdot \pi} \left[\ln \frac{2}{11} + \frac{1}{2} - \frac{0.1}{1.1} \right] = 20\,621 \, .$$

The component can be cleared until the next service interval because the number of cycles to failure is significantly larger than the interval time.

Solution 31:

a) We start by reading off the number of cycles to failure from the S-N diagram:

$$\sigma_{a,1} = 60\,\mathrm{MPa} \quad \Rightarrow \quad N_{f,1} = \infty \, ,$$
$$\sigma_{a,2} = 100\,\mathrm{MPa} \quad \Rightarrow \quad N_{f,2} = 1\,500\,000 \, ,$$
$$\sigma_{a,3} = 150\,\mathrm{MPa} \quad \Rightarrow \quad N_{f,3} = 45\,000 \, ,$$
$$\sigma_{a,4} = 200\,\mathrm{MPa} \quad \Rightarrow \quad N_{f,4} = 4000 \, .$$

We can use this to calculate the damage contributions:

$$D_1 = \frac{n_1}{N_{f,1}} = \frac{10\,000}{\infty} = 0 \, ,$$

$$D_2 = \frac{n_2}{N_{f,2}} = \frac{5000}{1\,500\,000} = 3.33 \times 10^{-3} \, ,$$

$$D_3 = \frac{n_3}{N_{f,3}} = \frac{2000}{45\,000} = 4.44 \times 10^{-2} \, ,$$

$$D_4 = \frac{n_4}{N_{f,4}} = \frac{200}{4000} = 5.00 \times 10^{-2} \,.$$

The total damage in a week is thus $D_{\text{week}} = \sum_{i=1}^{4} D_i = 9.77 \times 10^{-2}$. Fracture of the component is to be expected after n weeks when a damage value of $D = n \cdot D_{\text{week}} = 1$ has been reached:

$$n = D_{\text{week}}^{-1} = 10.23 \,.$$

The number of cycles per week is 17 200; 10.23 weeks thus correspond to about 176 000 cycles.

Solution 32:

a) We use equation (11.5) and solve for the unknown P and C at the two data points T_1, t_1 and T_2, t_2:

$$P = \frac{\ln(t_1/\text{h}) - \ln(t_2/\text{h})}{1/T_1 - 1/T_2} = 57.4 \times 10^3 \,\text{K} \,,$$

$$C = \frac{P}{T_1} - \ln\frac{t_1}{\text{h}} = 44.1 \,.$$

b) From the known stress value σ_2, we can determine the Larson-Miller parameter P_2 at this stress:

$$P_2 = T_3\left(\ln\frac{t_3}{\text{h}} + C\right) = 59.7 \times 10^3 \,\text{K} \,.$$

The same parameter value is reached at a service time of $t_4 = 100\,000$ hours at a temperature

$$T_3 = \frac{P_2}{\ln(t_4/\text{h}) + C} = 1075 \,\text{K} \,.$$

Solution 33:

a) Because the deformation is viscoplastic, but not viscoelastic, a spring and a dashpot model have to be connected in series.

b) At constant stress and time t, the creep strain is $\varepsilon_c = \dot{\varepsilon} t$. The total strain is thus

$$\varepsilon = A\sigma^n t + E\sigma \,.$$

c) Since we need an approximation only, we can use the stress and strain values in the middle of the tube. The elastic part of bending is negligible, so we can write $\varepsilon = \dot{\varepsilon} t$ because the strain rate is constant at constant stress, We thus find

$$h = \frac{\varepsilon l^2}{4d} = \frac{\dot{\varepsilon} l^2 t}{4d} = \frac{A\sigma l^2 t}{4d} = \frac{A\varrho g l^4 t}{32d^2} = 2 \times 10^{-4} \,\text{m} \,.$$

In each year, the middle of the tube is displaced by 0.2 mm.

Solution 34:

a) Heating causes thermal strains ε_{th} in the rod which are compensated for by elastic strains ε_{el} because of the clamping. In the course of time, the stress caused by the elastic strains is relaxed by a creep strain ε_{c}. We can write

$$\varepsilon = \varepsilon_{\text{th}} + \varepsilon_{\text{el}} + \varepsilon_{\text{c}} = 0\,.$$

The thermal strain is $\varepsilon_{\text{th}} = \alpha\Delta T$. We start by calculating the stress in the component:

$$\sigma = E\varepsilon_{\text{el}} = E(-\alpha\Delta T - \varepsilon_{\text{c}})\,.$$

Its time-dependence can be found by differentiating, noting that the first term is constant:

$$\dot{\sigma} = -E\,\dot{\varepsilon}_{\text{c}} = -EA\sigma^n\,.$$

This differential equation can be solved by separating variables (see exercise 25):

$$\frac{\mathrm{d}\sigma}{-AE\sigma^n} = \mathrm{d}t\,,$$

$$\frac{1}{-n+1}\sigma^{-n+1} = -AEt + C\,,$$

$$\sigma^{n-1} = \frac{1}{1-n}\cdot\frac{1}{-AEt + C}\,,$$

where C is a constant of integration yet to be determined.

For the case of interest, $n = 3$, this yields

$$\sigma^2 = \frac{1}{1-3}\cdot\frac{1}{-AEt + C} = \frac{1}{2(AEt - C)}\,. \tag{13.25}$$

Because the stress is negative (the rod is compressed), we have to use the negative sign when taking the square root on the right-hand side:

$$\sigma = -\frac{1}{\sqrt{2(AEt - C)}}\,.$$

The constant of integration can be found by using the condition $\sigma_0 = -E\alpha\Delta T = -2223\,\text{MPa}$ at time $t = 0$ in equation (13.25):

$$C = -\frac{1}{2\sigma_0^2} = -\frac{1}{2E^2\alpha^2\Delta T^2} = -1.01\times 10^{-7}\,\text{MPa}^{-2}\,.$$

After $100\,\text{s}$, the stress is only $\sigma_{\text{e}} = -113\,\text{MPa}$.

b) The elastic strain at the end of holding time is only $\varepsilon_{\text{e}} = \sigma_{\text{e}}/E = -8.70\times 10^{-4}$. This strain reduces to zero at a temperature difference of $\Delta T = \varepsilon_{\text{e}}/\alpha = -49.7\,\text{K}$. The rod will fall from the clamping at a temperature of about $950\,°\text{C}$.

A

Using tensors

In this chapter, we discuss the basics of how to calculate with tensors. For a more detailed study, the reader is referred to the technical literature e. g., *Holzapfel* [67].

A.1 Introduction

In general, a tensor is a physical quantity that is associated with coordinates. Although its numerical representation may change when switching to another reference or coordinate system, the tensor itself remains unchanged. As an example, consider the vector $\underline{a}$ in figure A.1. To state the value of the vector, a single number is not sufficient. Instead, its *components* have to be stated in a coordinate system. Usually, this is done by writing the coordinate values as a column vector. In the x_1-x_2 coordinate system shown in the figure, the vector's representation is

$$(a_i) = \begin{pmatrix} 1 \\ 2 \end{pmatrix} .$$

If we use the $x_{1'}$-$x_{2'}$ coordinate system, rotated by 45°, the components are

$$(a_{i'}) = \begin{pmatrix} 3\sqrt{2}/2 \\ \sqrt{2}/2 \end{pmatrix} .$$

The components are not the same in both systems, but the vector $\underline{a}$ itself is nevertheless the same as can be seen in figure A.1.

A.2 The order of a tensor

Tensors can be classified by their *order*, sometimes also called 'rank'. As a first example, we consider again a *vector*. As already explained, its components are

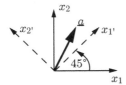

Fig. A.1. Vector $\underline{a}$ in two different coordinate systems

usually written one below the other, resulting in a *component matrix* that is a (3×1) matrix in three-dimensional space. It is thus one-dimensional, and the tensor is a tensor of first order. The components of the matrix can be characterised by a single index.

In a *scalar,* which is coordinate-independent, the component matrix is a single number. Since a scalar thus has dimension zero, it is a tensor of zeroth order. Using an index is thus not necessary.

If we now move on to a quantity with components written as a (3×3) matrix, we get a tensor of second order. In this case, we need two indices to specify a component of the tensor. One example for a second-order tensor is the stress tensor $\underline{\sigma}$.

Tensors of second order can be represented by a matrix in a specified coordinate system. In general, a matrix is simply a rectangular arrangement of numbers. Arbitrary matrices do not necessarily share the important tensor property of invariance: If a coordinate transformation is done, a tensor's components may change, but the tensor itself does not. Thus, many quantities that are usually called 'matrices' should better be denoted as second-order tensors.

A third-order tensor is represented by a 'coordinate cube' with $3 \times 3 \times 3 = 27$ components. This scheme can be extended to arbitrarily high orders. A fourth-order tensor, having $3^4 = 81$ components, cannot be imagined geometrically. Nevertheless, it is of great importance in material science (see section 2.4.2).

To summarise, it can be said that the order of a tensor specifies the dimension of the 'hypercube' of edge length three that contains the components in a certain coordinate system. In three-dimensional space, a tensor of order m has 3^m components.

A.3 Tensor notations

There are different ways to write down tensors and their components, whose usefulness depends on the context. If we talk about the tensor itself, independent of a coordinate system, we use the *symbolic notation.* Different typographical styles can be found in the literature. In this book, a first-order tensor is underscored once $(\underline{a}, \underline{b}, \dots)$, a second-order tensor twice (and usually denoted with a capital letter, $\underline{\underline{A}}, \underline{\underline{B}}, \dots$). Higher-order tensors get a tilde and the value of the order below ($\underset{3}{A}$, $\underset{4}{B}$, ...). Other typographic conventions frequently used elsewhere are bold letters ($\underline{a} \mathrel{\widehat{=}} \mathbf{a}$, $\underline{\underline{A}} \mathrel{\widehat{=}} \mathbf{A}$) or arrows ($\underline{a} \mathrel{\widehat{=}} \vec{a}$, $\underline{\underline{A}} \mathrel{\widehat{=}} \vec{A} \mathrel{\widehat{=}} \overrightarrow{A}$).

If we want to specify a component of a tensor in a certain coordinate system, we use the *index notation*. Each tensor order requires its own index. Usually, lowercase letters are used as indices, starting with i (a_i, A_{ij}, C_{ijkl}). Algebraic rules are frequently written down in index notation. Although the components themselves depend on the coordinate system, the rules are nevertheless valid in all systems.

Everywhere in this book, we use a right-handed Cartesian coordinate system with perpendicular coordinate axes and unit vectors of length 1. If this is not done, the notation becomes much more cumbersome. Nevertheless, this step has to be done sometimes e. g., when dealing with large deformations (see section 3.1).

If several coordinate systems are used, they are distinguished by adding primes to the indices. Thus, a representation in the x_1-x_2-x_3 coordinate system may be written as a_i, A_{ij}, and C_{ijkl}, changing to $a_{i'}$, $A_{i'j'}$, and $C_{i'j'k'l'}$ in the $x_{1'}$-$x_{2'}$-$x_{3'}$ coordinate system. It is important to add the prime to each index because tensors might be written using indices mixed from different systems e. g., $A_{ij'}$ or $A_{i'j}$.

If all components of a tensor are to be described, this is done by adding parentheses to the tensor written in index notation, for example, (a_i), (A_{ij}), (C_{ijkl}). Implicitly, it is assumed that each index runs from 1 to 3. In second-order tensors, the first index denotes the row and the second index denotes the column of the component in the matrix notation. The components of the tensor

$$(A_{ij}) = \begin{pmatrix} 1 & 2 & 3 \\ 4 & 5 & 6 \\ 7 & 8 & 9 \end{pmatrix}$$

are thus $A_{11} = 1$, $A_{12} = 2$, $A_{13} = 3$, $A_{21} = 4$, $A_{22} = 5$, $A_{23} = 6$, $A_{31} = 7$, $A_{32} = 8$, $A_{33} = 9$. Using this parenthesis notation, the connection to the symbolic notation can easily be made: $\underline{a} \mathrel{\widehat{=}} (a_i)$, $\underline{\underline{A}} \mathrel{\widehat{=}} (A_{ij})$.

A.4 Tensor operations and Einstein summation convention

One of the most important tensor operations is the *product*. We can write the so-called (single[1]) *contraction* of two tensors as

$$\underline{\underline{C}} = \underline{\underline{A}} \cdot \underline{\underline{B}} = \underline{\underline{A}}\,\underline{\underline{B}} \,.$$

---

[1] We will see below that there is also a double contraction.

The contraction of tensors is sometimes also referred to as their *scalar product,* *inner product,* or *dot product.* The resulting components of $\underline{\underline{C}}$ can be written as (using the rule 'column vector times row vector')

$$C_{ij} = \sum_{k=1}^{3} A_{ik} B_{kj} \, . \tag{A.1}$$

This equation holds for all 9 components of the tensor C_{ij}. We need not specify the range of the indices $i = 1 \ldots 3$, $j = 1 \ldots 3$ because they are always fixed by the space dimension. To further ease the notation, even the sum sign itself can be omitted, simplifying the equation to

$$C_{ij} = A_{ik} B_{kj} \, . \tag{A.2}$$

This notation is the so-called *Einstein summation convention,* stating that each index that occurs twice in an expression is summed over (with a range from 1 to 3). This index is called *summation index.* All other indices are called *free indices.* Thus, we can write

$$c = a_i \, b_i = \sum_{i=1}^{3} a_i \, b_i \, ,$$

$$C_{ik} = A_{ij} \, B_{jk} = \sum_{j=1}^{3} A_{ij} \, B_{jk} \, .$$

The summation convention is still used even if the double index occurs within the same tensor:

$$A_{ii} = \sum_{i=1}^{3} A_{ii} = A_{11} + A_{22} + A_{33} \, .$$

In the unusual case that a double index is not to be summed over, the indices are underscored: $A_{\underline{ii}}$. In this case, we really mean only one of the components, A_{11}, A_{22}, or A_{33}.

Because a tensor product is a scalar quantity in index notation, the commutative law can be used within the sums:

$$C_{ij} = A_{ik} \, B_{kj} = B_{kj} \, A_{ik} \, .$$

If we want to re-write this to calculate the components of $\underline{\underline{C}}$, the same indices have to be next to each other as shown in the left-hand part. The rule 'row times column' can be used only in this case. Therefore, the contraction itself is not commutative:

$$\underline{\underline{C}} = \underline{\underline{A}} \, \underline{\underline{B}} \neq \underline{\underline{B}} \, \underline{\underline{A}} \, .$$

If there are several double indices in a product, it is a *multiple contraction*. In this case, all summation indices are summed over, for example

$$c = A_{ij} B_{ji} = \sum_{i=1}^{3} \sum_{j=1}^{3} A_{ij} B_{ji} .$$

In the symbolic notation, a multiple contraction is denoted by several dots, as many as there are summation indices:

$$c = A_{ij} B_{ji} = \underline{\underline{A}} \cdot\cdot \underline{\underline{B}} . \tag{A.3}$$

In elasticity theory (section 2.4.2) , Hooke's law uses a double contraction between the elasticity tensor of order four and the strain tensor of order two:

$$\underline{\underline{\sigma}} = \underline{\underline{C}}_4 \cdot\cdot \underline{\underline{\varepsilon}}$$

or, in index notation,

$$\sigma_{ij} = C_{ijkl}\, \varepsilon_{kl} .$$

In this case, it doesn't matter that the indices of the strain tensor are in the wrong sequence because $\underline{\underline{\varepsilon}}$ is symmetric i. e., $\varepsilon_{kl} = \varepsilon_{lk}$.

The number of indices in the result of a contraction of tensors of arbitrary order is equal to the number of indices in the contraction that are not doubled, the free indices. Here are a few examples, whose symbolic notation is, in some cases, explained later:

$$
\begin{aligned}
a_i &= A_{ij}\, b_j , & \underline{a} &= \underline{\underline{A}} \cdot \underline{b} , \\
a_i &= b_j\, A_{ji} , & \underline{a} &= \underline{b}^{\mathrm{T}} \cdot \underline{\underline{A}} , \\
c &= a_i\, b_i , & c &= \underline{a}^{\mathrm{T}} \cdot \underline{b} = \underline{a} \cdot \underline{b} , \\
a &= A_{ij}\, B_{jk}\, C_{ki} , & a &= \mathrm{tr}\big(\underline{\underline{A}}\,\underline{\underline{B}}\,\underline{\underline{C}}\big) , \\
A_{ijk} &= B_{ijkl}\, c_l , & \underline{\underline{A}}_3 &= \underline{\underline{B}}_4 \cdot \underline{c} , \\
A_{ijkl} &= B_{ijkm}\, C_{ml} , & \underline{\underline{A}}_4 &= \underline{\underline{B}}_4 \cdot \underline{\underline{C}} , \\
A_{ij} &= B_{ijkl}\, C_{lk} , & \underline{\underline{A}} &= \underline{\underline{B}}_4 \cdot\cdot \underline{\underline{C}} .
\end{aligned}
$$

The result of a calculation must not depend on the coordinate system. Therefore, the result of a tensor operation must in itself be a tensor (of the appropriate order). All rules in this section fulfil this condition. A product definition of the form $c_i = a_i b_i$, directly multiplying the components, is not physically meaningful because the value of (c_i) would depend on the coordinate system.

A.5 Coordinate transformations

As already stated at the beginning of this chapter, a coordinate transformation does not change a tensor, but only its components. Thus, for the tensor itself, we can write $\underline{\underline{A}}^{(x_i)} = \underline{\underline{A}}^{(x_{i'})}$ (with the superscript denoting the coordinate system), but for the components, we usually find $A_{ij} \neq A_{i'j'}$. To transform from one coordinate system to another, we have to change the component matrix accordingly. The relation between the representation in the old and the new coordinate system can be stated using a matrix, the *transformation matrix* $(g_{i'i})$. Each index is multiplied with this matrix. It has to be noted that i and i' are different and not to be summed over. A first-order tensor (vector) is transformed like this:

$$a_{i'} = g_{i'i}\, a_i\,,\tag{A.4}$$

a second-order tensor like this:

$$A_{i'j'} = g_{i'i}\, g_{j'j}\, A_{ij}\,.\tag{A.5}$$

To re-write this transformation in the symbolic notation, we have to ensure that identical indices are next to each other. Thus, we have to exchange two indices of the matrix, a matrix operation known as *transposing* and denoted by a superscript 'T':

$$A_{ij}^{\mathrm{T}} = A_{ji}\,.$$

Thus, the transformation rule

$$(A_{i'j'}) = (g_{i'i})\,(A_{ij})\,(g_{j'j})^{\mathrm{T}}$$

follows. The components $g_{i'i}$ of the transformation matrix $\underline{\underline{g}} = (g_{i'i})$ can be calculated from the basis vectors $\underline{g}^{(i)}$ and $\underline{g}^{(i')}$ of the two coordinate systems. Each of its components is determined by the contraction of the basis vectors:

$$g_{i'i} = \underline{g}^{(i')} \cdot \underline{g}^{(i)}\,.\tag{A.6}$$

All basis vectors have to be written in the same coordinate system.

This can be explained most easily using an example. We again consider the two-dimensional system from section A.1. The basis vectors of the un-primed coordinate system (see figure A.1) are

$$\underline{g}^{(1)} = \begin{pmatrix} 1 \\ 0 \end{pmatrix} \qquad \text{and} \qquad \underline{g}^{(2)} = \begin{pmatrix} 0 \\ 1 \end{pmatrix}\,,$$

the basis vectors of the primed system, *written in the un-primed system*, are

$$\underline{g}^{(1')} = \begin{pmatrix} \sqrt{2}/2 \\ \sqrt{2}/2 \end{pmatrix} \qquad \text{and} \qquad \underline{g}^{(2')} = \begin{pmatrix} -\sqrt{2}/2 \\ \sqrt{2}/2 \end{pmatrix}\,.$$

Using equation (A.6), we can calculate the transformation matrix

$$(g_{i'i}) = \begin{pmatrix} \sqrt{2}/2 & \sqrt{2}/2 \\ -\sqrt{2}/2 & \sqrt{2}/2 \end{pmatrix}.$$

If we perform the coordinate transformation of $\underline{a}$, $(a_i) = \begin{pmatrix} 1 & 2 \end{pmatrix}^{\mathrm{T}}$, we find

$$(a_{i'}) = (g_{i'i})\,(a_i) = \begin{pmatrix} \sqrt{2}/2 & \sqrt{2}/2 \\ -\sqrt{2}/2 & \sqrt{2}/2 \end{pmatrix} \begin{pmatrix} 1 \\ 2 \end{pmatrix} = \begin{pmatrix} 3\sqrt{2}/2 \\ \sqrt{2}/2 \end{pmatrix}.$$

A.6 Important constants and tensor operations

In this section, some important rules, conventions, and constants are summarised.

The *Kronecker delta* δ_{ij} represents a tensor with invariant components that do not change in any coordinate rotation. It is defined as

$$\delta_{ij} = \begin{cases} 1 & \text{for } i = j, \\ 0 & \text{for } i \neq j. \end{cases} \tag{A.7}$$

Written in the component notation, the Kronecker delta is nothing but the *unit tensor:*

$$(\delta_{ij}) = \underline{\underline{1}} = \begin{pmatrix} 1 & 0 & 0 \\ 0 & 1 & 0 \\ 0 & 0 & 1 \end{pmatrix}.$$

A tensor of any order can be *transposed* by reverting the sequence of the indices:

$$(C_{ijkl})^{\mathrm{T}} = (C_{lkji}).$$

In a tensor of second order, this means exchanging the rows and columns.

The *trace* of a tensor of second order is the sum of its diagonal elements:

$$\operatorname{tr}\underline{\underline{A}} = A_{ii} = A_{11} + A_{22} + A_{33}. \tag{A.8}$$

The *determinant* $\det\underline{\underline{A}}$ of a tensor of second order is the determinant of its matrix representation. In three dimensions, it is

$$\det\underline{\underline{A}} = \quad A_{11}A_{22}A_{33} + A_{12}A_{23}A_{31} + A_{13}A_{21}A_{32}$$
$$- A_{11}A_{23}A_{32} - A_{12}A_{21}A_{33} - A_{13}A_{22}A_{31}.$$

The positive terms are formed by the components on the diagonals that point downwards and to the right, the negative by those pointing upwards and to the right.[2]

---

[2] This simple rule is only valid in three dimensions.

A.7 Invariants

Each tensor representation is characterised by some properties that are not changed by a coordinate transformation. The length of a vector, for example, is not changed by a coordinate transformation. It can be computed by the contraction of the vector with itself:

$$|\underline{a}|^2 = a_1^2 + a_2^2 + a_3^2 = a_{1'}^2 + a_{2'}^2 + a_{3'}^2 \,,$$

or, in index notation,

$$|\underline{a}|^2 = a_i \, a_i = a_{i'} \, a_{i'} \,. \tag{A.9}$$

Since $|\underline{a}|$ is defined by a contraction, it does not change during a coordinate transformation. The length of a vector is thus called an *invariant*. If we choose a special $x_{i''}$ coordinate system where the vector representation is

$$(a_{i''}) = \left(\begin{array}{ccc} |\underline{a}| & 0 & 0 \end{array}\right)^{\mathrm{T}} \,,$$

the vector is completely specified by stating its length and the coordinate system.

Like vectors, tensors of any order possess invariants, quantities that do not change during a coordinate transformation. A second-order tensor has three invariants, called its *eigenvalues* (from the German word 'eigen' meaning 'own, peculiar, particular') $\lambda^{(k)}$. For a tensor (A_{ij}), they can be calculated from the *characteristic equation*

$$\left(A_{ij} - \delta_{ij}\lambda^{(k)}\right) v_j^{(k)} = 0 \tag{A.10}$$

or

$$\left(\underline{\underline{A}} - \underline{\underline{1}}\lambda^{(k)}\right) \underline{v}^{(k)} = \underline{0} \,,$$

where the trivial solution $v_j^{(k)} = 0$ is ignored. This equation always has three (not necessarily distinct) solutions $\lambda^{(k)}$ with the associated *eigenvectors* $\underline{v}^{(k)}$. In general, the eigenvalues and -vectors calculated this way are complex and thus cannot be interpreted physically.

In the common case of a symmetric tensor, fulfilling $A_{ij} = A_{ji}$, the eigenvalues are real numbers and the eigenvectors are perpendicular to each other. This, for example, is the case in stress or strain states in a classical continuum. The eigenvalues in this case are called *principal stresses* and *principal strains*, respectively; the eigenvectors are the *principal directions* or *axes*.

Similar to a vector, a symmetric second-order tensor can be completely characterised by stating its eigenvalues and -vectors. To do so, we form a new $x_{i''}$ coordinate system of the eigenvectors, normalised to length 1. In this system, the tensor is diagonal:

$$\underline{g}_{i''} = \underline{v}^{(i'')},$$

$$(A_{i''j''}) = \delta_{\underline{i}''j''} \lambda^{(\underline{i}'')} = \begin{pmatrix} \lambda^{(1)} & 0 & 0 \\ 0 & \lambda^{(2)} & 0 \\ 0 & 0 & \lambda^{(3)} \end{pmatrix}.$$

In the representation using the principal axes, the component matrix contains only diagonal elements.

There is a special set of invariants, the *principal invariants* J_k, that can be formed from the eigenvalues:

$$J_1 = \lambda^{(1)} + \lambda^{(2)} + \lambda^{(3)},$$
$$J_2 = -\lambda^{(1)}\lambda^{(2)} - \lambda^{(1)}\lambda^{(3)} - \lambda^{(2)}\lambda^{(3)},$$
$$J_3 = \lambda^{(1)}\lambda^{(2)}\lambda^{(3)}.$$

Written for the tensor itself, they can be calculated as follows:

$$J_1 = A_{ii} \qquad\qquad = \operatorname{tr}\underline{\underline{A}},$$
$$J_2 = \frac{1}{2}\left[A_{ij}A_{ji} - A_{ii}A_{jj}\right] \quad = \frac{1}{2}\left[\operatorname{tr}(\underline{\underline{A}}\,\underline{\underline{A}}^{\mathrm{T}}) - (\operatorname{tr}\underline{\underline{A}})^2\right],$$
$$J_3 = \det(A_{ij}) \qquad\qquad = \det\underline{\underline{A}}.$$

These principal invariants are important in the context of yielding of materials and are used in section 3.3.1.

A.8 Derivations of tensor fields

Physical quantities are frequently not defined by a single tensor, but by assigning a tensor to each point in space, thus defining a *field*. Examples are a temperature field (a scalar field), where a temperature is specified at each point in space, a velocity field in a flowing fluid (a vector field), where the flow direction and speed are stated at each point, or the stress field in a material (a tensor field of second order), assigning a value of the stress tensor to each point.

Because the value of the field may differ at different points in space, fields can be derived in the different spatial directions. The derivative of a scalar field $f(\underline{x})$ with respect to $\underline{x}$ is a vector, calculated by the rule

$$\frac{\partial f(\underline{x})}{\partial \underline{x}} = \begin{pmatrix} \partial f(\underline{x})/\partial x_1 \\ \partial f(\underline{x})/\partial x_2 \\ \partial f(\underline{x})/\partial x_3 \end{pmatrix}.$$

Each component states how the scalar field changes in the respective spatial direction. This vector field is called the *gradient* of f.

The gradient of a vector field $\underline{v}(\underline{x})$ can be calculated in a similar way. It states how the vector field changes in each spatial direction and is thus a tensor field of second order. The rule to calculate this gradient can be most easily written in component notation:

$$\left(\frac{\partial \underline{v}(\underline{x})}{\partial \underline{x}} \right)_{ij} = \frac{\partial v_j}{\partial x_i} \, .$$

It is important to note that the first index of the gradient, i, is in the denominator, not in the numerator of the right-hand side. In the same way, higher-order tensors can be derived with respect to a vector, always using the first index from the denominator.

Occasionally, tensor fields have to be derived with respect to scalar quantities. To do so, each component of the field is simply derived separately, for example

$$\frac{d\underline{v}(\alpha)}{d\alpha} = \left(\begin{array}{ccc} \frac{dv_1(\alpha)}{d\alpha} & \frac{dv_2(\alpha)}{d\alpha} & \frac{dv_3(\alpha)}{d\alpha} \end{array} \right)^{\mathrm{T}} .$$

The order of the tensor remains unchanged in this operation.

B

Miller and Miller-Bravais indices

In many cases, it is necessary to specify directions and planes in a crystal lattice. It is most sensible to do so using a crystallographic coordinate system, with axes parallel to the edges of the chosen unit cell. All parallel directions and planes in a crystal are equivalent, rendering it unnecessary to state the origin of the direction vector or plane.

B.1 Miller indices

For specifying directions and planes in a crystal, the origin of the crystallographic coordinate system is positioned in a lattice point, and the axes are scaled so that the length of every edge of the unit cell is one. Thus, for non-cubic lattice types, this coordinate system is non-Cartesian.

Using the so-called *Miller indices,* a direction is specified by a straight line through the origin of the coordinate system. The coordinates describing the line are called the *indices* and written with square brackets: $[hkl]$. They result from the intersection of the line with the nearest lattice point e. g., $[112]$ in figure B.1(a). If negative values occur, they are denoted by a bar on top of the coordinate, for example $[1\bar{1}0]$. If we do not want to specify a certain direction, but all cristallographically equivalent directions, we use indices in angle brackets: $\langle hkl \rangle$. Cristallographically equivalent directions are those that can be transformed into each other by using a crystal symmetry. In a cubic crystal, for example, all space diagonals ($[111]$, $[\bar{1}11]$, $[1\bar{1}1]$, $[11\bar{1}]$) are equivalent.

A plane is also specified by three numbers. They are determined in the following way: Choose the origin of the coordinate at any point *not* in the plane considered. Determine the intersections m, n, and p of the plane with the coordinate axes as shown in figure B.1(b) (here: $m = 1$, $n = 1$, $p = 2$). If the plane is parallel to one of the axes, the intersection is assumed to occur at infinity. We now calculate the reciprocal values of the intersections: $\tilde{h} = m^{-1}$, $\tilde{k} = n^{-1}$, $\tilde{l} = p^{-1}$ (here: $\tilde{h} = 1$, $\tilde{k} = 1$, $\tilde{l} = 0.5$). Next, we form the smallest

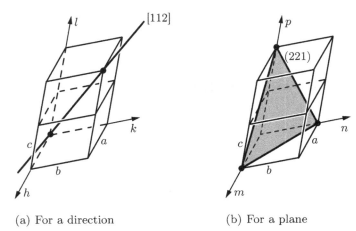

(a) For a direction   (b) For a plane

Fig. B.1. Determination of Miller indices

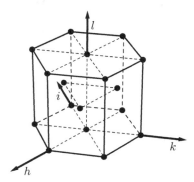

Fig. B.2. Coordinate system used for Miller-Bravais indices

triple consisting of integers, $h : k : l$, with the same ratios as $\tilde{h} : \tilde{k} : \tilde{l}$. This triple characterises the plane and is written in parentheses: (hkl), for example (221). If the set of all equivalent planes is to be specified, curly braces are used: $\{hkl\}$. In the case of a cubic lattice, the indices of a plane specify its normal vector.

B.2 Miller-Bravais indices

In hexagonal crystals, where the base plane exhibits a 120° symmetry, the *Miller-Bravais system* is used to specify planes and directions, using a coordinate system with four axes: Three of these, with angles of 120° between them, lie in the base plane and are equivalent, the fourth is perpendicular to the base plane: $[hkil]$ (figure B.2). In this way, the symmetry of the crystal is reflected in the indices. The first three indices obey the additional constraint

$$h + k + i = 0 .$$ (B.1)

Apart from this peculiar way of defining the coordinate system, the calculation of the indices of directions and planes is the same as for the Miller indices. However, the normal vector of a plane does *not* correspond to the plane's indices.

C

A crash course in thermodynamics

In this appendix, we will explain two important thermodynamic concepts needed in this book in different places: The first concept is thermal activation of processes, the second the concept of free energy. A detailed discussion of thermodynamics can be found, for example, in *Reif* [116].

C.1 Thermal activation

We are looking for the probability that a process needing an energy ΔE occurs in a certain system. The system may take this energy from its stored thermal energy. If the temperature of the system is above absolute zero, its components (for example, the atoms it consists of) are in permanent, irregular motion, the so-called *Brownian motion*. Slightly simplified, we can consider the thermal activation of a process as being caused by these random movements of the atoms acting together and enabling the process. The probability for such an event will become the larger, the higher the temperature.

To understand thermal activation in greater detail, we need one basic principle of thermodynamics. If a system can exist in a number of different states $\mathcal{Z}_1$, $\mathcal{Z}_2$, ... with energies E_1, E_2, ... and if it is in thermal equilibrium, *Boltzmann's law* states that the probability $P(\mathcal{Z}_i)$ to find the system in state $\mathcal{Z}_i$ is given by

$$P(\mathcal{Z}_i) \propto \exp\left(-\frac{E_i}{kT}\right). \tag{C.1}$$

Here T is the system temperature and k is *Boltzmann's constant* ($k = 1.38 \times 10^{-23}$ J/K). The constant of proportionality in the equation can be calculated from the fact that the sum over all probabilities must be 1.

From this law, the probability that the system changes its state by thermal activation to a state with an energy that is larger by ΔE is

$$P(\Delta E) \propto \exp\left(-\frac{\Delta E}{kT}\right). \tag{C.2}$$

As an example, we can estimate the density of vacancies in a metallic crystal. If a vacancy is formed, all atoms adjacent to the vacancy thus possess unsaturated bonds, thus increasing the energy. A typical value for the activation energy required for this process is $\Delta E \approx 10^{-19}$ J. If we put this number into equation (C.2),[1] we can calculate the probability of a vacancy being at a certain lattice site as 3×10^{-12} at 0°C and as 10^{-4} at 500°C. Due to the large number of atoms in a crystal, these probabilities also correspond to the vacancy density. As can be seen, the number of vacancies strongly increases with temperature.

Chemists frequently use a different form of Boltzmann's law, replacing Boltzmann's constant with the so-called *gas constant* $R = 8.314$ J/mol K. Using this, the energy in equations (C.1) and (C.2) has to be given per mole. The formula is then to be interpreted as giving the probability that the process of interest does not happen only once, but 6.022×10^{23} times. The activation energy in the example above then takes a value of 60 kJ/mol.

C.2 Free energy and free enthalpy

All of thermodynamics is based on the following two laws:

- First law of thermodynamics:
  The energy U of a closed system is constant.
- Second law of thermodynamics:
  The *entropy* S of a closed system takes its maximum value in thermal equilibrium.

A closed system in this context is a system with constant volume that can exchange neither heat nor particles with its environment. The statement of the first law is nothing but the law of the conservation of energy.

To understand the second law, we need to think about the meaning of *entropy*. The entropy of a system is a measure of the probability that the system is in a certain macroscopic state. In a closed system, in which the energy is constant, all microscopic states are equally probable according to Boltzmann's law. If we observe the system macroscopically, we will find that state most often that can be obtained by the largest number of distinct microscopic states. Therefore, it is highly improbable that all gas molecules inside a container will gather in one corner, leaving a vacuum everywhere else. The process is not impossible, but there are only a very small number of possibilities to arrange the gas molecules in the corner, compared to the number of possibilities to distribute them evenly over the whole volume. Simplifying, it can be said that

---

[1] In this calculation we assume that each lattice site can exist in one of two states, either occupied with an atom or vacant. As the probability for a vacancy is small, the probability of the site being occupied is close to one. The proportionality constant in equation (C.2) can therefore be set to 1 in very good approximation.

the entropy of a system is a measure of its disorder because an ordered state can only be created in a comparably small number of ways.

These considerations were made for a closed system. In practice, we usually have to deal with systems in contact with their environment. According to the previous section, the probability for a system to be in a certain state $\mathcal{Z}_i$ in thermal equilibrium at a temperature T is given by equation (C.1). The most probable state is therefore the state with the lowest energy, and the higher the energy of a state, the lower is the probability to find the system in this state.

> This statement seems to contradict every-day experience: If we consider again the example of the gas molecules in a container, now at a fixed temperature, it seems to imply that all gas molecules should lie at rest on the floor of the container because this would minimise their potential and kinetic energy. This, however, is not observed.
>
> This seeming contradiction can be resolved by considering the number of different states the gas molecules can be in to obtain a certain macroscopic state. There are only a small number of possibilities to produce the state of lowest energy described above, but a very large number of configurations in which the gas molecules are irregularly distributed everywhere in the container. For this reason we will almost certainly observe one of the irregular configurations.
>
> A simple example can serve to illustrate the distinction: A die is loaded so that it shows the number 6 with a probability of 25 %, each other number with a probability of only 15 %. If we throw the die ten times, the probability to throw ten sixes is larger than the probability of any other exactly specified sequence of numbers. Nevertheless, the probability for this event is only $(1/4)^{10} \approx 10^{-6}$, for there is only one possibility to get ten sixes, but, for example, already 50 ways to throw nine sixes together with another number.

If we consider a system $\mathcal{S}$ held at a certain temperature by bringing it in contact with a heat bath $\mathcal{W}$, the two can exchange energy. The entropy is maximised for the complete system i.e., for the system $\mathcal{S}$ and the heat bath $\mathcal{W}$ together. In this case, the entropy of $\mathcal{S}$ itself is not necessarily maximised because the complete system may increase its entropy by a process that diminishes the entropy of $\mathcal{S}$ but increases the entropy of $\mathcal{W}$ by a greater amount.

To describe the system $\mathcal{S}$, we introduce a new quantity, the *free energy F*. It is defined as

$$F = U - TS, \tag{C.3}$$

where U is the *internal energy* of the system $\mathcal{S}$ and S is its entropy. F is minimised when the system is in contact with a heat bath and has a fixed volume.[2]

---

[2] F is not maximised because the entropy enters with a minus sign.

If the system can also change its volume, being held under constant pressure p,[3] it can also change its inner energy by increasing or decreasing its volume V against the pressure. In this case, the *free enthalpy* or *Gibb's energy* G takes the place of F:

$$G = U - TS + pV \,. \tag{C.4}$$

In a system at constant temperature and pressure, the free enthalpy is minimised. If we are interested in whether a certain process will take place under these conditions, we have to look at changes in free enthalpy: If it decreases, the process can take place. One example for this is the investigation of nucleation in section 6.4.4. In solids, the distinction between free energy and free enthalpy is usually unnecessary because the volume change on changes in pressure is small. [4]

C.3 Phase transformations and phase diagrams

As explained in the previous section, a system with constant volume in contact with a heat bath (i.e., at constant temperature) minimises its free energy. At low temperatures, the influence of the entropy is small according to equation (C.3) so that the system tends to minimises its inner energy. With increasing temperature, the entropy becomes more and more important, and the system will not be in the state of lowest energy. This temperature dependence of the state is the reason for the occurrence of phase transformations as will be explained in the following.

As an example, consider a metal: At low temperatures, it is crystalline because this arrangement minimises the energy by ensuring strong bonds between the atoms (see section 1.2). The entropy of the crystal, however, is small due to its long-range order, for the positions of all atoms in the crystal are (almost) fixed. In the liquid state, on the other hand, the metal atoms are more weakly bound, resulting in a higher inner energy. The entropy is larger, however, because there is no long-range order and the atoms can move about freely.

Figure C.1 shows curves of the free energy for the liquid and solid state of a material. According to equation (C.3), both free energies depend linearly on temperature[5] so that the curves intersect at a certain temperature value. Below this value, the crystalline state is the state of lowest free energy, above this temperature, the liquid state is favoured. If we heat the system beyond

---

[3] The pressure p is the negative hydrostatic stress σ_{hyd} and is thus negative under tensile stresses.

[4] One counter-example is the transformation toughening of ceramics discussed in section 7.5.4.

[5] Here we make the simplifying assumption that neither the internal energy nor the entropy themselves are temperature-dependent.

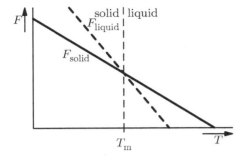

Fig. C.1. Free energy in the liquid and solid state as function of the temperature

this temperature, a phase transformation between the solid and the liquid state will take place. This temperature is thus the melting temperature.

In general, a phase transformation does not occur exactly at the transformation temperature. Directly above the melting temperature, the reduction in free energy that can be obtained by the transformation is small. Furthermore, some additional energy is needed to form a liquid phase because the interface between the two phases is in a high-energy state. This makes it possible to heat a system beyond its melting temperature without a phase transformation. The phase transformation can only occur if the increase in free energy is large enough to compensate for the additional interface energy. If the process occurs infinitely slow, however, the phase transformation can occur directly at the transformation temperature because there is always a finite probability of the additional interface energy being provided by thermal activation.

If we look at an alloy instead of a pure material, the situation becomes more complicated. Depending on the solubility of the alloying elements, the system can reduce its free energy by either mixing the components or by separating different phases. Consider a completely solid system made of two elements A and B. The entropy of the system is largest if both elements form a solid solution because separating the atoms in two phases would reduce the number of possibilities to arrange the atoms. At sufficiently high temperatures, we therefore expect complete solubility of the two elements. At low temperatures, the behaviour of the system depends on the strength of the bond between the elements A and B: If it is stronger than the bond between A–A and B–B, a solid solution is favoured at low temperatures as well. If the bond is weaker, it is better to separate the phases at low temperatures. The result is a *miscibility gap*, a region in the phase diagram (see below) where the elements are not completely soluble.

Figure C.2 shows a *phase diagram* of a system with a miscibility gap. In this kind of diagram, the concentration of the elements is put on the horizontal axis, the temperature on the vertical. Within the diagram, the regions consisting of different phases are marked to allow determining the phases of the system as function of temperature and concentration of the alloying elements. The regions are separated by boundary lines which denote a phase transformation.

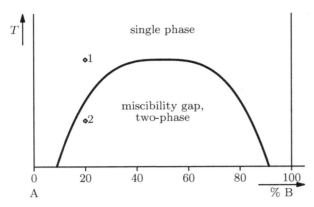

Fig. C.2. Sketch of a phase diagram of a system with a miscibility gap in the solid state. At low temperatures, two separate phases coexist (point 1), at high temperatures the elements mix in solid solution (point 2)

In some regions, the alloy is in a single phase; in others, it is in a two-phase state.[6] In phase diagrams of binary alloys, regions with one phase always adjoin to regions with two phases, except for single points. This rule is helpful in reading more complicated phase diagrams.

In the two-phase region, an A-rich and a B-rich phase coexist. The concentration within the two phases can be read off by drawing a horizontal line from the point characterising the actual system state. At the intersection of this line with the boundaries of the two-phase region, the concentration within the two phases can be found. The amount of the two phases can also be determined because the overall concentration must equal the known concentration in the alloy. If c denotes the total concentration of B, c_A the B concentration in the A-rich phase, and c_B the B concentration in the B-rich phase, the masses m_A and m_B of the A- and B-rich phase are given by the *lever rule*

$$\frac{m_A}{m_A + m_B} = \frac{c_B - c}{c_B - c_A},$$
$$\frac{m_B}{m_A + m_B} = \frac{c - c_A}{c_B - c_A}. \tag{C.5}$$

If we cool down a system with a miscibility gap from the high-temperature region, phases will separate similarly to the transition between solid and liquid phases. Again, the transformation will only occur exactly at the transformation temperature shown in the diagram if the cooling is extremely slow. To separate the two phases, diffusion processes in the solid have to occur which are slow at low temperatures. Thus, it is possible to supercool the system by

---

[6] In real systems, even at higher temperatures, a solid solution may not form because the melting temperature may be reached before the increase in entropy is large enough.

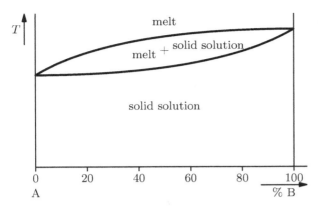

Fig. C.3. Schematic phase diagram of a binary system with complete solubility of the components

cooling down so fast that no diffusion can occur. If we keep the system at low temperatures, the diffusion coefficient may be so small that the system state is practically stable, although it is not the thermodynamically favoured state of lowest free energy. This is called a *metastable state.*

As already explained, a solid will transform to a liquid at high temperatures. The liquid state can also be drawn in the phase diagram. As an example, figure C.3 shows the phase diagram of a binary alloy with complete solubility in the liquid and the solid state. At low temperatures, a solid solution forms; at high temperatures, the system is liquid. If the system is cooled down from the liquid phase, a two-phase region is encountered in which melt and solid phase coexist. As we can see by drawing a horizontal line, the solid regions are richer in the component with the higher melting point, whereas the concentration of the lower-melting component is larger in the melt. With decreasing temperature, the concentrations in the melt and the solid change until the single-phase solid solution forms when the second boundary line is crossed.

Frequently, the components A and B are only partially soluble in the solid state. In this case, the phase diagram is more complicated. One possible diagram is shown in figure C.4. At high temperatures, in the liquid state, both components are completely soluble, but at low temperatures, the solubility is small so that two phases can coexist, one A-rich phase with dissolved B atoms, usually called the α *phase*, and one B-rich phase with dissolved A atoms, usually called the β *phase*. The lower part of the phase diagram is similar to figure C.2, for in this case there is also a miscibility gap with a two-phase region in the solid state. However, the gap ends at a certain temperature because the material starts to melt. At higher temperatures, a mixture of melt and α or β phase forms, depending on the concentration. There is a certain concentration, called the *eutectic concentration*, where the solid transforms directly into a melt and becomes liquid at a certain temperature, not in a temperature

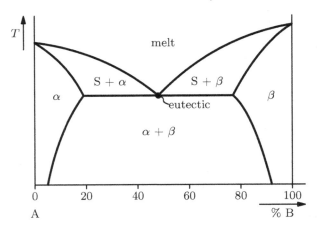

Fig. C.4. Schematic phase diagram of a binary system with limited solubility of the components

regime. In this case, the temperature is below the melting temperature of both components.

Depending on the binding energy between the atoms and on the possibility of additional phases that may form, real phase diagrams can be much more complex than the simple examples shown here. One example is the phase diagram of the iron-carbon system in figure 6.50.

D

The J integral

In this appendix, the J integral, introduced in chapter 5, will be derived and discussed in detail. The derivation starts by introducing some concepts of vector calculus. It is based on *Gross / Seelig* [58] and *Rice* [118].

D.1 Discontinuities, singularities, and Gauss' theorem

Gauss' theorem relates the surface integral and the volume integral of a vector-valued function $\underline{F}(\underline{x})$:

$$\iiint_V (\underline{\nabla} \cdot \underline{F}(\underline{x})) \, d\underline{x} = \iint_S \underline{F}(\underline{x}) \cdot \underline{n} \, dS \,. \tag{D.1}$$

V is the integration volume with surface S, and $\underline{n}$ is the *normal vector* on this surface.

Gauss' theorem can be interpreted as follows: A vector field $\underline{F}(\underline{x})$ can be visualised as consisting of flux lines filling space as shown in figure D.1. The *divergence* $\underline{\nabla} \cdot \underline{F}(\underline{x})$ of a vector field is a measure of its source strength. If the divergence is zero in a region of space, the number of flux lines entering and leaving the volume is the same (region 1 in figure D.1). If this is the case in the whole space, the flux lines do not end or begin anywhere; they are closed. If the divergence is non-zero, flux lines begin at this point (a source) or they end there (a sink), as in region 2 of figure D.1. This is exactly the statement of Gauss' theorem: The volume integral is a measure of the total source strength within V, the surface integral counts the flux lines entering and leaving the volume. The normal vector in the surface integral ensures that ingoing and outgoing flux lines are counted with opposite signs.

Gauss' theorem is not universally valid, but only if the function $\underline{F}(\underline{x})$ is continuously differentiable with continuous derivative. If, however, the function possesses a singularity (i.e., if its value approaches infinity), Gauss' theorem cannot be used and the left-hand and the right-hand side of equation (D.1) are not identical anymore.

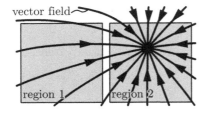

vector field

region 1   region 2

Fig. D.1. Plot of a vector field $\underline{F}(\underline{x})$. In region 1, $\underline{\nabla} \cdot \underline{F}(\underline{x}) = 0$ holds, in region 2 $\underline{\nabla} \cdot \underline{F}(\underline{x}) \neq 0$

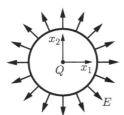

Fig. D.2. Electrical field on the surface of a sphere surrounding a point charge

This fact can be exploited by using Gauss' theorem to check whether a function with divergence zero contains a discontinuity or singularity in a certain region. If the surface integral $\iint_S \underline{F}(\underline{x}) \cdot \underline{n} \, dS$ of a function $\underline{F}(\underline{x})$ with $\underline{\nabla} \cdot \underline{F}(\underline{x}) = 0$ yields a non-zero value, there must be a discontinuity or singularity in the volume V enclosed by S.

This can be illustrated using an example from electrostatics: The divergence of the electrical field $\underline{E}(\underline{x})$ vanishes in vacuum, $\underline{\nabla} \cdot \underline{E}(\underline{x}) = 0$, implying that the flux lines of the field do not end in free space. If, however, there is an electrical charge, it acts – depending on its sign – as a source or sink of flux lines. The electrical field of a small sphere with charge Q situated in the origin of our coordinate system (figure D.2) fulfils the equation

$$\underline{E}(\underline{x}) = \frac{Q}{4\pi\varepsilon_0} \frac{\underline{x}}{|\underline{x}|^3} \tag{D.2}$$

away from the charge. A simple calculation shows that the divergence of this function vanishes.

To prove this, we write $\underline{E}(\underline{x})$ in Cartesian coordinates for each component:

$$E_i(\underline{x}) = \frac{Q}{4\pi\varepsilon_0} \cdot \frac{x_i}{(x_1^2 + x_2^2 + x_3^2)^{3/2}} \,.$$

In Cartesian coordinates, the divergence operator is

$$\underline{\nabla} = (\partial/\partial x_1, \partial/\partial x_2, \partial/\partial x_3) \,.$$

Applying this to the field, we find

$$\underline{\nabla} \cdot \underline{E}(\underline{x}) = \frac{\partial E_i}{\partial x_i} = \frac{\partial E_1}{\partial x_1} + \frac{\partial E_2}{\partial x_2} + \frac{\partial E_3}{\partial x_3} = 0$$

for $\underline{x} \neq \underline{0}$.

The function $\underline{E}(\underline{x})$ contains a singularity at $\underline{x} = 0$. If we integrate the electrical field over the surface S with radius R, we find

$$\iint_S \underline{E}(\underline{x}) \cdot \underline{n} \, \mathrm{d}S = \frac{Q}{4\pi\varepsilon_0} \iint_S \frac{\underline{x}\,\underline{n}}{|\underline{x}|^3} \, \mathrm{d}S = \frac{Q}{4\pi\varepsilon_0} \iint_S \frac{R}{R^3} \, \mathrm{d}S$$
$$= \frac{Q}{4\pi\varepsilon_0} \frac{1}{R^2} \iint_S \mathrm{d}S = \frac{Q}{4\pi\varepsilon_0} \frac{1}{R^2} \cdot 4\pi R^2$$
$$= \frac{Q}{\varepsilon_0}.$$

As can be seen, the surface integral does not vanish although the divergence of the field vanishes at the integration surface. Thus, the surface integral can be used to probe for charges in a volume. This is especially useful if the charge is a point charge because in this case integrating over the volume would be problematic because of the singularity.

A similar problem occurs in media containing cracks. In this case, there is a singularity (for example in the stress field) at the crack tip. Furthermore, the detailed conditions near the crack tip may be unknown, although they may be known at some distance. Thus, if we can find a quantity with normally vanishing divergence that contains a singularity or discontinuity at the crack tip, this quantity can be used to gain information on the crack by integrating over a surface far away. This idea is pursued in the following.

D.2 Energy-momentum tensor

Our task is to find a physical quantity that can be defined for any elastic-plastic material, that has a vanishing divergence, and that becomes discontinuous or singular at a crack tip.

The most obvious choice would be the stress tensor $\underline{\sigma}$. If there are no volume forces, the stress tensor fulfils the equation $\underline{\nabla} \cdot \underline{\sigma} = \underline{0}$ because stresses are generated only where forces act. Using the stress tensor, however, has a severe disadvantage because the stress is frequently prescribed by the external load. This is illustrated in the following example:

Consider a tensile test specimen made of two materials with different Young's moduli E_1 and E_2 (figure D.3). The specimen is loaded with constant force at its end. In this case, the stress is constant anywhere within the specimen so that any surface integral over the stress vanishes within the material.[1] The discontinuity in the material properties thus cannot be found using such an integral.

Another quantity whose values differ in both halves of the specimen is the *energy density* $w = \int \sigma_{ij} \, \mathrm{d}\varepsilon_{ij}$. At a given stress, the strain is larger in

---

[1] For simplicity, transversal contraction is neglected here.

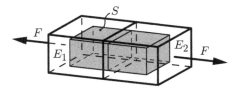

Fig. D.3. Elastic medium with a discontinuous change in Young's modulus. Also shown is an integration surface S

the region with smaller Young's modulus, corresponding to a larger energy density. The energy density itself is not a suitable quantity, however, because Gauss' theorem requires a vector-valued function. Nevertheless, it is a good idea to use a quantity containing the energy density. One such quantity is the *energy-momentum tensor* $\underline{\underline{T}}$, defined by

$$T_{ij} = w \cdot \delta_{ij} - \sigma_{jk} \frac{\partial u_k}{\partial x_i} . \tag{D.3}$$

Here, w is the energy density, $\underline{\underline{\sigma}}$ the stress tensor, $\underline{u}$ the displacement vector, and δ_{ij} is the Kronecker delta introduced in appendix A.6.

> The derivation of the energy-momentum tensor is way beyond the scope of this book. The name stems from classical field theory which deals with arbitrary physical fields like the electromagnetic field, the velocity field in a fluid, or the strain field in an elastic medium. In field theory, a generalised tensor is used, with some components describing the energy and momentum density of the system. In the context of elasticity theory, the name energy-momentum tensor is misleading because none of its components are the energy or momentum density. A detailed, but mathematically involved, introduction to the subject can be found in *Landau / Lifschitz* [86].

The energy-momentum tensor has the desired property of a vanishing divergence:

$$\left(\underline{\nabla} \cdot \underline{\underline{T}}\right)_i = \frac{\partial}{\partial x_j} T_{ij} = 0 \tag{D.4}$$

for $j = 1 \ldots 3$. This is easy to show if we take the equation $\partial \sigma_{ij}/\partial x_j = 0$ into account that is valid if there are no volume forces. We thus get not only one, but three quantities to be used in Gauss' theorem to test for discontinuities or singularities.

D.3 J integral

To further study the energy-momentum tensor, we take another look at the example of the medium with two different Young's moduli (see figure D.3). The stress in the specimen is everywhere equal to the external stress σ; the strain is constant in each half of the specimen, with 11-component $\varepsilon^{(n)} = \sigma/E^{(n)}$, where the superscript 'n' denotes the two halves. Using the energy density of a linear-elastic material, $w = \sigma\varepsilon/2$, and the Poisson's ratio ν (assumed to be the same in both parts), the energy-momentum tensor is

$$T_{ij}^{(n)} = w^{(n)}\delta_{ij} - \sigma_{jk}^{(n)}\frac{\partial u_k^{(n)}}{\partial x_i} \text{ i.e.,}$$

$$
\begin{aligned}
\underline{\underline{T}}^{(n)} &= \frac{\sigma^2}{2E^{(n)}}\begin{pmatrix} 1 & 0 & 0 \\ 0 & 1 & 0 \\ 0 & 0 & 1 \end{pmatrix} - \begin{pmatrix} \sigma & 0 & 0 \\ 0 & 0 & 0 \\ 0 & 0 & 0 \end{pmatrix} \cdot \begin{pmatrix} \varepsilon^{(n)} & 0 & 0 \\ 0 & -\nu\varepsilon^{(n)} & 0 \\ 0 & 0 & -\nu\varepsilon^{(n)} \end{pmatrix} \\
&= \frac{\sigma^2}{2E^{(n)}}\begin{pmatrix} 1 & 0 & 0 \\ 0 & 1 & 0 \\ 0 & 0 & 1 \end{pmatrix} - \begin{pmatrix} \sigma & 0 & 0 \\ 0 & 0 & 0 \\ 0 & 0 & 0 \end{pmatrix} \cdot \frac{\sigma}{E^{(n)}} \cdot \begin{pmatrix} 1 & 0 & 0 \\ 0 & -\nu & 0 \\ 0 & 0 & -\nu \end{pmatrix} \\
&= \frac{\sigma^2}{2E^{(n)}}\begin{pmatrix} 1 & 0 & 0 \\ 0 & 1 & 0 \\ 0 & 0 & 1 \end{pmatrix} - \frac{\sigma^2}{E^{(n)}}\begin{pmatrix} 1 & 0 & 0 \\ 0 & 0 & 0 \\ 0 & 0 & 0 \end{pmatrix} \\
&= \frac{\sigma^2}{2E^{(n)}} \cdot \begin{pmatrix} -1 & 0 & 0 \\ 0 & 1 & 0 \\ 0 & 0 & 1 \end{pmatrix}.
\end{aligned}
$$

The energy-momentum tensor is thus rather simple and is constant in each half of the material.

What happens if we apply Gauss' theorem? As already stated, we have three possibilities, one for each column of the matrix representation of the energy-momentum tensor. We consider the surface S in figure D.3 and define the three quantities J_1, J_2, and J_3, called the J integrals,

$$J_i = \iint_S T_{ij} \cdot n_j \, \mathrm{d}A, \tag{D.5}$$

where $\underline{n}$ is the normal vector on the surface [118].

Since the tensor is constant in each half of the material, the four side faces of the integration volume cancel (for each face, there is an opposite face with opposite normal vector). We have to look at the two end faces with surface area A only. This results in

$$
\begin{aligned}
J_1 &= \iint_{A_1} \frac{-\sigma^2}{2E^{(1)}} \cdot n_1 \, \mathrm{d}A + \iint_{A_2} \frac{-\sigma^2}{2E^{(2)}} \cdot n_2 \, \mathrm{d}A \\
&= \frac{\sigma^2 A}{2}\left(-\frac{1}{E^{(1)}} + \frac{1}{E^{(2)}}\right),
\end{aligned} \tag{D.6}
$$

$$J_2 = 0,$$
$$J_3 = 0.$$

As can be seen, two of the three J values vanish, and only J_1 is non-zero. Because the discontinuity surface has a normal vector in the 1-direction, this is a plausible result.

The J integral can also be interpreted directly: Looking at its unit, we see that it has the unit of a force, but it is not too obvious what kind of force it is and what it is applied to. A better interpretation can be found if we multiply the J integral with a distance $\mathrm{d}x_1$. The resulting quantity $\mathrm{d}\Pi = J_1\mathrm{d}x_1$ is an energy. If we remember that the energy density in the elastic media is $w = \sigma^2/2E^{(n)}$, we see that $\mathrm{d}\Pi$ is the energy that would be released if we were to shift the interface in 1 direction by a distance $\mathrm{d}x_1$. The same interpretation is valid for the other two J integrals: If we shift the interface in the 2 or 3 direction, nothing changes, and the energy release is zero. The J integral can thus be interpreted as the differential energy release, the change in energy when the discontinuity in the system is shifted by an infinitesimal distance.

So far, we considered the J integral only for the case of a medium with an elastic discontinuity. The results, however, are universally applicable. Thus, if S is an arbitrary surface with normal vector $\underline{n}$ enclosing a discontinuity or singularity of the system, each component of the J integral $\underline{J}$ is defined as the surface integral

$$J_i = \iint_S T_{ij} \cdot n_j \, \mathrm{d}A \,. \tag{D.7}$$

The differential energy release during an infinitesimal displacement $\mathrm{d}\underline{x}$ of the position of the 'disturbance' is $d\Pi = \underline{J} \cdot \mathrm{d}\underline{x}$. This simple interpretation of this quantity as energy release is only valid in elastic (linear) media. In elastic-plastic media, $\mathrm{d}\Pi$ is the difference of the energy of two systems in which the position of the disturbance differs by $\mathrm{d}\underline{x}$, but it is not always ensured that this energy would in fact be released if the disturbance would actually move by this distance (for example, when a crack propagates). This will be explained in more detail below.

One important property of the J integral, following directly from what we saw so far, is that it is independent of the integration surface. As long as it encloses the disturbance, the exact choice of the surface is irrelevant. This was already obvious in the example of the surface integral containing a point charge: As long as the charge is enclosed, the value of the integral is independent of the radius of the sphere and always equals Q/ε_0. The same holds for the medium with elastic discontinuity; again, the exact shape of the surface does not matter, and only the enclosed surface of the disruption is important. This property is important because it allows to perform numerical calculations (for example, using finite elements) without calculating the detailed conditions near the crack tip – it is sufficient if the stress and displacement fields in some distance from the crack tip are correct.

Fig. D.4. Simple configuration to evaluate the *J* integral

D.4 *J* integral at a crack tip

In the following, we will use the *J* integral to understand material behaviour at a crack tip. The considerations are limited to the two-dimensional case i. e., a system in a state of plane strain or stress. In this case, the integration surface is chosen to have the same cross section at any x_3 position. If we fix an arbitrary x_3 value, the integral is not a surface integral anymore, but only a path integral[2].

In two dimensions, the *J* integral in i direction is thus defined as

$$J_i = \int_C T_{ij} \cdot n_j \, \mathrm{d}s \,. \tag{D.8}$$

Here, C denotes the integration path (or contour), $\underline{n}$ is the normal vector on this path, and $\mathrm{d}s$ is an infinitesimal path element along the curve. The path of integration has to enclose the crack tip. Apart from that, it is arbitrary as explained above.

This path independence explains why the *J* integral can be used to characterise a crack and is independent of other aspects of the material's state. If we choose the integration path close to the crack tip, it is plausible that the value of the *J* integral depends only on the state there. Due to the path independence, the value of the integral does not change if we continuously move the path away from the crack tip, allowing us to use a distant integration path.

We now want to study the *J* integral using the simple example from figure D.4. The figure shows a crack in x_1 direction in a material that is infinite in the x_1 and x_3 direction and has a height h. The material is clamped at the upper and lower end, with constantly prescribed displacement $\underline{u}$ on both boundaries. The integration path C is chosen as shown in the figure. It starts at one crack surface and ends at the other. In principle, the contour has to be closed but because there is no material within the crack, the energy-momentum tensor vanishes there. Since the crack is in the x_1 direction, we choose the J_1 integral

$$J = J_1 = \int_C \left[w \, \mathrm{d}x_2 - \left(\underline{\underline{\sigma}} \cdot \frac{\partial \underline{u}}{\partial x_1} \right) \cdot \underline{n} \, \mathrm{d}s \right] \,. \tag{D.9}$$

---

[2] Care has to be taken in evaluating this path integral: In most path integrals occurring in mathematics or physics, the integration variable is a vector tangential to the curve describing the integration path. This is different here: because the path integral is in fact a dimensionally reduced surface integral, the quantity $\mathrm{d}s$ is a scalar and the vector $n_j \, \mathrm{d}s$ is perpendicular to the curve.

If we move the x_1 coordinate of the vertical parts of C to $\pm\infty$, it is easy to calculate the value of the J integral. The two vertical parts above and below the crack at $x_1 = -\infty$ do not contribute to the integral because the energy density and the derivative of the displacement are zero. The integration along the clamped ends does not contribute as well because the integration over the energy density vanishes (dx_2 is zero along a path in x_1 direction) and because the second term also vanishes due to $\partial\underline{u}/\partial x_1 = 0$. The only remaining part is the path at $+\infty$. In this region, $\partial\underline{u}/\partial x_1 = 0$ holds, so it is only the energy density w_∞ that contributes to the integral:

$$J_1 = w_\infty h\,. \tag{D.10}$$

It is somewhat problematic to use the energy interpretation of the J integral in an infinitely extended system because the total energy of the system is infinite. We can try to argue as follows: If we shift the crack tip in x_1 direction by dx_1, some energy is released, and since the configuration after the shift is the same as before (only displaced by dx_1), this energy release can be written as $w_\infty h\, dx_1$. This argument is a bit obscure, however, because if the configuration is the same, its energy must be the same as well. The problem is that we consider the difference between two infinite quantities. Nevertheless, this simple example shows that the J integral may serve to characterise a crack in an elastic material.

According to the statements made above, the J integral can only be non-zero if there is a singularity in the energy-momentum tensor. This is indeed the case if there is a crack tip in an elastic material As shown in section 5.2.1, stresses and strains become infinite if we approach the crack tip, see equation (5.1):[3]

$$\sigma \propto \frac{1}{\sqrt{r}}\,,$$
$$\varepsilon \propto \frac{1}{\sqrt{r}}\,. \tag{D.11}$$

r denotes the distance from the crack tip.

Such a singularity may seem unphysical. However, this is not the case because stresses, strains, and energy densities are only relative quantities that cannot be measured directly. The strain, for example, is the normalised difference of the displacements at two points. The displacement itself cannot take infinite values, but its change may, if normalised to an infinitesimally small distance. The stress, defined as force per unit area, may become singular as long as the forces in the medium stay finite.[4] The energy density may also

---

[3] A simplified argument for this is given below.

[4] Strictly speaking, continuum mechanics cannot be used at the crack tip because we must not neglect the fact that matter consists of atoms. This is discussed in exercise 13.

become infinite provided the energy of the system (the integral over the energy density) stays finite. According to equation (D.11), the energy density $w = \int \sigma \, d\varepsilon$ is proportional to $1/r$. In cylindrical coordinates, we can write $w(r, \phi) = w_\phi(\phi)/r$. If we integrate the energy density within a circle C with radius R around the crack tip, we find

$$\iint_C w(\underline{x}) \, dA = \int_0^\pi \int_0^R w(r, \phi) r \, dr \, d\phi = \int_0^\pi \int_0^R \frac{w_\phi(\phi)}{r} r \, dr \, d\phi$$
$$= \int_0^\pi w_\phi(\phi) \, d\phi \cdot \int_0^R dr = R \int_0^\pi w_\phi(\phi) \, d\phi \, .$$

The energy stored in the region near the crack tip is thus finite.

D.5 Plasticity at the crack tip

The considerations made so far were valid in an elastic material. The J integral is also to be used if the material is yielding close to the crack tip. Again, stresses, strains, and energy densities become singular in this case.

The kind of singularity can be analysed – in a slightly simplified argument – as follows: If we integrate the J integral along a circular path with radius R, we find

$$J_1 = \int_{-\pi}^\pi \left[w n_1 - \left(\underline{\underline{\sigma}} \, \frac{\partial \underline{u}}{\partial x_1} \right) \underline{n} \right] R \, d\phi \, . \tag{D.12}$$

The J integral must be path-independent as we saw above and thus have the same value for all values of R. If we consider a small vicinity of the crack tip, it is plausible to assume that the influence of the stress state far away from the crack tip (i. e., the influence of the exact geometry of the component) becomes less and less important and everything is determined by the geometry of the crack (and the mode of loading).[5]

Furthermore, we assume that the energy-momentum tensor can be written as $\underline{\underline{T}}(r, \phi) = \underline{\underline{T}}_r(r) \underline{\underline{T}}_\phi(\phi)$. The independence of the integral on the radius of the integration path thus means that the value of the energy-momentum tensor must be proportional to $1/R$. Thus, we find $\underline{\underline{T}}_r(r) = 1/r$. The energy density and the product $\sigma_{ij} \varepsilon_{ij}$ are each proportional to $1/r$. In the linear-elastic case, where $\sigma \propto \varepsilon$, this implies that σ and ε are proportional to $1/\sqrt{r}$ (see equation (D.11)).

For the case of a plastic material, we can assume the following relation between stress and strain, using equation (3.16),

$$\varepsilon = K^{-1/n} \sigma^{1/n} \, , \tag{D.13}$$

---

[5] This assumption is justified with hindsight by the fact that we find a singularity in the energy-momentum tensor, showing that external influences become negligible if we approach the crack tip.

if elastic parts of the strain are small compared to the plastic parts. Using equation (D.13) and $\sigma\varepsilon \propto 1/r$ yields[6]

$$\underline{\sigma}(r, \phi) \propto \frac{1}{r^{n/(n+1)}} \underline{\sigma}_\phi(\phi)\,, \tag{D.14}$$

$$\underline{\underline{\varepsilon}}(r, \phi) \propto \frac{1}{r^{1/(n+1)}} \underline{\underline{\varepsilon}}_\phi(\phi) \tag{D.15}$$

for the singularity near the crack tip. The smaller the exponent n becomes, the stronger is the singularity in the strain and the weaker is the singularity in the stress.

What is the role of the J integral in this context? Because of the independence of the stress- and strain fields from external influences far away from the crack tip, all stress fields that we can find for a given crack geometry and mode of loading (mode I, mode II, or mode III) are similar. They just differ by a factor specifying the amount of loading. This factor is nothing but the J integral. Thus, we can write $\underline{\underline{T}}(r, \phi) = (J/r)\underline{\underline{T}}_\phi(\phi)$. If we raise the external stress, the value of the energy-momentum tensor changes accordingly. The equations for the stress and strain are thus

$$\underline{\underline{\sigma}}(r, \phi) = c_\sigma \left(\frac{J_1}{r}\right)^{n/(n+1)} \underline{\underline{\sigma}}_\phi(\phi)\,, \tag{D.16}$$

$$\underline{\underline{\varepsilon}}(r, \phi) = c_\varepsilon \left(\frac{J_1}{r}\right)^{n/(n+1)} \underline{\underline{\varepsilon}}_\phi(\phi)\,. \tag{D.17}$$

The constants c_σ and c_ε depend on the geometry, the exponent n, and the mode of loading, but not on the load strength. The J integral is thus a measure quantifying the stress and strain field. The same is true for the linear-elastic case from the previous section.

D.6 Energy interpretation of the J integral

As already detailed in section D.3, the J integral can be interpreted as specific energy difference between two systems that differ by an infinitesimal displacement of the singularity or discontinuity.[7] This will be proven here.

To do so, we consider two three-dimensional bodies with an axially symmetric cavity. The task is to calculate the energy difference between the bodies (see figure D.5). Both bodies are assumed to be absolutely identical except for the cavity being larger in body B than in body A by an infinitesimal volume ΔV. We assume both bodies to be elastic, but not necessarily linear-elastic. This assumption is important because the work done by a deformation on an elastic body is independent of the deformation history. Furthermore, we

---

[6] Frequently, an exponent $m = 1/n$ is used instead of n in these equations.

[7] This may be, for instance, due to crack propagation.

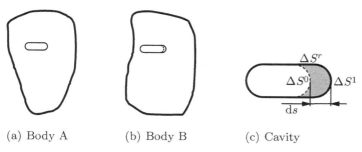

(a) Body A   (b) Body B   (c) Cavity

Fig. D.5. Comparison of two bodies with cavities that differ by a volume element ΔV. Further explanations in the text

assume all volume forces to be zero to ensure that the divergence of the energy-momentum tensor vanishes.

To further simplify matters, we assume that the displacements on the surface of both bodies are prescribed. Thus, no work is done by external forces when material is removed from one of the bodies.[8] The cavity within the material is force-free as well.

The volume element to be removed is bounded by the following surfaces: ΔS^0, ΔS^1, and ΔS^r (see figure D.5). ΔS^1 can be generated by an infinitesimal displacement of ΔS^0 by a distance $\mathrm{d}s$. In three dimensions, ΔS^r is the outer surface of a cylinder with height $\mathrm{d}s$. The two parts generated by the displacement, ΔS^1 and ΔS^r, are united as $\Delta S = \Delta S^1 + \Delta S^r$.

We are looking for the difference of the energy stored in both bodies. To calculate this energy, we can imagine to transform one body into the other in two steps. We start by removing the volume element ΔV, replacing the forces the material in the volume ΔV exerts onto the surface ΔS^1 by external forces. Thus, no work is done on the remaining volume of the material because the removal of the volume is not seen by it. The energy $\mathrm{d}W_V$ released in this step is given by the integral of the energy density w over the volume ΔV:

$$
\begin{aligned}
\mathrm{d}W_V &= \int_{\Delta V} w(\underline{x})\,\mathrm{d}V \\
&= \int_{\Delta S^0} w(\underline{x})\,\mathrm{d}s\,\mathrm{d}S \\
&= \int_{\Delta S^1} w(\underline{x})\,\mathrm{d}s\,\mathrm{d}S \,,
\end{aligned}
\tag{D.18}
$$

where the volume integral can be transformed to a surface integral because the energy density within the infinitesimal volume can be considered to be constant.

---

[8] This assumption is not necessary in proving the energy interpretation of the J integral, but it considerably simplifies the proof.

In the second step, we now have to remove the surface forces that kept the material in its old state so far. These surface forces $\underline{F}$ are determined by the stress within body A on the surfaces ΔS^1 and ΔS^r. If $\underline{n}$ is the normal vector on the surface, we can write $F_i = \sigma_{ij} n_j$. These forces are now reduced to zero. In this step, the systems does work $\mathrm{d}W_S$ that is given by

$$\mathrm{d}W_S = \int_{\Delta S} F_i \Delta u_i \, \mathrm{d}S \, . \tag{D.19}$$

Here, $\Delta\underline{u}$ is the displacement resulting from reducing the forces. If the displacement of the surface $\mathrm{d}s$ is small, we can assume that the displacement field on the surface ΔS^1 in body B is equal to the field on the surface ΔS^0 in body A. In the case of a crack, where stresses and strains near the crack tip are determined completely by the crack tip geometry, this assumption is particularly plausible. The change in the displacement $\Delta\underline{u}(\underline{x})$ on the surface ΔS^1 is thus determined by the difference between the displacement field at a point $\underline{x}^1$ on ΔS^1 in body B and the displacement field of the corresponding point $\underline{x}^0$ on ΔS^0 in body A:

$$\Delta\underline{u}(\underline{x}^1) = \underline{u}(\underline{x}^0) - \underline{u}(\underline{x}^1) = \frac{\partial\underline{u}}{\partial\underline{x}}\mathrm{d}s \, . \tag{D.20}$$

Furthermore, the work $\mathrm{d}W_S$ contains a further term from the cylinder surface ΔS^r. This contribution is small because the height of the cylinder is infinitesimal. Thus, the total work done upon unloading of the surface is

$$\mathrm{d}W_S = \int_{\Delta S^1} \sigma_{ij} n_j \frac{\partial u_i}{\partial x_k}\mathrm{d}s_k \, \mathrm{d}S \, . \tag{D.21}$$

The total change in the energy is thus

$$\mathrm{d}W = \int_{\Delta S^1} \left(w(\underline{x})\mathrm{d}s - \sigma_{ij} n_j \frac{\partial u_i}{\partial x_k}\mathrm{d}s_k \right) \mathrm{d}S \, . \tag{D.22}$$

The term in parentheses is the J integral multiplied by the displacement $\mathrm{d}s$ of the surface. This proves the energy interpretation of the J integral.

During the derivation, we used the fact that the result is independent of how we obtained the final state of the system (i. e., body B) from the initial state. Strictly speaking, this is valid only for elastic bodies. If plastic deformations occur, the energy interpretation has to be handled with care. It is still valid if the so-called theory of deformation plasticity is used, which assumes that there is no unloading within the material. This condition is frequently met so that the J integral can used even in plasticity in many cases.

References

1. 3M Corporation: *Aluminum matrix composites.* http://www.3m.com/mmc/.
2. R. B. Abernethy: *The New Weibull Handbook.* Published personally, 4th edition, 2000.
3. R. Abinger, F. Hammer, and J. Leopold: *Großschaden an einem 300-MW-Dampfturbosatz.* Der Maschinenschaden, 61(2):58–60, 1988.
4. D. Altenpohl: *aluminium von innen.* Aluminium-Verlag, Düsseldorf, 5th edition, 1994.
5. Aluminium-Zentrale (editor): *Aluminium-Taschenbuch.* Aluminium-Verlag, Düsseldorf, 14th edition, 1983.
6. B. F. Antolovic: *Fatigue and fracture of nickel-base superalloys.* In S. R. Lampman (editor): *Fatigue and Fracture*, volume 19 of ASM Handbook, pages 854–868. ASM International, 1996.
7. M. F. Ashby: *Materials Selection in Mechanical Design.* Butterworth-Heinemann, Oxford, 1992.
8. M. F. Ashby and D. R. H. Jones: *Engineering Materials 1: An Introduction to their Properties and Applications.* Pergamon Press, Oxford, 1980.
9. M. F. Ashby and D. R. H. Jones: *Engineering Materials 2: An Introduction to Microstructure, Processing and Design.* Pergamon Press, Oxford, 1986.
10. N. W. Ashcroft and N. D. Mermin: *Solid State Physics.* ITPS Thomson Learning, 2000.
11. D. R. Askeland and P. P. Phleé: *The Science and Engineering of Materials.* Thomson-Engineering, 5th edition, 2005.
12. K. Autumn, Y. A. Liang, S. T. Hsieh, W. Zesch, P. C. Wai, T. W. Kenny, R. Fearing, and R. J. Full: *Adhesive force of a single gecko foot-hair.* Nature, 405:681–685, 2000.
13. A. M. M. Baker and C. M. F. Barry: *Effects of composition, processing, and structure on properties of engineering plastics.* In G. E. Dieter (editor): *Materials Selection and Design*, volume 20 of ASM Handbook, pages 434–456. ASM International, 1997.
14. O. H. Basquin: *The exponential law of endurance tests.* Proceedings of the ASTM, 10:625–630, 1910.
15. K. J. Bathe: *Finite Element Procedures.* Pearson Higher Education, 1995.
16. E. Becker and W. Bürger: *Kontinuumsmechanik*, volume 20 of *Leitfäden der angewandten Mathematik und Mechanik.* B. G. Teubner, Stuttgart, 1975.

486 References

17. A. Beiser: *Concepts of Modern Physics*. McGraw Hill, New York, 2002.
18. W. Beitz and K. H. Küttner (editors): *Dubbel – Handbook of Mechanical Engineering*. Springer, Berlin, 1994.
19. W. Bergmann: *Werkstofftechnik, Teil 1: Grundlagen*. Hanser Verlag, 3rd edition, 2000.
20. G. Blugan, M. Hadad, J. Janczak-Rusch, J. Kuebler, and T. Graule: *Fractography, mechanical properties, and microstructure of commercial silicon nitride-titanium nitride composites*. Journal of the American Ceramical Society, 88:926–933, 2005.
21. H. Blumenauer and G. Pusch: *Technische Bruchmechanik*. Deutscher Verlag für Grundstoffindustrie, Stuttgart, 1993.
22. J. E. Bramfitt and R. L. Hess: *A novel heat activated recoverable temporary stent (hearts system)*. In *Proceedings of SMST-94*, pages 435–442, California, 1994.
23. D. Broek: *The Practical Use of Fracture Mechanics*. Kluwer Academic, 1989.
24. I. N. Bronštein and K. A. Semendjaev: *Handbook of Mathematics*. Springer, Berlin, 2004.
25. R. J. Brook (editor): *Concise Encyclopedia of Advanced Ceramic Materials*. Pergamon Press, Oxford, 1991.
26. R. Bürgel: *Handbuch Hochtemperatur-Werkstofftechnik*. Vieweg Verlag, Braunschweig, 1998.
27. W. D. Callister: *Materials Science and Engineering*. John Wiley & Sons, 5th edition, 2000.
28. K. K. Chawla: *Ceramic Matrix Composites*. Chapman & Hall, London, 1993.
29. K. K. Chawla: *Composite Materials*. Springer, New York, 2nd edition, 1998.
30. R. E. Clegg and G. D. Paterson: *Ductile particle toughening of hydroxyapatite ceramics using platinum particles*. In A. Atrens, J. N. Boland, R. Clegg, and J. R. Griffiths (editors): *Structural Integrity and Fracture, Proceedings of SIF 2004*, Brisbane, 2004. http://eprint.uq.edu.au/archive/00000836/.
31. L. L. Clements: *Polymer science for engineers*. In J. N. Epel (editor): *Engineering Plastics*, volume 2 of *Engineered Materials Handbook*, pages 48–62. ASM International, 1988.
32. L. F. Coffin and j. Schenectady, N. Y.: *A study of the effects of cyclic thermal stresses on a ductile metal*. Transactions of the ASME, 76:931–950, August 1954.
33. *The copper page*. http://www.copper.org.
34. A. H. Cottrell: *Dislocations and plastic flow in crystals*. Clarendon Press, Oxford, 1965.
35. T. H. Courtney: *Mechanical Behaviour of Materials*. McGraw-Hill, New York, 1990.
36. J. D. Currey: *Bones: Structure and Mechanics*. Princeton University Press, Princeton, 2002.
37. M. Dankert: *Ermüdungsrißwachstum in Kerben – ein einheitliches Konzept zur Berechnung von Anriß- und Rißfortschrittslebensdauern*. PhD thesis, Institut für Stahlbau und Werkstoffmechanik, Technische Universität Darmstadt, 1999.
38. R. H. Dauskardt, R. O. Ritchie, and B. N. Cox: *Fatigue of brittle materials*. In S. R. Lampman (editor): *Fatigue and Fracture*, volume 19 of *ASM Handbook*, pages 936–945. ASM International, 1996.
39. J. R. Davis (editor): *Stainless Steels*. ASM Specialty Handbook. ASM International, 1994.

40. G. E. Dieter: *Mechanical Metallurgy*. McGraw-Hill, New York, 1988.
41. H. Domininghaus: *Plastics for Engineers. Materials, Properties, Applications*. Hanser Verlag, München, 1999.
42. W. Domke: *Werkstoffkunde und Werkstoffprüfung*. Cornelsen Verlag, Berlin, 10th edition, 1994.
43. N. E. Dowling: *Mechanical Behavior of Materials*. Pearson Education, 2007.
44. G. W. Ehrenstein: *Polymer-Werkstoffe*. Hanser Verlag, München, 2nd edition, 1999.
45. G. Erhard: *Konstruieren mit Kunststoffen*. Carl-Hanser-Verlag, München, 1993.
46. A. G. Evans and R. M. Cannon: *Toughening of brittle solids by martensitic transformations*. Acta metallurgica, 34:761–800, 1986.
47. R. P. Feynman, R. B. Leighton, and M. Sands: *The Feynman Lectures on Physics, Vol I, II & III*. Pearson Higher Education, 1989.
48. U. Fischer, M. Heinzler, R. Kilgus, F. Näher, H. Paetzold, W. Röhrer, K. Schilling, and A. Stephan: *Tabellenbuch Metall*. Verlag Europa-Lehrmittel, Haan-Gruiten, 41st edition, 1999.
49. J. V. Foltz: *Metal-matrix composites*. In J. R. Davis (editor): *Nonferrous Alloys and Special Purpose Materials*, volume 2 of ASM *Handbook: Properties and Selection*, pages 903–912. ASM International, 1990.
50. P. G. Forrest: *Fatigue of Metals*. Pergamon Press, Oxford, 1962.
51. H. J. Frost and M. F. Ashby: *Deformation-Mechanism Maps*. Pergamon Press, 1982.
52. F. E. Fujita (editor): *Physics of New Materials*, volume 27 of *Springer Series in Materials Science*. Springer-Verlag, Berlin, 2nd edition, 1998.
53. C. J. Gilbert and R. O. Ritchie: *On the quantification of bridging tractions during subcritical crack growth under monotonic and cyclic fatigue loading in a grain-bridging silicon carbide ceramic*. Acta materialia, 46:609–616, 1998.
54. M. B. Goddard, P. D. Burke, D. E. Kizer, R. Bacon, and W. C. Harrigan Jr.: *Continuous graphite fiber* MMCs. In T. J. Reinhart (editor): *Composites*, volume 1 of *Engineered Materials Handbook*, pages 867–873. ASM International, 1987.
55. G. Gottstein: *Physikalische Grundlagen der Materialkunde*. Springer-Verlag, Berlin, 1998.
56. *Grafil Inc.* http://www.grafil.com/pyrofil.html.
57. J. K. Gregory: *Fatigue crack growth of titanium alloys*. In S. R. Lampman (editor): *Fatigue and Fracture*, volume 19 of ASM *Handbook*, pages 845–853. ASM International, 1996.
58. D. Gross and T. Seelig: *Fracture Mechanics with an Introduction to Micromechanics*. Springer, Berlin, 2006.
59. W. Guo, C. H. Wang, and L. R. F. Rose: *Elastoplastic analysis of notch-tip fields in strain hardening materials*. Technical Report AR-010-065, Aeronautical and Maritime Research Laboratory, Melbourne, 1998.
60. S. Hampshire: *Engineering properties of nitrides*. In S. J. Schneider (editor): *Ceramics and Glasses*, volume 4 of *Engineered Materials Handbook*, pages 812–820. ASM International, 1991.
61. J. Harder: *Simulation lokaler Fließvorgänge in Polykristallen*. PhD thesis, Mechanik-Zentrum, TU Braunschweig, 1997.

488   References

62. W. C. Harrigan: *Discontinuous silicon fiber* MMCs. In T. J. Reinhart (editor): *Composites*, volume 1 of *Engineered Materials Handbook*, pages 889–895. ASM International, 1987.
63. D. J. Henwood and J. Bonet: *Finite Elements*. Palgrave Macmillan, 1996.
64. F. Hild: *Probabilistic Approach to Fracture: The Weibull Model*, pages 558–565. In J. Lemaitre [92], 2001.
65. R. Hill: *The Mathematical Theory of Plasticity*. Oxford University Press, Oxford, 1998.
66. M. Hoffman, Y. W. Mai, S. Wakayama, M. Kawahara, and T. Kishi: *Crack-tip degradation processes observed during* in situ *cyclic fatigue of partially stabilized zirconia*. Journal of the American Ceramic Society, 78:2801–2810, 1995.
67. G. A. Holzapfel: *Nonlinear Solid Mechanics*. John Wiley & Sons, Chichester, 2000.
68. R. W. K. Honeycombe: *Steels – Microstructure and Properties*. Edward Arnold, London, 2000.
69. S. Horibe and R. Hirahara: *Cyclic fatigue of ceramic materials: Influence of crack path and fatigue mechanisms*. Acta metallurgica et materialia, 39:1309–1317, 1991.
70. E. Hornbogen: *Werkstoffe*. Springer-Verlag, Berlin, 6th edition, 1994.
71. T. J. R. Hughes and J. Winget: *Finite rotation effects in numerical integration of rate constitutive equations arising in large-deformation analysis*. International Journal for Numerical Methods in Engineering, 15:1862–1867, 1980.
72. E. A. Humphreys and B. W. Rosen: *Properties analysis of laminates*. In T. J. Reinhart (editor): *Composites*, volume 1 of *Engineered Materials Handbook*, pages 218–235. ASM International, 1987.
73. A. Hunsche and P. Neumann: *Quantitative measurement of persistent slip band profiles and crack initiation*. Acta metallurgica, 34(2):207–217, 1986.
74. B. Ilschner and R. F. Singer: *Werkstoffwissenschaften und Fertigungstechnik*. Springer-Verlag, Berlin, 2002.
75. G. R. Irwin: *Analysis of stresses and strains near the end of a crack traversing a plate*. Journal of Applied Mechanics, pages 361–364, September 1957.
76. L. Issler, H. Ruoß, and P. Häfele: *Festigkeitslehre – Grundlagen*. Springer-Verlag, Berlin, 2nd edition, 1997.
77. R. Janda (editor): *Kunststoffverbundsysteme*. VCH Verlagsgesellschaft, Weinheim, 1990.
78. Z. D. Jastrzebski: *The Nature and Properties of Engineering Materials*. John Wiley & Sons, 1976.
79. A. D. Jenkins (editor): *Polymer Science*, volume 1. North-Holland Publishing Company, Amsterdam, 1972.
80. M. Jirasek and Z. P. Bazant: *Inelastic Analysis of Structures*. John Wiley & Sons, 2001.
81. S. Kaliszky: *Plastizitätslehre*. VDI-Verlag, Düsseldorf, 1984.
82. H. H. Kausch (editor): *Crazing in Polymers, Vol. 1*, volume 52/53 of *Advances in Polymer Science*. Springer-Verlag, Berlin, 1983.
83. H. H. Kausch (editor): *Crazing in Polymers, Vol. 2*, volume 91/92 of *Advances in Polymer Science*. Springer-Verlag, 1990.
84. C. Kittel: *Solid State Physics*. Wiley, New York, 8th edition, 2004.
85. D. P. Knight and F. Vollrath: *Liquid crystals and flow elongation in a spider's silk production line*. Proc. R. Soc. Lond., 266:519–523, 1999.

86. L. D. Landau and E. M. Lifschitz: *Course of Theoretical Physics*, volume 2: The Classical Theory of Fields. Butterworth-Heinemann, 4th edition, 1987.

87. R. W. Landgraf: *Fracture mechanics of steel fatigue*. In S. R. Lampman (editor): *Fatigue and Fracture*, volume 19 of ASM *Handbook*, pages 632–644. ASM International, 1996.

88. G. Lange: *Vereinfachte Ermittlung der Fließkurve metallischer Werkstoffe im Zugversuch während der Einschnürung der Proben*. Archiv für das Eisenhüttenwesen, 45(11):809–812, November 1974.

89. G. Lange (editor): *Systematic Analysis of Technical Failures*. Ir Publications, 1985.

90. G. Lange (editor): *Systematische Beurteilung technischer Schadensfälle*. Wiley-VCH, Weinheim, 5th edition, 2001. An English translation of a previous edition is available [89].

91. R. L. Lehman: *Overview of ceramic design and process engineering*. In S. J. Schneider (editor): *Ceramics and Glasses*, volume 4 of *Engineered Materials Handbook*, pages 29–37. ASM International, 1991.

92. J. Lemaitre (editor): *Handbook of Materials Behavior Models*. Academic Press, San Diego, 2001.

93. P. Lukáš and L. Kunz: *Specific features of high-cycle and ultra-high-cycle fatigue*. Fatigue and fracture of engineering materials and structures, 25(8):747–753, 2002.

94. S. S. Manson: *Behavior of materials under conditions of thermal stress*. In *Heat Transfer Symposium*, pages 9–76. University of Michigan Engineering Research Institute, 1953.

95. S. S. Manson: *Behavior of materials under conditions of thermal stress*. NACA report 1170, National Advisory Committee for Aeronautics, 1954.

96. C. Matthek: *Design in der Natur: der Baum als Lehrmeister*. Rombach Wissenschaften, Reihe Ökologie, 1997.

97. N. G. McCrum, C. P. Buckley, and C. B. Bucknall: *Principles of Polymer Engineering*. Oxford University Press, 1988.

98. M. Merkel and K. H. Thomas: *Taschenbuch der Werkstoffe*. Fachbuchverlag Leipzig, 5th edition, 2000.

99. M. A. Miner: *Cumulative damage in fatigue*. Journal of Applied Mechanics, pages A-159–A-164, September 1945.

100. *Mitsubishi Chemical Corporation*. http://www.mitsubishichemical.com/MCFPAmerica.html.

101. M. Miyayama, K. Koumoto, and H. Yanagida: *Engineering properties of single oxides*. In S. J. Schneider (editor): *Ceramics and Glasses*, volume 4 of *Engineered Materials Handbook*, pages 748–757. ASM International, 1991.

102. A. Moet and H. Aglan: *Fatigue failure*. In J. N. Epel (editor): *Engineering Plastics*, volume 2 of *Engineered Materials Handbook*, pages 741–750. ASM International, 1988.

103. D. Munz and T. Fett: *Mechanisches Verhalten keramischer Werkstoffe*. Springer-Verlag, Berlin, 1989.

104. D. Munz and T. Fett: *Ceramics*, volume 36 of *Materials Science*. Springer-Verlag, Berlin, 1998.

105. E. M. Nadgornyi: *Dislocation dynamics and mechanical properties of crystals*, volume 31 of *Progress in materials science*. Pergamon Press, Oxford, 1988.

106. H. Neuber: *Kerbspannungslehre*. Springer-Verlag, Berlin, 4th edition, 2001.

490   References

107. T. A. Osswald and G. Menge: *Materials Science of Polymers for Engineers*. Hanser Verlag, München, 2003.
108. H. Parisch: *Festkörperkontinuumsmechanik*. B. G. Teubner, Stuttgart, 2003.
109. W. D. Pilkey: *Peterson's Stress Concentration Factors*. John Wiley & Sons, New York, 2nd edition, 1997.
110. M. de Podesta: *Understanding the Properties of Matter*. UCL Press, London, 1996.
111. I. J. Polmear: *Light Alloys*. Metallurgy and Materials Science Series. Arnold, London, 3rd edition, 1995.
112. W. Prager: *Einführung in die Kontinuumsmechanik*, volume 20 of *Lehr- und Handbücher der Ingenieurwissenschaften*. Birkhäuser Verlag, Basel, 1961.
113. D. Radaj: *Ermüdungsfestigkeit*. Springer-Verlag, Berlin, 2nd edition, 2003.
114. S. Rawal: *Metal-matrix composites for space applications*. Journal of Metals, 53:14–17, 2001.
115. K. A. Reckling: *Plastizitätstheorie und ihre Anwendung auf Festigkeitsprobleme*. Springer-Verlag, Berlin, 1967.
116. F. Reif: *Fundamentals of statistical and thermal physics*. McGraw Hill, 1987.
117. T. J. Reinhart (editor): *Composites*, volume 1 of *Engineered Materials Handbook*. ASM International, 1987.
118. J. R. Rice: *A path independent integral and the approximate analysis of strain concentration by notches and cracks*. Journal of Applied Mechanics, pages 379–386, June 1968.
119. H. Riedel: *Fracture at High Temperatures*. Materials Research and Engineering. Springer-Verlag, Berlin, 1987.
120. R. O. Ritchie: *Mechanisms of fatigue-crack propagation in ductile and brittle solids*. International Journal of Fracture, 100:55–83, 1999.
121. J. C. Romine: *Continuous aluminum oxide fiber MMCs*. In T. J. Reinhart (editor): *Composites*, volume 1 of *Engineered Materials Handbook*, pages 874–877. ASM International, 1987.
122. B. W. Rosen: *Analysis of material properties*. In T. J. Reinhart (editor): *Composites*, volume 1 of *Engineered Materials Handbook*, pages 185–205. ASM International, 1987.
123. J. Rösler, H. Harders, and M. Bäker: *Mechanisches Verhalten der Werkstoffe*. B. G. Teubner, Wiesbaden, 2nd edition, 2006.
124. J. Rösler, R. Joos, and E. Arzt: *Microstructure and creep properties of dispersion-strengthened aluminium alloys*. Metallurgical Transactions A, 23 A:1521–1539, May 1992.
125. J. Ruge and H. Wohlfahrt: *Technologie der Werkstoffe*. Vieweg Verlag, Braunschweig, 6th edition, 2001.
126. H. Salmang and H. Scholze: *Keramik, Teil 2*. Springer-Verlag, Berlin, 6th edition, 1983.
127. V. Saß: *Untersuchung der Anisotropie im Kriechverhalten der einkristallinen Nickelbasis-Superlegierung CMSX-4*. PhD thesis, TU Berlin, 1997.
128. R. Schirrer: *Damage mechanisms in amorphous glassy polymers: Crazing*, pages 488–499. In J. Lemaitre [92], 2001.
129. K. Schneider: *Advanced blading*. In E. Bauchelet et al. (editors): *High Temperature Materials for Power Engineering 1990 – Part II*, pages 935–954, Dordrecht, 1990. Kluwer Academic Publishers.
130. G. Schott (editor): *Werkstoffermüdung – Ermüdungsfestigkeit*. Deutscher Verlag für Grundstoffindustrie, Stuttgart, 4th edition, 1997.

131. K. Schulte: *Faserverbundwerkstoffe mit Polymermatrix – Aufbau und mechanische Eigenschaften*. DLR-FB 92-28, Deutsche Forschungsanstalt für Luft- und Raumfahrt, 1992.

132. J. M. Schultz: *Properties of solid polymeric materials – Part B*, volume 10 of *Treatise on materials science and technology*. Academic Press, New York, 1977.

133. K. H. Schwalbe: *Bruchmechanik metallischer Werkstoffe*. Carl Hanser Verlag, München, 1980.

134. SGL *Carbon Group*. http://www.sglcarbon.com/sgl_t/fibers/sigra_c.html.

135. K. Shiozawa and L. Lu: *Very high-cycle fatigue behaviour of shot-peened high-carbon-chromium bearing steel*. Fatigue and fracture of engineering materials and structures, 25(8):813–833, 2002.

136. N. J. A. Sloane: *Kepler's conjecture confirmed*. Nature, 395:435–436, 1998.

137. R. Stevens: *Engineering properties of zirconia*. In S. J. Schneider (editor): *Ceramics and Glasses*, volume 4 of *Engineered Materials Handbook*, pages 775–786. ASM International, 1991.

138. R. A. Storer et al. (editors): *Metals Test Methods and Analytical Procedures*, volume 03.01 of *1995 Annual Book of ASTM Standards*, chapter E 399: Standard Test Method for Plane-Strain Fracture Toughness of Metallic Materials, pages 412–442. American Society for Testing and Materials, Philadelphia, 1995.

139. K. Tanaka and Y. Akiniwa: *Fatigue crack propagation behaviour derived from S-N data in very high cycle regime*. Fatigue and fracture of engineering materials and structures, 25(8):775–784, 2002.

140. M. W. Toaz: *Discontinuous ceramic fiber* MMCs. In T. J. Reinhart (editor): *Composites*, volume 1 of *Engineered Materials Handbook*, pages 903–910. ASM International, 1987.

141. *Toray Industries, Inc.* http://www.torayca.com.

142. Verband der Keramischen Industrie e. V. (editor): *Breviary Technical Ceramics*. Fahner Verlag, Lauf, 2004. http://www.keramverband.de/brevier_engl/brevier.htm.

143. Verlag Stahlschlüssel Wegst GmbH, Marbach/N.: *Stahlschlüssel-Taschenbuch*, 1992.

144. J. F. V. Vincent and J. D. Currey (editors): *The mechanical properties of biological materials*. Cambrige University Press, 1980.

145. R. Z. Wang, A. G. Evans, N. Yao, and I. A. Aksay: *Deformation mechanisms in nacre*. Journal of Materials Research, 16:2485–2493, 2001.

146. W. Weibull: *A statistical distribution function of wide applicability*. Journal of Applied Mechanics, 18:293–297, 1951.

147. K. Wellinger and H. Dietmann: *Festigkeitsberechnung: Grundlagen und technische Anwendung*. Alfred Kröner Verlag, Stuttgart, 3rd edition, 1976.

148. H. von Wieding: *Entwicklung eines Programmpaketes zur Berechnung der ertragbaren Spannungen bzw. der Lebensdauer bei mehrachsiger Betriebsbeanspruchung auf der Basis der modifizierten Oktaederschubspannungshypothese*. Student's thesis, Institut für Werkstoffe, TU Braunschweig, 1991.

149. G. Ziegler: *Engineering properties of carbon-carbon and ceramic-matrix composites*. In S. J. Schneider (editor): *Ceramics and Glasses*, volume 4 of *Engineered Materials Handbook*, pages 835–844. ASM International, 1991.

List of symbols

Scalars

α	angle in lattice	μ	Lamé's elastic constant
α	(mean) coefficient of thermal expansion	ν	Poisson's ratio
		ϱ	density
α_{k}	stress concentration factor, we use K_{t}	ϱ	notch radius
		ϱ	dislocation density
β	angle in lattice	σ	normal stress
β_{k}	fatigue notch factor, we use K_{f}	$\sigma_{\mathrm{I}}, \sigma_{\mathrm{II}}, \sigma_{\mathrm{III}}$	principal stresses, sorted by their value ($\sigma_{\mathrm{I}} \geq \sigma_{\mathrm{II}} \geq \sigma_{\mathrm{III}}$)
γ	angle in lattice		
γ	shear strain	σ_0	stress with largest density of failure probability
γ, γ_0	specific surface energy		
Γ_0	surface energy	$\sigma_1, \sigma_2, \sigma_3$	principal stresses, unsorted
δ	thickness of a grain boundary	σ_{a}	stress amplitude
δ	displacement of loading points	σ_{B}	rupture strength of a polymer
δ_{t}	crack tip opening displacement	σ_{b}	buckling stress
ε	(nominal) normal strain	σ_{C}	cleavage strength
ε_{B}	rupture strain of a polymer	σ_{c}	critical stress
ε_{M}	strain at the tensile stress of a polymer	σ_{c}	inert strength
		σ_{E}	fatigue limit
$\varepsilon_{\mathrm{tB}}$	rupture strain of a polymer	$\sigma_{\mathrm{E,nss}}$	fatigue strength of a notched specimen (net-section stress at the notch root)
ε_{Y}	yield strain of a polymer		
$\varepsilon_{\mathrm{eq}}^{(\mathrm{pl})}$	equivalent plastic stress		
$\dot{\varepsilon}_{\mathrm{eq}}^{(\mathrm{pl})}$	equivalent plastic strain rate	σ_{eq}	equivalent stress
θ	angle	$\sigma_{\mathrm{eq,cM}}$	equivalent stress of the conically modified yield criterion
λ	half distance of obstacles		
λ	Lamé's elastic constant	$\sigma_{\mathrm{eq,pM}}$	equivalent stress of the parabolically modified yield criterion
λ	compliance		
λ	angle	$\sigma_{\mathrm{eq,M}}$	von Mises equivalent stress
$\dot{\lambda}$	proportionality factor in the flow rule	$\sigma_{\mathrm{eq,T}}$	Tresca equivalent stress
		σ_{F}	plastic flow stress

σ_f	stress in fibre	a^*	critical crack length
$\sigma_{f,B}$	fracture strength of a fibre	$\mathrm{d}a/\mathrm{d}N$	crack propagation per cycle
σ_{hyd}	hydrostatic stress	Δa	crack propagation
σ_l	lower limit stress (Weibull)	A	anisotropy factor
σ_{limit}	maximum allowed stress	A	short name of the elongation
σ_m	stress in matrix		after fracture $A_{5.65}$
σ_m	mean stress	A, A^*	material parameters of subcriti-
$\sigma_{m,F}$	yield strength of the matrix		cal crack growth
σ_{max}	maximum stress in the notch	A	(cross-sectional) area
	root	$A_{5.65}$	elongation after fracture for
σ_{max}	maximum stress		$L_0/\sqrt{D_0} = 5.65$
σ_{min}	minimum stress	$A_{11.3}$	elongation after fracture for
σ_M	tensile strength of a polymer		$L_0/\sqrt{D_0} = 11.3$
σ_{nss}	net-section stress at the notch	A_C	material constant of Coble
	root		creep
σ_p	proof stress in a proof test	A_{GBS}	material parameter of grain
σ_t	true stress		boundary sliding
σ_Y	yield strength of a polymer	A_{NH}	material parameter of Nabarro-
$\Delta\sigma$	stress range		Herring creep
$\Delta\sigma_d$	amount of work hardening	b	fatigue ductility exponent
$\Delta\sigma_{gbs}$	amount of grain boundary	b	lattice constant
	strengthening	b	length of the Burgers vector
$\Delta\sigma_{ps}$	amount of particle strengthen-	B, B^*	material parameters of subcriti-
	ing		cal crack growth
$\Delta\sigma_{sss}$	amount of solid solution	B	material parameter of power-
	strengthening		law creep
τ	shear stress	c	lattic constant
τ^*	effective shear stress	c	concentration
τ_{crit}	critical shear stress	C	Larson-Miller constant
τ_F	shear flow stress	C	constant in the Paris equation
$\tau^{(ss)}$	shear stress in a slip system	C	hardening parameter
τ_i	interfacial shear stress	d	diameter
τ_i	frictional stress	d	grain size
τ_{rel}	relaxation time	d^*	width of an obstacle
τ_{ret}	retardation time	d_0	initial diameter in a tensile test
φ	true strain	D	damage
φ	angle coordinate	D	diffusion constant
φ_{neck}	strain at which necking of the	D_0	diffusion constant
	specimen sets	D_{GB}	diffusion coefficient of self-
χ	stress gradient at the notch		diffusion
	root	D_V	diffusion coefficient of volume
χ^*	relative stress gradient at the		diffusion
	notch root	E	energy
Ω	volume of a vacancy	E	Young's modulus
a	crack length of surface cracks,	E_c	creep modulus
	half crack length of interal	E_r	relaxation modulus
	cracks	f	yield function
a	lattice constant	f	probability density
a	fatigue strength exponent	f_f	volume fraction of the fibres

$f_{\rm m}$	volume fraction of the matrix	l	crystallographic index
$f_{\rm V}$	volume fraction	l	length
F	free energy	l_0	initial length
F	force	$l_{\rm c}$	critical length in fibre composites
$F_{\rm i}$	binding force		
$F_{\rm Q}$	critical force	L	current gauge length in the tensile test
g	yield function		
g	probability density of the proof test	L_0	initial gauge length in the tensile test
G	free enthalpy	m	ratio of tensile and compressive yield strength in polymers
G	shear modulus		
$G_{\rm f}$	failure probability during the proof test	m	Weibull modulus
		m^*	Weibull modulus of the life time
$\mathcal{G}, \mathcal{G}_{\rm I}, \mathcal{G}_{\rm II}, \mathcal{G}_{\rm III}$ energy release rate			
$\mathcal{G}_{\rm Ic}$	critical energy release rate	M	Taylor factor
h	crystallographic index	n	exponent of the Paris equation
H	hardening coefficient	n	exponent of the power law of subcritical crack growth
HB	Brinell hardness		
j	vacancy current density	n	creep exponent
J	vacancy current	n	hardening exponent
J	J integral	n_χ	notch support factor
J_1, J_2, J_3 principal invariants		N	count
k	Boltzmann's constant	N	number of stress cycles
k	spring stiffness	$N_{\rm E}$	limiting number of cycles
k	crystallographic index	$N_{\rm f}$	number of cycles to failure
k	hardening parameter	p	pressure
$k_{\rm d}$	work hardening constant	P	Larson-Miller parameter
$k_{\rm F}$	critical stress for the von Mises yield criterion	P	probability
		$P_{\rm f}$	failure probability
$k_{\rm HP}$	Hall-Petch constant	$P_{\rm s}$	probability of survival
k_l	hardening parameter	Q	activation energy
$k_{\rm ps}$	particle strengthening constant	$Q_{\rm ex}$	activation energy for vacancy migration
K	bulk modulus		
$K, K_{\rm I}, K_{\rm II}, K_{\rm III}, K_{\rm Q}$ stress intensity factor		$Q_{\rm V}$	enthalpy required to create a vacancy
$K_{\rm I0}$	limiting value of the stress intensity factor of subcritical crack growth	$Q_{\rm V}$	energy difference to an oversaturated phase
		r	atomic distance
$K_{\rm Ic}$	fracture toughness	r	reciprocal transition time (Garofalo equation)
$K_{\rm IR}$	crack-growth resistance		
$K_{\rm f}$	fatigue notch factor	r	radial coordinate
$K_{\rm m}$	mean stress intensity factor	r	particle radius
$K_{\rm op}$	crack opening stress intensity factor	r^*	critical particle radius
		r_0	stable atomic distance
$K_{\rm t}$	stress concentration factor	R	gas constant
ΔK	cyclic stress intensity factor	R	radius
$\Delta K_{\rm eff}$	effective cyclic stress intensity factor	R	stress ratio (R ratio)
$\Delta K_{\rm th}$	fatigue-crack-growth threshold	$R_{\rm c}$	compressive yield strength of ductile materials

R_{cm}	compressive strength of brittle materials	T	line tension of a dislocation
R_{eH}	upper yield strength of materials with an apparent yield point	T	period in fatigue
		T	temperature
R_{eL}	lower yield strength of materials with an apparent yield point	T_g	glass transition temprature
		T_m	melting temperature
R_m	tensile strength	T/T_m	homologous temperature
R_p	yield strength of materials without an apparent yield point without specifying the plastic deformation	U	energy, potential
		$U^{(el)}$	stored elastic strain energy
		v_0	displacement of the crack surfaces (half crack opening)
$R_{p0.2}$	0.2%-yield strength of materials without an apparent yield point	V	volume
		V^*	activation volume
S	current cross-sectional area in a tensile test	V_0	reference volume
		W	work
S	entropy	w	energy density, work density
S_0	initial cross-sectional area in a tensile test	$w^{(el)}$	elastic energy density
		$\dot{w}^{(pl)}$	power dissipated during plastic deformation
S_i	partial damage	x	coordinate
t	notch depth	X	position of the notch root
t	thickness of a specimen	Y	geometry factor (fracture mechanics)
t	time		
t_f	failure time		

Vectors

In this list, the vectors are printed in the index notation. The corresponding symbol notation can be built using the scheme $(a_i) \cong \underline{a}$. Unless stated otherwise, the indices run from 1 to 3.

ε_α	strain vector (Voigt notation, $\alpha = 1 \dots 6$)	m_i	slip direction
		n_i	normal vector
ξ_i	coordinate in an undeformed system	t_i	line vector of a dislocation
		u_i	displacement
σ_α	stress vector (Voigt notation, $\alpha = 1 \dots 6$)	v_i	direction of movement of a dislocation
b_i	Burgers vector	x_i	coordinate

Matrices and tensors

In this list, the matrices and tensors are printed in the index notation. The corresponding symbol notation can be built using the scheme $(A_{ij}) \mathrel{\widehat=} \underline{\underline{A}}$ and $(A_{ijkl}) \mathrel{\widehat=} \underline{\underline{A}}_{4}$, respectively. Unless stated otherwise, the indices run from 1 to 3.

δ_{ij}	Kronecker delta	C_{ijkl}	elasticity tensor
ε_{ij}	strain tensor	F_{ij}	deformation gradient
$\dot{\varepsilon}_{ij}^{(\mathrm{pl})}$	plastic strain rate	G_{ij}	Green's strain tensor
σ_{ij}	stress tensor	R_{ij}	rotation tensor
σ_{ij}'	deviatoric stress tensor	$S_{\alpha\beta}$	compliance matrix (Voigt notation, $\alpha, \beta = 1 \ldots 6$)
$\sigma_{ij}^{(\mathrm{eff})}$	effective stress		
$\sigma_{ij}^{(\mathrm{kin})}$	kinematic backstress	S_{ijkl}	compliance tensor
$C_{\alpha\beta}$	elasticity matrix (Voigt notation, $\alpha, \beta = 1 \ldots 6$)	T_{ij}	energy-momentum tensor
		U_{ij}	right stretch tensor

Indices and operators

α, β	tensor calculus: indices in the Voigt notation, possible values: 1 to 6	i, j, k, l	tensor calculus: running index for the tensor components, possible values: 1 to 3
$\underline{\nabla}$	divergence	m	mean value in cyclic loads e. g., mean stress $\sigma_{\mathrm{m}} = (\sigma_{\max} + \sigma_{\min})/2$
Δ	range in cyclic loads e. g., $\Delta\sigma = \sigma_{\max} - \sigma_{\min}$		
$1, 2, 3$	unsorted principal values of stress tensors	max	maximum value in cyclic loads e. g., maximum stress $\sigma_{\max}$
$\mathrm{I, II, III}$	sorted principal values of stress tensors	min	minimum value in cyclic loads e. g., minimum stress $\sigma_{\min}$

Index